The GMO Handbook

The GMO Handbook

Genetically Modified Animals, Microbes, and Plants in Biotechnology

Edited by

***Sarad R. Parekh,* PhD**

Dow AgroSciences, Indianapolis, IN

HUMANA PRESS TOTOWA, NEW JERSEY

999 Riverview Drive, Suite 208
Totowa, New Jersey 07512

humanapress.com

This publication is printed on acid-free paper. ∞
ANSI Z39.48-1984 (American Standards Institute)
Permanence of Paper for Printed Library Materials.

Production Editor: Wendy S. Kopf.

Cover illustrations: Sheep and corn photographs by Corbis Images (Corbis.com). Background DNA figure and foreground is the microscopic photo of *Escherichia coli* cells. The photograph was provided by Sarad R. Parekh.

Cover design by Sarad R. Parekh and Patricia F. Cleary.

For additional copies, pricing for bulk purchases, and/or information about other Humana titles, contact Humana at the above address or at any of the following numbers: Tel.: 973-256-1699; Fax: 973-256-8341; E-mail: humana@humanapr.com or visit or website: humanapress.com

Printed in the United States of America. 10 9 8 7 6 5 4 3 2 1
e-ISBN: 1-59259-801-3
Library of Congress Cataloging in Publication Data

The GMO handbook : genetically modified animals, microbes, and plants in
biotechnology / edited by Sarad R. Parekh.
p. ; cm.
Includes bibliographical references and index.
ISBN 1-58829-307-6 (alk. paper)
1. Transgenic organisms--Handbooks, manuals, etc.
[DNLM: 1. Organisms, Genetically Modified. 2. Genetic
Engineering--methods. 3. Genetic Engineering--standards. QH 442.6 G569
2004] I. Parekh, Sarad R., 1953-
QH442.6.G657 2004
660.6--dc22

2003026980

Preface

The 21st century began in the midst of a biotechnology revolution. To many scientists and biotechnologists, the amount of information concerning new and novel methodologies and their impact is overwhelming. The development and application of biotechnology is forging ahead, bringing with it rapid changes in everyday life. Because all living organisms depend on molecular activity for survival and reproduction, research on molecular biotechnology and the utility of genetically altered cells has acquired a more dominant role in multidisciplinary activities. In addition, the biosafety and economic feasibility of genetically modified crops, microbes, and animals—often referred to as genetically modified organisms (GMOs)—have become paramount issues for consumers, farmers, and governments, as well as the agriculture, chemical, food, and pharmaceutical industries. An intense global debate is focused on ethical issues and the potential benefits and risks of genetic engineering technology.

Although there is general agreement that attention must be focused on educating the public on the potential and technical challenges of GMOs, it is equally important to recognize the complexities that arise from a heightened public awareness of the safety and environmental impacts of biotechnology. If GMO technology is to succeed, all concerned parties require a greater understanding of this special research.

Updating one's knowledge of GMO biotechnology, the most recent advances, the accepted guidelines for commercial application, and the new opportunities GMO research offers to the benefit of humankind is a continual challenge. *The GMO Handbook* introduces and explains the fundamentals of molecular biotechnology as a scientific discipline, provides an understanding of how current GMO research is conducted, discusses the problems that have arisen from genetic technology, explains the tools needed to address and resolve conflicts on GMO issues, and provides in-depth discussions on how GMO-derived technology may impact our lives in the future.

The GMO Handbook is divided into three sections. In each section, special emphasis is placed on explaining the wide range of current technologies and new strategies employed from cell biology, molecular biology, and biochemistry in constructing GMOs. In addition, large scale expression and production of recombinant products in cultured cells is described. Furthermore, the significance of public acceptance of GMOs and the economic benefits for the agricultural and pharmaceutical markets is also discussed. Finally, the unknown challenges and ethical concerns associated with GMOs are raised.

In most instances, *The GMO Handbook* addresses special topics by first reviewing the molecular details and then explaining their broader applications. Whereas many books on GMOs address registration procedures, regulation policies, environmental impacts on the food chain, or biosafety concerns with deliberate release, many fail to explain the basic principles, methodologies, and concepts behind the methods of construction of GMOs. They rarely provide the reader with critical information on the limitations of the technology, the pitfalls, and the necessary guidelines for commercialization. This book intends to fill these gaps. An extensive glossary is provided to help the reader understand the terminology employed in GMO biotechnology.

This book is the culmination of the efforts of many. The chapters were written by scientists who have made significant contributions to GMO research and can address some of the latest developments in the field. *The GMO Handbook* presents the work of scientists from a broad range of disciplines including botany, entomology, plant pathology, chemical engineering, microbiology, virology, genetics, public health, and government policy.

The GMO Handbook gives a cross section of current accomplishments in GMO research and provide insight into its future. The handbook targets a unique range of scientists. Young researchers beginning their undergraduate- and graduate-level studies will benefit from the ability to see the full range of techniques and applications used to culture and analyze animal, microbial, and plant GMOs. The chapters are written so as to facilitate their use as a teaching aid. Experienced scientists, biotechnology business managers, and manufacturing managers engaged in major GMO technologies are challenged to contemplate the potential of these technologies. The hope is that the book will illuminate the collaboration among scientists of many differing disciplines that has resulted in the invention of state-of-the-art techniques for developing GMOs.

Each individual chapter presents the opinions and views of its author(s). Although every effort has been made by the editor to provide a consistent style, the results presented, views expressed, and final content of the chapters remain the sole responsibility of the chapter authors. It is our hope that the information presented here will prove to be a useful contribution not only to science, but also to the general public interest.

I wish to express my appreciation to all of the contributing authors. I am especially grateful to the members of the international advisory board and Anne Gregg, information scientist, at Dow AgroSciences. Anne provided critical help in editing and organizing this book. Thanks are also owed to many associate friends for their advice and collaboration, as well as to Humana Press and their team members for the opportunity to publish this book.

Finally, with love and affection, I am indebted to my wife, Ranna, and my daughter, Puja, for their patience and understanding.

Sarad R. Parekh

Contents

Contributors

W. Michael Ainley • *Plant Transformation and Gene Expression, Dow AgroSciences, Indianapolis, IN*

Raymond Anthony • *Department of Philosophy and Religious Studies, Iowa State University, IA*

Steven L. Evans • *Discovery Research, Dow AgroSciences, Indianapolis, IN*

Yong Gao • *Biotechnology Regulatory Science, Dow AgroSciences, Indianapolis, IN*

Anne Gregg • *Information Management Center, Dow AgroSciences, Indianapolis, IN*

William F. Goure • *Business Development, Mendel Biotechnology Inc., Hayward, CA*

Manju Gupta • *Plant Trait Discovery Research, Dow AgroSciences, Indianapolis, IN*

Lei Han • *Supply Research and Development, Dow AgroSciences, Indianapolis, IN*

Siva P. Kumpatla • *Plant Biotechnology, Dow AgroSciences, Indianapolis, IN*

Sarad R. Parekh • *Supply Research and Development, Dow AgroSciences, Indianapolis, IN*

Raghav Ram • *Plant Biotechnology, Dow AgroSciences, Indianapolis, IN*

Douglas J. Stemke • *Department of Biology, Elon University, Elon, NC*

Paul B. Thompson • *Food, Agricultural, and Community Ethics, Michigan State University, East Lansing, MI*

William Whitford • *Research and Product Development, HyClone, Logan, Utah*

I

INTRODUCTION

1

Introduction

Sarad R. Parekh and Anne Gregg

1. Introduction

The rationale for accelerated research in biotechnology has several thrusts; two are to understand the living biological system for advancement of science and to find newer applications for commercialization. Equally motivating factors include satisfying human curiosity about the nature of cell cultures and living things and exploiting the opportunities that biotechnology offers to eliminate long-threatening diseases, improve crop productivity to maintain an adequate supply of food for humans and animals, and generally improve the quality of life *(1)*.

Until discovery of the microscope in 1670, the basic studies in biology consisted of simple and understandable observations. With technological advancements, research in biology has become highly sophisticated and more expensive. As a result, not only academic but also government institutions are engaged in the research to the extent that consequences of this biological research, and its impact on human life, now require scrutiny for justification of the expenditure of tax and investment dollars. With such support mounting to billions of dollars annually worldwide, it is no wonder that public interest in applied research and biotechnology has been created *(2)*.

Although new breakthroughs and advances in biological research continue rapidly, the complexity of biotechnology is not clearly understood by the public. The new breakthroughs and technological advances in the biological sciences that will serve and benefit humankind in many cases are either ignored or misinterpreted *(3)*.

As early as the 1980s, attention was focused on recombinant deoxyribonucleic acid (DNA) research, in which specific genes were manipulated in a more elegant way to create new entities of living microbes, the so-called genetically modified organisms (GMOs). The interest primarily came from scientists who recognized the potential opportunities for this technology and its future applications *(4)*. The benefits and new potentials continued to be realized and reviewed at scientific meetings, and the opportunities associated with the implementation of the technology continued to be discussed in scientific journals; the public community, however, in many instances remained unaware of the developments and could not begin to contemplate the potential values. Furthermore, genuine concern for public safety, coupled with the memories of World War II,

From: *The GMO Handbook: Genetically Modified Animals, Microbes, and Plants in Biotechnology*
Edited by: S. R. Parekh © Humana Press Inc., Totowa, NJ

have caused some work on GMOs to become a subject of controversy, sometimes within the scientific community and sometimes throughout nations worldwide *(5)*.

These issues have been long debated in public places through panel discussions involving subject matter experts, reviews in the newspapers, programs on television, and, in a few instances, government policy hearings *(6)*. Therefore, it is paramount that both the scientific community and those with limited understanding about GMOs appreciate and interpret correctly the application of GMOs in biotechnology. It is critical to communicate clearly and openly with the general public, not only about the benefits and the rational for embracing GMO technologies, but also about the basic principles of the technology, including its evolution, essential protocols, environmental risks, and level of predictability *(5)*. Equally important is the review of biosafety issues—the scientific, economic, political, sociological, and public health issues that exist when considering risk assessment and management procedures *(7)*. Once a comprehensive perspective emerges from open communication, coupled with evidence and experience in handling GMOs, it will become simpler for any person to make decisions independently and to harness the potentials of GMOs.

2. Historical Perspectives and GMO Technology

To understand GMOs fully, including their impact and the promises they hold, it is essential to trace the history of how the technology of manipulating genetic material in organisms has evolved and how this past research has laid the foundation of today's biotechnology industries. The utility and practice of using GMOs is not new. Since the dawn of civilization, agricultural biotechnology practices have guaranteed food supply, sources of income, and quality of life *(2)*. Domestication of plants and animals, combined with gradual, long-term changes in their quality and quantity, can be traced as far back as 10,000 years ago. A continuous effort to improve plant and animal yields has been documented in the early writings of many civilizations in both the Old and New Worlds. The documented history of all cradles of civilization has contained detailed descriptions of primitive methods to improve plant and animal productivity.

Later, domestication of plants and animals, followed by food shortages, population growth, and industrialization of various nations, coincided most probably with the growth of microbiology *(8)*. Thus evolved the classical food fermentation industry, the earliest known application of biotechnology, which employed selective microbes for culturing and deriving fermented foods. The techniques are well documented for successful beer and wine production and bread making in various archaeological relics of the ancient civilizations. Cheese, yogurt, vinegar, soy sauce, and curd are other examples of the application of selective microbes used in old biotechnology and commercial applications *(9)*.

With time came a gradual improvement in agricultural techniques for the domestication of plants and animals, including step-by-step selection of better performing and more adaptive genotypes along with intuitive breeding. Mendel's discovery of genetics in the 1860s led to the beginning of applied science and controlled breeding methods *(8)*. Breeding, referred to here as *old biotechnology*, was an efficient tool to make better hybrids and genotypes of crops and diversified animal stocks. This significantly set the pace for establishment of a "green" revolution *(9)*.

Soon, this old biotechnology practice was propelled by another round of biological revolution toward the mid-20th century with the discovery of the structure and function of DNA. Thereafter, scientists exploited the capability of changing organisms by altering DNA. This altering was done initially by mutagenesis. Using a specific chemical agent, scientists could modify the DNA sequence, and by this process of mutation, they could induce inheritable change in the structure of DNA (genetic material).The organism that inherited the new or modified DNA sequence was called a *mutant strain*. The technique of genetic alteration by recombination, so-called crossing over, was then used to create a diverse organism carrying two mutations from two cells, each carrying single but completely different DNA sequences (mutations). This dogma was an extremely important one and led to the understanding of physiology, metabolism in the biological process, and research *(8)*.

These tools and techniques worked, however, only when the two cell variants were of similar species. Because the biological system as it exists in nature is very complex, with thousands of gene variations, numerous genetic alterations by mutations and recombination events are required to result in the formation of a new organism that then can be used to yield valuable information in simple experiments. Such experiments provided insight into the role of the newly formed inheritable DNA and thus assisted in the characterization of the mutant organism.

The formation of mutant organisms provided the foundations for studying genetic regulation as done by Jacob and Monod in their studies of the regulation of lactose metabolism in *Escherichia coli (8)*. They created several thousand variants of the parent population by mutations and new combinations of the DNA sequence. By studying the properties of such variant strains in detail, they were able to dissect the information. In this way, Jacob and Monod uncovered many mysteries and principles that govern gene expression. Their research was rewarded with a shared Nobel Prize *(8)*. Their techniques laid the foundations for construction of variant bacterial strains adopted by thousands of laboratories across the world and became the standard procedure to create new varieties of constructed *E. coli* strains having limitless combination of properties already present in the wild *E. coli (4)*.

In the early 1960s, application of similar technology for modified strain construction was performed in recombining DNA of two distinct species of bacteria, *E. coli* and *Serratia marcescens*. Extension of this work resulted in the construction of hybrid species carrying mixtures of genes, and thus hybrid bacterial strains with mixed properties from each parent strain were derived.

Although some advances were made in inducing genetic exchanges with diverse species, the major impact of the power of genetic modification no doubt was realized in the mid-1970s with the discovery of restriction and ligation enzymes. These enzymes could break isolated DNA molecules into fragments that could be reassembled in any desired sequence. This method provided a quantum leap in molecular techniques and made possible the ability of any isolated DNA molecule to be spliced together in large blocks of genes, either from the same or different organism *(9)*. Shortly thereafter, experiments by Stanley and Cohen *(8)* used these techniques to transfer genes from one organism to another (i.e., from *E. coli* to frog).

This work paved the way for the next wave in biotechnology, the demonstration that modified genes can be manipulated and inserted from bacteria into plant and animal cells. This breakthrough in producing recombinant DNA molecules gave birth to genetic engineering or "cloning," and it opened up the new science of engineering genes and creating GMOs and new transgenic living cells. These tools greatly increased the ability of the molecular biotechnologist to intervene in the hereditary process of plants, animals, and microbes. Some individuals saw this breakthrough as an increase in both qualitative and quantitative changes in natural selection and life processes, affecting not only the amount of intervening processes, but also the kind of effects that can be produced *(9)*. Regardless of the date, at the dawn of GMO biotechnology, the utility and application of GMOs opened unimaginable possibilities and resulted in the establishment of the "new biotechnology" industry *(10)*.

The commercial development and practice of GMOs became routine in research laboratories, in which scientists initially used the techniques to study the structure, function, and organization of genetic material. As technologies advanced, it became possible to investigate and probe precise questions on the physiological system and interaction and network of biochemical pathways *(11)*. Eventually, with the introduction of computer models and bioinformatics tools, more powerful and precise methods resulted for the construction of GMOs with newer applications for various industries *(12)*. The roots of these new applications in industry ultimately found their way into practices of cell culture fermentation, agriculture, and human husbandry *(13)*. To this age, the old practice of biotechnology adds not only the application of recombinant DNA technology and cell culture, but also a host of applications using the diversity of living organisms to make commercial products and therapeutics (Table 1). These GMO-based technologies raise promises to reshape the future of humankind by revolutionizing the agrochemical and biopharmaceutical industries and medical diagnosis and treatment *(14–16)*.

3. Why Use GMOs?

The main drivers for accepting and integrating GMOs include the economics and novelty of products and services this technology offers and its ability to help meet the demand of food supply and quality of life of the world's increasing population *(17)*. For example, conventional agricultural practice and animal farming suffer from serious limitations in facing new changes in international markets *(7)*. Also, the advancements in computer information and communication technology have simplified and globalized world markets. Local pricing policies, political pressure, and profitability are no longer effective, especially in keeping pace with the world's population demands and the international trade volume for novel agricultural products *(18–20)*.

Furthermore, the agricultural and food-based industries (beef, poultry, fruit, and vegetables) that depend on natural resources are facing new conditions of reduced land and water availability. The rapid industrial revolution coupled with changes in global climatic conditions have caused water and air quality to deteriorate from pollution and soil erosion.

Table 1
Approximate Time-Scale for Successful Introduction of Commercial Products (GMOs) From Expressed Transgene Event (Protein)

Transgenic animals (farm animals)	2–3 years
Transgenic crops (tobacco and corn)	0.5–1 years
Transgenic mammalian cell culture	6 months
Insect cells culture	1–2 months
Fungi (*Aspergillus* yeast)	2 weeks
Bacteria (*E. coli*)	1 weeks

These changes have an impact not only on developed countries, but also on those areas where rapid industrialization and urbanization cause severe competition for land *(19)*. Regions that were employed once for agriculture are turning into real estate targets for modernization.

Hence, with the steady increase in world population, the needs of the hungry cannot be met or sustained with today's limited food supply and primitive animal and crop farming practices and therapeutics *(20)*. Therefore, needs are critical and urgent to accelerate technologies that will not only meet the supply of food and marketable products at competitive pricing, but also provide better health care. In addition, evidence of medical breakthroughs from the human genome projects exists, and the development of novel biotechnology products is becoming cost competitive *(21)*. All these factors compel the adoption and application of new biotechnology, including recombinant DNA practices and products derived from GMOs (Table 2).

4. New Opportunities With GMOs

GMOs as discussed in this book refer to the use of recombinant DNA and the biological techniques that use living organisms, cells, or parts of cells to make modified products or improved biological capabilities of microbes, plants, or animals for specific purposes *(22,23)*. The three major areas in which GMOs are currently being utilized in biotechnology are as (1) tools in classical breeding; (2) a means of generating transgenic plants, animals, and other microbes; and (3) a means of integrating GMOs into various agricultural, pharmaceutical, or industrial production systems (Table 2).

The agricultural potential of classical breeding has not been fully exploited to improve plant and animal productivity or to modify food production. In fact, classical techniques of plant and animal cell line advancement now rely on the use of newly developed recombinant DNA biotechnology tools to detect precise genotypes and to select specialty crops. Newly developed recombinant DNA technologies now are used to develop transgenics. These tools include (1) microscopic visualization of chromosomes in tissue culture; (2) staining by fluorescent *in situ* hybridization (FISH) for fluorescent labeling of markers in cells; (3) restriction fragment length polymorphism (RFLP); (4) development of simple sequence repeat (SSR); and (5) markers for trait development in plant cells. All of these new tools are an integral part of classical breeding and the selection approach for accelerating selective trait development.

Table 2
Products Made by Microbes, Yeast, Insect Cells, Mammalian Cells, Transgenic Animals, and Transgenic Plants for Other Applications by Recombinant DNA Technology and Selective Breeding

Bacteria	Yeast or Fungi	Insect Cells	Mammalian	Transgenic Animals	Transgenic Plants
Antibiotics	Beverages	Viral vaccines and proteins	Tissue plasminogen activator	α-Antitypic	Human lysosomic enzymes
Insulin	Vaccines	Rabies	Glycoprotein factor 8	Bile sale lipase	Human glucocerebrosidase
Interleukins2 and 3	Streptokinase	Glycoproteins	Hybridomas	Superoxide dismutase	Avidin
α,γ-Interferon	Hirudin	Tissue plasminogen activator	Monoclonal antibodies	Lymphotoxin	*Bacillus thuringiensis* proteins
Vitamin C	Aprotinin	β,α-Interferon	OKT-3	Epidermal growth factor	Aprotinin
Bacterial vaccines	γ-Interferon	Erythropoietin	B growth factor	Human serum albumin	Vaccines
Biopolymers	Interleukins 3	Interleukin 2	CD4 receptor	Calcitonin	Pesticide, viral herbicide resistance
Amino acids	Industrial	Adenosine deaminase	Parasitic vaccines	Tissue cells	
Bioremediation	Enzymes HIV-1 antigens	Diagnostic proteins	Growth hormone	Fibrinogen, collagen	
Indigo-chemicals			Hepatitis B antigen	Antithrombin	
Enzymes			Herceptin	Malaria vaccines	
Insecticide			Diagnostics	Stem cell therapy	
Proteins *(Bt)*			Antibodies		
N and H fixation			Urokinase		

5. GMO Applications

GMOs have diversified applications, multitechnology approaches, and a plethora of products. Some of these are briefly mentioned here.

The pharmaceutical industry felt the earliest effects of GMO technology. Products difficult to isolate or produce, such as growth hormones *(7,12,20)*, insulin, and others effecting human health, could now be produced in significant amounts from *E. coli* and yeast (Table 2). A major and logical extension of these technologies eventually gained application in biofarming, including the use of transgenics to produce valuable human proteins in farm animals, which act as drug manufacturing production lines *(4,22)*. Later, the agricultural sector felt a new wave of impact from GMOs, primarily from the planned introduction of GMO seeds and derivatives to serve precise purposes. The application of GMOs in agriculture can now be envisioned for the production of value-added crops such as transgenic plants resistant to disease, drought, or frost or capable of metabolizing toxic waste *(9,22,23)*.

Crops with protection traits seem to increase yields significantly for farmers in parts of the world where pest pressure is high and chemical pesticides are prohibitively expensive. If concerns about contamination of food supplies can be overcome, plant-based plastics, polymers and films, lubricants, inks, enzymes, and synthetic fibers could be made, which would then begin to make inroads into a $60 billion US market dominated by petrochemicals *(21)*.

Although industrial use of new biotechnology and developing GMOs continues to progress, the current international market is limiting the introduction of GMO agricultural products. Limitations result from the "free market" economy, volume, international trade barriers, pricing, and regulatory policies. Despite these difficulties, biotechnology-based industry has continued to grow and is now the world's largest business *(8)* because of the prospect of reduced costs of plant and animal stocks with desired traits (Table 3).

In the last decade, a number of GMO-derived products introduced into the market have driven the evolution of these new industries *(7,16,23)*. Plant-based pharmaceuticals could present a multi-billion dollar market opportunity, with contained farmland acting as a production factory (e.g., molecular farming). The chemical companies Dow, Epicyte, Prodigene, and Monsanto are using crops to make nutraceuticals and for other applications *(17)*. To name a few products, Monsanto's Polsilac version of the bovine growth hormone is now used in 20% of US dairy cows. Monsanto has developed transgenic soybeans resistant to their premier herbicide, RoundUp™, with other herbicide-resistant crops such as cotton, corn, potato, and rapeseed also in the market. China has marketed a virus-resistant tomato *(17)*.

Calgene, one of the early leaders in the agricultural biotechnology industry, and Mycogen, Ecogen, Concept, and Syntro are all among the American industries that have launched several agricultural-based GMOs . A Wall Street report suggested that the annual growth rate of agricultural biotechnology companies is anticipated to exceed the annual growth rate of other industries, indicating that investment in GMO-based technologies will finally pay off *(8,17)*.

Further benefits of GMOs can be realized with changes in perception of GMO products (Tables 4 and 5). However, changes are slow in coming. For example, the Euro-

Table 3
Characteristics and Attributes Desired in Transgenic Cultures

Cell Biology	Product Expression	Recovery and Isolation	Safety and Quality
• Rapid growth • Few growth factors • Serum-free or protein free medium • Protein processing and transport close to human	• In high levels • Stable productivity • Efficient and accurate post-translation modifications • Control of product expression	• In supernatant • Thermostable • No interfering proteins • Few steps for recovery • Less sensitive to degradation	• No endogenous viral sequences • No virus materials • No oncogenic DNA • No adventitious agents (mycoplasma, fungi, bacteria)

Table 4
Advantages and Disadvantages With GMO Cell Types

GMO	Advantages	Disadvantages
Bacteria	• Established regulatory tack record • Inexpensive media • High expression level • Fast growing • Easy to characterize	• Certain proteins not secreted • Contains endotoxin • No posttranslation modification • Possibility of incorrect folding • Harvest can damage proteins
Insect cells	• Product is secreted • High expression level • Baculovirus are harmless to humans • Proper posttranslation modification	• Minimum regulatory record • Short history of use • Few incorrect protein glycosylations • Expensive media Mammalian virus can infect cells
Mammalian	• Can secrete proteins • Correct folding • Posttranslation modification • Good camp track record	• Expensive media • May contain allergens and component of bovine source • Requires extensive characterization • May contain adventitious material
Yeast	• Recognized as "safe" • Long history of use • High expression level • No endotoxin • Fast growth • Inexpensive media • Protein mostly folded correctly	• Overglycosylation can alter bioactivity • May contain immunogens • Nonreactive proteins not folded correctly

Table 5
Opportunities With Transgenic Animals and Plants

	Animals	Plants
Advantages	• High expression levels • Can process complex proteins • Correctly fold proteins • Easy scale up • Low-cost production	• Shorter cycle than animals • Easy storage of seed bank • Well-understood genetics • No plant virus known to infect humans • Low-cost production
Disadvantages	• Little regulatory experience • Variable expression level • Long time-scale • Complicated definition of lots and batches • Unresolved public image	• May contain allergens • Different posttranslation modification made by animals • Potential source of herbicide, pesticide impurity • Unresolved public issues

pean Regulatory Council recently agreed to limit the labeling of GMO food products, and Britain approved several products, including GMO-based tomato paste (Zeneca), rapeseed oil from Plant Genetic System (Belgium), and soy products from Monsanto *(17)*. However, based on the fear of presumed long-term adverse health effects, these approvals have not come easily as there have been countless demonstrations and strong public resistance in Europe against the use of transgenic products *(24)*.

The agricultural biotechnology industry seems different from biopharmaceuticals. For example, Genentech, Eli Lilly, Amgen, Merck, and Glaxo Smith Kline Beecham all have had accelerated Food and Drug Administration approvals and significant commercial successes for health care products derived from GMOs with little public objection *(25)*. Finally, with the recent Cleaning in Process Act enforced by the European Community to ensure the safety of commercial products, there has been an alteration in political perception, and the market demand for GMO derived products for medical use has surged *(26,27)*.

5.1. Transgenic Plants

Recent publications indicate that over 3500 field trials on transgenic crops in more than 40 countries are ongoing. This research reflects the dramatic and effective contribution that transgenic plants with unique traits can offer in agriculture and horticulture *(28–30)*. The major transgenic crops approved for commercial production include tomato with delayed ripening, cotton and corn with insect or herbicide resistance, soybean with herbicide resistance, and canola with enhanced oil quality *(17)*.

5.2. Somatic Cell and Micropropagation

The most significant contribution of in vitro recombinant methods to plant biotechnology has been somatic cell generation, especially haploid production of cell lines and somatic hybridization. Regeneration of haploid cells and plants from microspores is important for production of homozygous offspring and for further breeding. This has been demonstrated in barley, rice, rapeseed, potato, asparagus, and other plants. Somatic hybridization by protoplast fusion has offered an efficient solution to overcome interspecific crossing barriers.

Plant propagation in tissue culture (micropropagation) now has been adopted to develop top-quality clones. These clones are selected for their unique horticulture traits, pest resistance, or environmental stress conditioning *(4,31)*. The main advantage offered is the use of combining rapid, large-scale propagation of a new genotype with the use of small germplasm- and pathogen-free propagules. Virus- and bacteria-free commercial production of pyrethrum (a natural insecticide) is derived from this technology; it has a significant export value for countries such as those in East Africa *(20)*.

5.3. Bioremediation

Areas such as specialty chemicals, food additives, commodity chemicals, enzymes in food processing industries, waste disposal, mining, and energy production also have found the utility of GMOs. Microbes are genetically designed to assist detoxification of contaminated water. Environmental cleanup is another extended use; it involves the enrichment of nutrients such as nitrogen and phosphorus to augment the efficacy of microflora in assimilating environmental waste *(17)*. Fungal strains have been developed that degrade soil pollutants such as DDT, polychlorinated biphenyl (PCB), and cyanide *(5)*. Newly constructed GMOs have also been used in accelerated decomposition of organic matter to clean oil-contaminated shorelines. Certain transgenic plants and seaweed have been designed as "biofilters" for effluent detoxification, water salinization, and waste treatment; this is referred to as *phytomediation (18,21)*.

5.4. Biological Control of Pests, Pathogens, and Plant and Animal Diseases

Use of biological agents, or biopesticides, for the control of plant pests and pathogens has a long history *(28–31)*. The global market for chemical pesticides is becoming saturated and shows no signs of future growth. On the other hand, use of biopesticides is growing annually at the rate of 7% per year *(17)*. Biopesticides are considered more pest specific with fewer negative effects on humans, farm animals, and the environment. Sales of *Bacillus thuringiensis* alone have mounted to over $120 million *(32)*. The accepted use of baculoviruses for the protection of vegetables and cotton is estimated as valued at $2 billion *(17)*.

Dow AgroSciences received speedy approval and an Environmental Protection Agency (EPA) Green Chemistry award for the launch of spinosad, a biopesticide from a microbial source, as an environmentally friendly insect control agent *(32)*. Ecoscience has developed fungi to combat cockroaches and biofungicides to protect spoilage in citrus fruits. Biofertilizers (nitrogen-fixing bacteria) have also been a special niche *(30)*. The EPA and US Department of Agriculture have also approved several products for the control of nematodes and phytopathogenic fungi *(33)*.

In addition, recent mergers of biocontrol companies have resulted in the required critical mass to become dominant players in the biocontrol industry. The formation of such consortia is noted as a major movement in the biotechnology industry.

Although there are fears that repeated use and exposure of biopesticides may result in some pests developing resistance, this fear is counterbalanced by the fact that certain organophosphates and bromo- and chloro-derivatives of pesticides will be phased out because of overexposure and health risks. Furthermore, some of the global population has resorted to use of natural pesticides for organic farming; this population consumes only food and vegetables not exposed to synthetic chemical treatment.

Use of GMOs is improving the health of farm animals. New vaccines from GMOs are being developed and used against brucella, encephalitis, and hepatitis (Table 3). These new vaccines have become a dominant method for proper maintenance of farm animals and poultry. Increased regulation that requires germ-free fish and meat products and fresh fruits and vegetables free of pesticides has opened a new area of diagnostic biotechnology for animal and plant farming *(34–36)*. In this new area of diagnostic biotechnology, tools of new DNA biotechnology are used for the early monitoring and detection of diseases, pests, and chemical residues and for managing health care and life expectancy of plant and animal stocks to ensure the marketability of products *(37–39)*.

5.5. Gene Bank and Germplasm

Embryos in reproductive animals and germplasm (i.e., self-contained units of propagation in plant seed and vegetative propagules) are, in fact, a capsule that contains a package of genetic blueprint. Commercialization of unique germplasm or embryos with a specific GMO trait is one of the fastest growing sectors of the biotechnology industry. This includes new biotechnologies of hybrid and artificial seeds, germplasm bank cryopreservation, in vitro fertilization, and embryo implantation *(17)*.

Efforts for improving reproduction and the quality of animal in vitro biotechnology are growing steadily in the cattle and poultry industries. This includes integrating and assisting superovulation, in vitro fertilization, and embryo implantation in a surrogate mother. Such GMO practices allow breeders to produce multiple embryos with unique or most desired qualities *(40)*. With the success of gender selection and embryo manipulation, major efforts are now under way in livestock agriculture. The outcome is increased meat and milk production that is seven to eight times that of three decades ago.

Germplasm banks and germplasm cryopreservation of plant seeds, vegetative tissues, and sperm oocytes are the offshoots of GMO technology development and applications. This development has provided the required platform for breeding programs of quality germplasm. Sperm cryopreservation and selection of favorable subspecies can manipulate, for example, breeding of fish for coloration and shape *(41)*.

In agricultural biotechnology, transgenic seed use is developing rapidly. In certain GMO plants, somatic hybridization and mutagenesis have led to the identification and diagnosis of the role of mitochondria, chloroplast, and DNA elements in cell growth, fertility, and differentiation. Understanding these roles has led to the development of specific GMO plants with unique traits that induce high rates of meiotic embryogenesis. This method of genetic modification has led to the development of miniaturized propagules with such favorable traits as high yield and disease resistance *(28)*.

Encapsulation and coating of somatic embryo biotechnology have resulted in the evolution of artificial or synthetic seeds for use in automatic planting processes and mass production. The quantity of seeds developed through these technologies is estimated to reach millions of tons by 2005, still only representing a fraction of the $37 billion in sales of hybrid seeds worldwide. It should be emphasized that the rewards for improving seed quality and developing elite cell lines for agriculture is enormous. Examples of seed input represent only 5–10% of the total grower's investment. Thus, the reward for improved or GMO seeds with elite quality are highly profitable. For example, the market value of 1 kg of elite tomato seeds can reach $20,000 compared to $4 to $7 per kilogram of fresh tomatoes in season *(40)*.

6. Potential Issues With Commercial Application of GMO

Although commercial GMO-based biotechnology has emerged as a major force behind the marketing of innovative products in the global food and pharmaceutical industries, its introduction and future application have become enmeshed in a complex set of economic and environmental controversies among scientists, educators, and the public *(42,43)*. At one end of the spectrum are those who recognize the potential benefits of GMO technology to humankind. These benefits may include ensuring success in maintaining and evolving new life forms, improving old tools or producing new ones, and applying the new technology to allow human activities to be more benign to the ecosystem than occurs with traditional chemical technologies. However, others espouse an opinion that GMOs introduced into the environment may have a chance of surviving and multiplying with undesirable consequences *(44)*.

Ecosystems are complex, and environmental conditions are unpredictable; these elements of uncertainty have led some scientists and public officials to express concern about the use of GMOs *(43–46)*. In one sense, this controversy has been compared to the similar issue that National Aeronautic and Space Administration faced about possibly introducing an alien bug from another planet and the consequences on our surroundings that would occur if the alien bug could not be controlled or traced.

Thus, in the present society, new technologies are considered in a far more critical way than they would have been several decades ago. Biological experiments and genetic manipulations in particular have seemed to evoke fear and resistance from some *(43,44)*. However, in most cases, GMOs are not new to the environment in which they are used. Almost invariably, either the same organism or one closely related already exists in the ecosystem in which the proposed biotechnology application would occur. The major difference between the native and GMO species lies in the addition or alteration of specific genes or sets of genes regulating aspects of the organism's biochemical pathway. However, certain critical issues differ from application to application and in laboratories or in the open environment *(47)*. Therefore, policymakers have to rely on sound scientific review and weigh carefully any potential risks against anticipated benefits. In addition, public perception of GMOs is likely to influence current and future developments. In fact, after the use of GMOs became more widely accepted, new policies were introduced and agreed to address and manage biosafety *(26)*.

In 1976, the National Institutes of Health issued the first formal guidelines for recombinant research. As scientists and industries learned more about the safety of GMOs, initial fears proved excessive, and the guidelines were repeatedly revised, with controls on GMO applications relaxed somewhat. Appropriate questions were then raised by the agencies that regulated GMO experiments and field trials about how to assess potential risks and coordinate the framework for commercial biotechnology and the jurisdiction of GMOs.

After 1986, the White House Office of Science and Technology Policy published a more comprehensive document identifying the critical role of specific agencies for approving GMO use. For example, the EPA and US Department of Agriculture govern regulatory policies, the Food and Drug Administration typically regulates drugs from GMOs, and the National Institutes of Health and National Science Foundation lay the research polices to ensure biosafety. The purpose of the framework was to enable the agencies to "operate" in an integrated and coordinated fashion to cover the full range of issues related to plant-, animal-, and microbial-derived GMO technology with the purpose of addressing biosafety, work practices, and risk prevention *(26)*.

One critical consideration for groups relying on GMO-based technologies is the need to develop more sophisticated strategies to address the interests of stakeholders with the power to influence consumer choice, politics, regulation, and trade *(17)*. Thus, as consumers in industrialized countries are more affected by changing prices, food sources, substitutes, and generic competition, this issue has become critical for the consumer-driven agricultural biotechnology and pharmaceutical sectors *(43)*. Although there are inherent advantages of the ability to improve the availability of food and health care for an ever-growing population, many developing and underdeveloped countries that face food and medical shortages may not be in a position to afford expensive biotechnology-based products. Education, explanation, desensitization, lower prices, and industry promotion are all factors that develop awareness and may eventually foster public confidence and acceptance *(40)*.

Planned introduction of GMOs for commercial practice in daily life may be accompanied by the following questions from the public: What are the potential risks and the outcome with continued use of GMOs? Are they beneficial, harmful, or have no relevance? How can they be tracked and traced? If harmful effects of GMOs ever manifest, can they be managed, contained, and neutralized? Special considerations exist for each case of GMO product development. Public acceptance and confidence needs to be built and supported through data that address the risks, rewards, and safety.

Although these issues are beyond the scope of this book, this section only summarized and highlighted specific and practical issues relevant to research and commercial practices for planned introduction of GMOs for human benefit. In this way, readers obtain a perspective on the foundation and framework of the technology for personal consideration. More perspectives on this area can be obtained in refs. *3, 8, 17,* and *21*.

7. Intragenetic Horizontal Transfers

The issue of intragenetic horizontal transfers is on the top of every GMO-based technology. Gene transfer between species can take place via a limited number of means, including hybridization or sexual crossing between plants and closely related animals. Transformation in bacteria is possible in the environment. Some of these mechanisms are shared between similar classes of organisms; whereas others are unique to particular groups. Gene transfer technology seems easy in some groups of bacterial species, but more challenging than cloning plants and animals.

More information pertaining to gene transfer technology comes from laboratory experience and well-studied experiments. However, this represents only a small sample of what really occurs naturally. Although the frequency of transfer may be low, rapid reproduction and large population growth means genes transferred into some bacterial populations could quite rapidly exert selection advantages *(47)*.

Among insects and animals, there is evidence that transfer can occur via transposable elements and viruses. Gene transfer among plants does occur rarely, but it has not been studied in detail. It therefore seems that, although horizontal transfer between closely related bacterial species occurs, there is a significant natural obstacle blocking gene transfer and flow between species.

It seems logical that natural bacterial populations experience lower rates of gene transfer than laboratory populations *(39)*. Several questions normally asked are as follows: How often does gene transfer occur within GMOs and by what mechanics? What environmental conditions favor the transfer, its integration on chromosomes, and expression at target sites? Do natural populations limit the inclusion of genetic material of foreign DNA into a new entity? Finally, if the new DNA is transferred and expressed, what environmental and ecological consequences occur from the transfer to nontargeted species?

Overall, no matter the outcome of the transfer, two critical questions remain to be addressed: How frequently is the genetic material likely to be transferred to nontargeted organisms? What is the estimated degree of distance for the genetic relationship between the original and the nontargeted species? Horizontal transfer cannot be assessed by simple information about host, gene vector, and final construct. Other factors influence and explain the outcome of genes inserted into a GMO. The outcome, the natural history, and the expression of the gene under various environmental conditions can be studied. The behavior of the vector in the host at various stages is fully evaluated to understand the horizontal transfer of the target gene *(47)*. Summarized below are some of the factors one needs to consider and address appropriately for successful development of a GMO-based platform.

7.1. GMO Host

It is important that the life cycles and physiology of microbes, plants, fungi, and animals used as hosts for genetic modification are studied and understood to facilitate selection of GMO product targets. For example, various means by which the potential gene can be transferred must be first characterized in laboratory experiments. Although the genetic exchanges and specific functions of some bacteria are well understood in controlled laboratory experiments, their population structure and specific functions in normal environmental conditions must also be considered. Therefore, a well-developed natural history of the host GMO supports the modeling and evaluation of the genetic and ecological implications of the planned introduction of the GMO product *(39)*.

7.2. Genes Inserted Into GMOs

The planned use of GMOs involves genes with characteristic attributes well understood by scientists (Table 3). Extensive characterization assists in determining the potential new interactions among genes, their environment, and methods of applications for novel expression phenomenon. Through experience, more specific understanding of GMO use is gained.

The toxin of *B. thuringiensis* offers a good example of a gene from an organism that is well characterized and for which the material from the bacterium has been used for decades *(17)*. There are now over 500 different *B. thuringiensis* toxin products available in over 20 formulations (e.g., powder, pellet, solution, emulsion) for the agricultural and forestry settings. Tons of the toxin from this bacterium have been applied to agriculture and forestry land. Intensive research has provided no evidence to date that the endotoxin gene, its product expressed in other microbes, or the toxin from the strain (var. *kurotsaki*) has any negative effect on organisms other than *Lepidoptera* and closely related insects *(24)*.

7.3. Vectors and Constructs

Just as it is essential to have well-characterized hosts and genes for GMO development, it is equally important to use vectors with precise gene sequences. Important properties to be considered here include the vector's ability to integrate independently in the strain, its special need to exchange information, and its degree of mobility and host range.

Another important factor affecting whether the gene inserted in a GMO has probability of movement from the altered host to a nontargeted organism is the final configuration of the gene–host, that is, the DNA structure at the site of gene insert. For example, inserting a gene into a chromosome minimizes subsequent gene transfer compared to insertion on a plasmid. The sequence of the regulatory gene controlling the expression of the inserted gene is also critical and plays a major role in limiting the transfer of an active gene to nontarget recipients *(47)*.

7.4. Ecological and Extrinsic Factors

The environment in which the GMO is introduced also influences the mobility and transfer of the gene. The expression of the trait, the surrounding environment, and its interaction with other GMOs it encounters must all be evaluated and analyzed to determine the impact and the fate of the targeted GMO.

7.5. Survival and Population Density

One critical component of the potential capability of GMOs for horizontal transfer is the ability of the GMO to establish and reproduce in its new habitat and express the engineered trait. Other factors that have a direct impact on gene transfer between cells are the absolute density of the GMOs in the surrounding ecosystem. Population densities can also be affected by the method used to introduce the GMO into the environment. In addition, timing of the planned introduction of the GMO has an effect on the densities of both the GMO and surrounding nontarget recipients.

7.6. Nontarget Recipients

There is a potential for gene transfer in organisms genetically similar to the GMO host. The probability of gene transfer declines as evolutionary relatedness decreases. The restriction enzyme system that exists in all cells recognizes and thereby degrades evolutionarily unrelated "foreign" DNA.

Horizontal transfer of genetic material between higher organisms (animals) is less likely than between simple ones (bacteria). Such events, however, are minimized if the engineered gene is already present in the recipient environment, and the concern about the transfer beyond the intended host is reduced. However, genetic transfer via sexual recombination among higher and lower forms could be an important consideration. Particularly genetic mobility via natural sexual transfer from crop plants (e.g., transgenic herbicide-resistant plants) to related weedy species must be evaluated. Chapters 8 and 9 explicitly describe measures to handle and prevent these transfers.

7.7. Selection Pressure

Selection pressure is a major variable in assessing the probability that new genetic material will persist, be expressed in a GMO, and be transferred to a nontarget population. Selection pressure is determined by a combination of factors, including encoded trait, the engineered gene, the compatibility of the recipient cell, and the value of the trait in its environment. Because the ecosystem is dynamic and the conditions in most cases are harsh and stringent (inadequate nutrients for growth, suboptimum temperature), selection pressure is critical, especially for gene and DNA replication.

7.8. Technologies for Measuring, Tracking, and Monitoring Gene Transfer

Central to the successful commercial introduction and implementation of new GMO products is the development of technologies for the economical, convenient, and effective tracking of the GMO or the engineered genes. Currently, two approaches are employed: selective markers and biochemical tests *(17)*.

The selective marker procedure is based on marking the chromosome with a resistance gene (mostly antibiotics or nutritional markers) that confers a selective advantage for identifying the GMO within a mixed population by isolating under selective conditions. For example, when exposed to selection pressure exerted by antibiotics, GMOs carrying a resistance gene survive against the nontargeted recipients and are easily detected. The antibiotic resistance strategy for tracking GMOs puts a resistance marker near the gene of interest. The antibiotic channel is used to recover any GMO that contains the resistance gene. Typically, the gene of interest is integrated with the resistance marker, thus obtaining and quantifying the resistant GMO cells. Use of a suitable marker for screening purposes should be considered because those that confer unintended resistance to whole families of antibiotics could lead to problems. The kanamycin-resistant gene, for example, could confer an ability to neutralize similar antibiotics.

Another selection approach relies on nutritional markers with a metabolite gene (e.g., those encoding for enzyme or metabolic activity) not associated with the host. Monsanto researchers have produced such a system via the insertion of genes for metabolizing sugar lactose into soil bacterium *(17)*.

Biochemical methods rely on gene probes made with recombinant technology. In most cases, a segment of DNA complimentary to the gene DNA sequence is employed as the probe. It is labeled with radioactivity or a dye that can be visibly detected with high precision. A gene probe will track a gene even if it is separated from the tightly linked selecting marker or inserted in an organism that cannot be isolated and cultured.

Precise identification is also possible with an antibody probe analysis derived from monoclonal antibody technology. But, quantifying gene transfer with this approach can require many tests and a large sample size. Processing a large volume of samples can be difficult and expensive.

With bacterial cells, the population is selected from the environment, their DNA is extracted, and the specific gene is probed and assayed. With plants, leaves or other segment are used; similarly, the DNA is extracted and tested for the presence or absence of the gene *(17)*.

The EPA universally accepts the use of the gene probe concept. In the case of soil organisms, extracted DNA must be purified to a certain degree to meet laboratory conditions for test precision and accuracy. For detection in water, enormous volumes must be processed to obtain an adequate sample size *(40)*. In a related biochemical test for plant GMO, the luciferase gene derived from fireflies has been used *(48)*. This gene codes for a light-emitting protein, luciferase, which when inserted and expressed in the plant, causes it to glow in dark. This probe offers a rapid method of tracking GMOs *(17)*. However, because it is highly energy intensive for the host cells to express, it does have limited use. With modern computer technology, other sophisticated imaging technologies are now being designed for tracking and monitoring GMOs (photography and X-ray film and three-dimensional computer analysis).

7.9. Biosafety Measures and Quality Control

Paramount to GMOs are biosafety assurance, quality assurance, and exposure risk analysis (Tables 4 and 5). These areas are legitimate concerns in the planned use of transgenics *(49)*. Several technologies are emerging that will prevent environmental transfer of genetic material by inactivating the GMO after its construction and application *(17)*. Also, advances in nucleotide chemistry have opened new opportunities for vector construction and custom gene designs that are immobilized and restricting, making an escape-proof host. Reducing or eliminating the use of a mobile plasmid and transposon in constructing GMOs helps minimize the risk of transfer. In some GMO construction, a "disarmed" transposon is typically engineered that can no longer separate or move independently from the chromosome in which it is inserted. This approach of crippling the vector has been used successfully in transferring genes from *B. thuringiensis* to *Pseudomonas fluorescens (31).*

7.10. Intellectual Property and Legal Issues Facing GMOs

The application of biotechnology for the benefit of humankind has been with us since farmers first began to "artificially" select plants, animals, and microbes to enhance their production *(50).* Only in the last century, by integrating Mendelian genetics, has it become possible to direct and focus the activities of biological matter to transcend the barrier between species and craft genes in forms that would have been impossible to bring about through natural breeding or would have required several generations *(51).* The advent of recombinant DNA technology has propelled a second, and even more radical, revolution in the bioengineering of plants, animals, and microbes *(22,48).*

Because of the mammoth investment and effort engaged to advance GMO technology, it is essential that inventors have some way of preventing others from unjustly reaping the fruits of their labor. Intellectual property law provides that protection by allowing those who can demonstrate the true invention to monopolize its commercial application temporarily, but not its every use. Not surprisingly, biotechnology-companies engaged in GMO-related technologies consider intellectual property protection the "lifeblood" or cornerstone of their industry *(51).*

Patent law was fundamental to the inventiveness of the Industrial Revolution; however, biological matter did not fit well into the categories that evolved during the Industrial Revolution. Legal changes generally have lagged behind the advances of biotechnology and therefore have resulted in controversies. There has been a long political debate *(40)* in the most recent move to allow patents on cloning. In addition to extending full patent protection to a living system, the current patent system poses the danger of overprotection by denying subsequent inventors the use of the genetic material to make further advances, in effect robbing the storehouse of knowledge of the "basic tools of science" needed for the technology to work.

For a patent to be granted for a GMO, it must be qualified as an "invention," and the subject matter claimed in the patent application must meet three basic requirements: novelty (new), utility (useful), and subject matter eligibility (composition of matter). However, GMOs claiming biological subject matter face tremendous obstacles in establishing eligibility of the inventions for patent protection. The US Patent and Trademark Office had rejected any claims on living matter, no matter how much ingenuity might have been involved in the creation of the new entity. The Supreme Court's landmark decision in *Diamond vs Chackrabartty* finally changed all that in 1980, when a patent application claiming a GMO with the unique ability to break down crude oil completely into a compound edible to aquatic life was granted *(17).* The usefulness of this organism in cleaning up oil spills withstood the legal appeal and paved the way to the marketing and commercialization of GMOs.

Since then, several new patents on GMOs have been issued to a number of biotechnology companies. Some of the newly patent protected inventions include development of transgenic plants for combatting pests; development of varieties of transgenic plants with different outputs (cotton, corn, soybean, alfalfa, rapeseed); enzymes; nitrogen-fixing bacteria; animal and human viral vaccines; growth hormones; antibiotics; human therapeutics; and medicines *(52)*.

8. Economic Challenges

The key economic issues facing the GMO industry include the impact of uncertain consumer adoption; midstream storage, processing, and transport of GMOs; reallocation of value within the food and pharmaceutical chains; and stiff competitive advantages between countries and trade flows *(53)*. Furthermore, financial analysts anticipate that GMO-derived crops and animals that are more productive will further place downward pressures on the price of food and drugs; thus, GMOs will leave potentially little gain for those producing compounds. Developments that signal the hurdles ahead are highlighted:

1. In 2003, an industrial working grouping in Canada established a new and rigorous set of guidelines for assessing the cost and benefits of introducing a new GMO in the supply chain.
2. Several US industry groups, including the Grocery Manufacturing and Food Marketing Institute, enforced stricter regulation of biopharma compliance.
3. The US National Organic Program has imposed limits on GMOs. Public opposition in Europe appears to be hardening. Europe plans to require labeling of products derived from GMOs *(54)*.

However, when the risk-to-benefit ratio is measured and evaluated, it seems that ample product opportunities exist for GMO based platforms. Successful integration and acceptance will require effective marketing strategies to gain product acceptance coupled with demonstrated leadership in addressing evolving issues facing GMO applications (Tables 4 and 5). In addition, industries will need to influence public opinion and decision makers in the global economy by carefully managing their brand identities to build trust on the part of regulatory agencies; in most cases, they must meet or exceed market performance. Furthermore, new and emerging technologies in agriculture, food, and medicine are anticipated to open new opportunities for GMOs in the future (Table 5).

9. About the Book

The chapters of this book are crafted to provide a how-to handbook and provide the layperson with up-to-date information on recent developments in GMOs and their future directions. The chapters address questions on how the structure of the industry is changing, which driving forces are shaping the economic future, which new technologies are emerging that will have a significant impact on the GMO industry, and which groups (public health, trade associations, etc.) are interested and which roles they play.

The book is divided into three sections, covering transgenic microbes, animals, and plants. In the first part of each section, fundamental aspects and the biological concepts are presented for understanding of how the DNA molecules are cut, aligned, and inserted into cells to result in the formation of a GMO. Recombinant techniques that lead to cloning of specific genes are discussed, and important laboratory procedures are explained. Furthermore, appropriate procedures to isolate desired genes to ensure that the genes function properly in the host GMO cells are presented. A section on optimizing the expression of cloned genes in the GMO is also included. How genes determine specific traits, are introduced into plant and animal GMOs, and are passed from one generation to the next are further examined. These topics provide the fundamental techniques and framework for understanding the application of GMO.

9.1. Microbial GMOs

The section on microbial GMOs explains the utility of microbial GMO species for carrying out specific biological processes. It further explains how genetically modified microbial cells are designed to act as biological engines and factories by altering metabolic pathways to create new and unique production methods for chemicals, pharmaceutical proteins, antibiotics, vaccines, and diagnostics tools (Tables 4 and 5). Because it is often assumed that growth of large quantities of GMOs is routine, the procedures and examples for managing large-scale production of recombinant proteins are discussed. The various parameters that must be controlled during the production and purification and isolation phases to ensure high yield from transgenic cells line *(55)* are emphasized.

9.2. Plant GMOs

Traditionally, novel high-yielding strains of different crops are developed through selective crosses and breeding. Now, this time-consuming process has been superseded. The development of methods for selective cloning and engineering of specific genes and the impact of gene silencing are discussed in this section. The section further highlights the various transgenic plants engineered for insecticide activity, viral infection tolerance, herbicide resistance, altered fruit ripening, altered flowering pigments, and improved nutritional quality. Also included is the use of plant cell suspensions and large-scale production of bioactive molecules.

9.3. Animal GMOs and Biotechnology

The various transgenic animal studies that are still in early discovery stages offer new challenges and hurdles to the prediction of their effect in the recipient species. The section on animal GMOs and biotechnology discusses the use of the mouse model to mimic cancer induction, cystic fibrosis, and Alzheimer's disease; the impact of animal cell lines; and the formation of transgenic animals for human therapeutics. Furthermore, the use of genetic technology and animal GMOs for additional diagnostic applications for human disease characterization and the limitless opportunities for efficient and cost-effective therapies is discussed.

For the paragraph "In each chapter," add a separate head to distinguish it from the discussion of animal GMOs and biotechnology in the preceding section.

In each chapter, all authors include a thorough explanation of GMO concepts and present them a format that is easy to read and understand. All chapters address the risks and rewards, the pros and cons, the environmental impacts and put into proper perspective the moral, bioethical, as well as legal issues facing introduction of GMOs without bias. All topics are articulated with the firm belief that education on GMOs is warranted and absolutely necessary so the public can acquire the facts needed to appreciate and evaluate this technology appropriately. The reader can rationally evaluate what is considered right and proper use of GMO technologies. The well-informed reader will have the option to embrace the new revolution of GMO biotechnology and its benefits without fear and with enhanced reassurance.

References

1. Pace, N. (1997) A molecular news of microbial diversity and biosphere. *Science* **276**, 734.
2. Drewes, J. (1993) Into the 21st century. *Biotechnology (NY)* **11**, S16–S20.
3. Hielman, I. (2000) Biotech regulation under attack. *Chem. Eng. News* May 22, p. 28.
4. Primrose, S. B. (1986) The use of genetically engineered microorganism in the production of drugs. *J. Appl. Bacteriol.* **61**, 99–116.
5. Raveria, P., Parker, I., and Apsual, M. (1996). Can we use experimental methods in predicting the GMO. *Ecology* **77**, 1670.
6. Conference of the Parties to the Convention on Biological Diversity, Cartagena Protocol on Biosafety, Dec. 19, 2003. Available at www.biodiv.org/biosafety/default.aspx. Accessed January 20, 2004.
7. Snow, A. and Moram-Palma, P. (1997) Commercialization of transgenic plants: potential ecological risk. *Bioscience* **47**, 86s.
8. Budd, R. (1993). 100 years of biotechnology. *Biotechnology (NY)* **11**, S14–S15.
9. Budd. R. (1991) Biotechnology in the 20th century. *Soc. Stud. Sci.* **21**, 415–457.
10. Anderson, W. (1992) Human gene therapy. *Science* **256**, 808–813.
11. Bailey, J. E. (1991) Towards a science of metabolic engineering. *Science* **272**, 1668–1675.
12. Wilcox, G. and Studnicka, G. (1998) Expression of foreign proteins in microbes. *Biotechnol. Appl. Biochem.* **10**, 500–509.
13. Johnson, I. S. (1983) Human insulin from recombinant DNA technology. *Science* **219**, 632–637.
14. Ratafia, M. (1987) Worldwide opportunities in genetically engineered vaccines. *Biotechnology* **5**, 1155–1158.
15. Pastan, I. and FitzGerald, D. (1991) Recombinant toxin for cancer treatment. *Science* **254**, 1173–1177.
16. Hodgson, J. (1991) Making monoclonal antibodies in microbes. *Biotechnology (NY)* **9**, 421–425.
17. Meiri, H. and Altman, A. (1998) Agricultural and agricultural biotechnology development towards the 21st century. In *Agricultural Biotechnology* (Altman, A., ed.). Marcel Dekker, New York, pp. 1–17.
18. Wilcox, G. and Studnicka, G. (1988) Expression of foreign proteins in microbes. *Biotech Appl. Biochem.* **10**, 500–509.
19. Murray, F. and Gaiannakas, K. (2001) Agricultural biotechnology and industry structure. *AgBio Forum* **4**, 137–151.
20. James, C. (2002) Global status of transgenic crops. *ISAAA Briefs No. 27* Available at www.Isaaa.org. Accessed January 19, 2004.
21. Raybould, A. F. (1999) Transgenic and agriculture going with the flow. *Trends Plant Sci.* **4,** 247.
22. Talbot, W. (1990) Manufacture of biopharmaceutical proteins by mammalian cell culture system. *Biotechnol. Adv.* **8**, 729–739.
23. Jaenisch, R. (1988) Transgenic animals. *Science* **240**, 1468–1474.
24. Duvick, N. D. (1999) How much caution in the fields. *Science* **286**, 418.
25. Liu, M. A. (1998) Vaccine development. *Nat. Med.* **4**, 515.
26 Jefrey, M. (1999) Immunotherapy of cancer. *Ann. NY Acad. Sci.* **866**, 67.
27. UN Environment Program. (1995) *UNEP International Technological Guidelines for Safety in Biotechnology*. Available at www.unep.org/unep/program/natres/biodiv/irp/unepgds.htm. Accessed January 20, 2004.

28. Qaim, M and Zilberman, D. (2003). Yield effects of genetically modified crops in developing countries. *Science* **299**, 900–1001.
29. Potrykus, I. (1991) Gene transfers in plants: an assessment. *Biotechnology (NY)* **8**, 535–542.
30. Sprent, J. (1986) Benefits of *Rhizobium* to agriculture. *Trends Biotechnol.* **4**, 124–129.
31. Meesun, R. and Warren, G. (1989) Insect control with genetically engineered crops. *Annu. Rev. Entomol.* **34**, 372–381.
32. Pollack, A. (2000) Opportunity for agricultural biotech. *Science* **288**, 615.
33. Kelly, P. D. (1992) Are isolated genes useful? *Biotechnology (NY)* **10**, 52–55.
34. Waldmann, T. (1991) Monoclonal antibodies in diagnosis and therapy. *Science* **252**, 1657–1662.
35. Fuxa, J. (1991) Insect control with baculovirus. *Biotechnol. Adv.* **9**, 425–442.
36. Graham, F. L. (1990) Adenoviruses as expression vectors and recombinant vaccines. *Trends Biotechnol.* **8**, 85–87.
37. Drahos, D. J. (1991) Field-testing of genetically engineered microbes. *Biotechnol. Adv.* **9**, 157–171.
38. Glick, B. and Skof, Y. (1986) Environmental implications of recombinant DNA technology. *Biotechnol. Adv.* **4**, 261–277.
39. Mulligan, R. C. (1993) The basic science of gene therapy. *Science* **260**, 926–936.
40. First, N., Schell, J., and Vasil, I. (1998) Prospects and limitations of agricultural biotechnologies: an update. In *Agricultural Biotechnology* (Altman, A., ed.). Marcel Dekker, New York, pp. 743–748.
41. Chen, T. and Powers, D. (1990) Transgenic fish. *Trends Biotechnol.* **8**, 209–215.
42. Carver, A. (1996) Transgenic on trial. *Scrip. Mag.* **11**, 51–53.
43. Jayraman, K. S. (1982) Foreign labs shut. *Nature* **296**, 104.
44. Wadman, M. (1999) US group sues over approval of *Bt* crops. *Nature* **397**, 636.
45. Macer, D. (1994) Perception of risks and benefits of vitro fertilization genetic engineering and biotechnology. *Soc. Science. Med.* **38**, 23–33.
46. Losey, J., Rayyor, L., and Carton, M. (1999) Transgenic pollen harms monarch butterfly. *Nature* **399**, 214.
47. Morrison, M. (1996) Do ruminal bacteria exchange genetic material? *Dairy Sci.* **79**, 1476.
48. Larkin, S. (1996) Immunoassay usage in ag and food safety sectors. *Genet. Eng. News* **16**, 12–13.
49. Clarke, T. (2003) Corn could make cotton pest *Bt* resistant. *Nature Sci. Update Dec 4, 2002.* Available at: www.nature.com/nsu/021202/021202-2.html. Accessed January 21, 2004.
50. Jaenisch, R. (1988) Transgenic animals. *Science* **240**, 1468–1474.
51. Barton, J. H. (1991) Patenting life. *Sci. Am.* **264**, 40–46.
52. Gregory, G., Raiser, G., and Small, A. (2003) Agricultural Biotechnology's Complementary Intellectual Assets Review of Economics and Statistics. Rev. Econ. Stat. **85**, 349–363.
53. Dunhill, J. and Paul E. (1990) Impact of genetically modified crops. *Agriculture* **9**, 280.
54. Kestrel, D., Taylor, M., Maryanski, J., Flamm, E., and Kahl, L. (1992) The safety of foods developed by biotechnology. *Science* **256**, 1747–1749.
55. Van Brunt, J. (1986) Fermentation economics. *Biotechnology (NY)* **4**, 395–401.

II

Microbial GMOs

2

Genetically Modified Microorganisms

Development and Applications

Lei Han

1. Introduction

With the development of recombinant deoxyribonucleic acid (DNA) technology, the metabolic potentials of microorganisms are being explored and harnessed in a variety of new ways. Today, genetically modified microorganisms (GMMs) have found applications in human health, agriculture, and bioremediation and in industries such as food, paper, and textiles. Genetic engineering offers the advantages over traditional methods of increasing molecular diversity and improving chemical selectivity. In addition, genetic engineering offers sufficient supplies of desired products, cheaper product production, and safe handling of otherwise dangerous agents. This chapter delineates several molecular tools and strategies to engineer microorganisms; the advantages and limitations of the methods are addressed. The final part of this chapter reviews and evaluates several applications of GMMs currently employed in commercial ventures.

2. Molecular Tools for Genetic Engineering of Microorganisms

A number of molecular tools are needed to manipulate microorganisms for the expression of desired traits. These include (1) gene transfer methods to deliver the selected genes into desired hosts; (2) cloning vectors; (3) promoters to control the expression of the desired genes; and (4) selectable marker genes to identify recombinant microorganisms.

2.1. Gene Transfer Methods

Table 1 *(1–23)* lists the gene transfer methods commonly used to introduce DNA into commercially important microorganisms. The most frequently used method is *transformation*. In this process, uptake of plasmid DNA by recipient microorganisms is accomplished when they are in a physiological stage of competence, which usually occurs at a specific growth stage *(24)*. However, DNA uptake based on naturally occurring competence is usually inefficient. Competence can be induced by treating bacterial cells with chemicals to facilitate DNA uptake. For *Escherichia coli*, an organism used commonly as a cloning host and a "bioreactor" for the commercial production

From: *The GMO Handbook: Genetically Modified Animals, Microbes, and Plants in Biotechnology*
Edited by: S. R. Parekh © Humana Press Inc., Totowa, NJ

Table 1
Gene Transfer Methods Used With Several Commercially Important Microorganisms

Type of Organism	Industrial Applications	Gene Transfer Methods	Reference
Aspergillus	Food fermentations	Protoplast transformation	*1*
		Electroporation	*1*
		Biolistic transformation	*1*
Yeasts	Food and beverage fermentations	Protoplast transformation	*2*
		Electroporation	*2*
Bacillus	Industrial enzymes	Transformation of competent cells	*3*
	Fine chemicals	Protoplast transformation	*4*
	Antibiotics	Electroporation	*5,6*
	Insecticides		
Corynebacterium	Amino acids	Protoplast transformation	*7*
		Conjugation	*8*
		Electroporation	*9*
Escherichia coli	Therapeutic protein production	Transformation of competent cells	*10*
	Biodegradable plastics		
Lactic acid bacteria	Food fermentations	Electroporation	*11*
	Organic acids	Protoplast transformation	*11*
Pseudomonas	Plant biological control agents	Electroporation	*12,13*
	Bioremediation	Conjugation	*14*
Streptomyces	Antibiotics, antitumor and antiparasitic agents	Protoplast transformation	*15*
		Electroporation	*16–18*
	Herbicides	Conjugation	*19–21*
			22,23

of numerous therapeutic proteins, the uptake of plasmid DNA is achieved when cells are first treated with calcium chloride or rubidium chloride *(10)*.

For many microorganisms, such as the antibiotic producing *Streptomyces*, transformation of plasmid DNA is a more complicated process. For these organisms, transformation involves preparation of protoplasts using lysozyme to remove most of the cell wall. Protoplasts are mixed with plasmid DNA in the presence of polyethylene glycol to promote the uptake of DNA. Growth medium, growth phase, ionic composition of transformation buffers, and polyethylene glycol molecular weight, concentration, and treatment time are variables that must be studied to identify the optimum conditions for protoplast formation and regeneration.

Electroporation is an alternative method to transform DNA into microorganisms. This method, originally used to transform eukaryotic cells, relies on brief high-voltage pulses to make recipient cells electrocompetent *(25)*. Transient pores are formed in the cell membrane as a result of an electroshock, thereby allowing DNA uptake. Growth phase, cell density, growth medium, and electroporation parameters must be optimized to achieve desirable efficiency. The main advantage of this method is that it bypasses the need to develop conditions for protoplast formation and regeneration of cell wall. Electroporation is often used when the efficiency of protoplast transformation is insufficient or ineffective. Several reports have documented the application of this method to industrially important *Streptomyces (16–18)*, *Corynebacterium (9)*, and *Bacillus (5,6)*. Electroporation is also the primary method of choice for transferring DNA into lactic acid bacteria *(11,26,27)*. In addition to using purified DNA for electroporation procedures, methods have been developed to transfer DNA directly from DNA-harboring cells into a recipient without DNA isolation *(28)*.

Conjugation is another method used to introduce plasmid DNA into microorganisms. This method involves a donor strain that contains both the gene of interest and the origin of transfer (*oriT*) on a plasmid and the genes encoding transfer functions on the chromosome *(29)*. Upon brief contact between donor and recipient, DNA transfer occurs. After conjugation takes place, donor cells are eliminated with an antibiotic to which the recipient cells are resistant. Recipient cells containing the transferred plasmid are identified based on the selectable marker gene carried by the plasmid. One advantage of this method is that it does not rely on the development of procedures for protoplast formation and regeneration of cell wall. In addition, this method offers the possibility of bypassing restriction barriers by transferring single-stranded plasmid DNA *(21)*. Introducing DNA by conjugation from donor *E. coli* has proven useful with *Streptomyces* and *Corynebacterium (8,19–23)*.

2.2. Vectors

Selection of a cloning vector to carry out genetic modifications depends on the choice of the gene transfer method, the desired outcome of the modification, and the application of the modified microorganism. Several classes of vectors exist, and the choice of which to use must be made carefully. Replicating vectors of high or low copy numbers are commonly used to express the desired genes in heterologous hosts for manufacturing expressed proteins. Replicating vectors are also used to increase the dosage of the rate-limiting gene of a biosynthetic pathway, such as that used for an amino acid, to enhance the production of the metabolite. Cosmid and bacterial artificial chromosome vectors, which accept DNA fragments as large as 100 kb, are necessary when cloning a large piece of DNA into a heterologous host for manipulation and high-level metabolite production *(30,31)*. Conjugal vectors facilitate gene transfer from an easily manipulated organism such as *E. coli* into a desired organism that is usually more difficult to transform. Gene replacement vectors allow stable integration of the gene of interest. Food-grade vectors differ from the conventional cloning vectors in that they do not carry antibiotic resistance marker genes.

Special consideration must be given when constructing GMMs for industrial applications. If a GMM is to be released into the environment as a biological control agent, conjugal vectors should be avoided to prevent the horizontal transfer of the vectors and the genes into indigenous microorganisms. If a GMM is used as a starter culture for food fermentation, conjugal vectors should also be avoided *(32)*, and food-grade vectors should be developed and used for genetic manipulation *(27,33)*.

2.3. Promoters

A *promoter* is a segment of DNA that regulates the expression of the gene under its control. Constitutive promoters are continuously active; inducible promoters become activated only when certain conditions, such as the presence of an inducer, are met. It is important to select an appropriate promoter to optimize the expression of the target genes for desired timing and level of expression. A strong constitutive promoter is used when continuous expression of a target gene is desirable. For example, constitutive promoters were used to drive the expression of selectable marker genes to achieve complete elimination of nontransformed cells *(34)*. However, inducible promoters are often chosen when it is necessary to control the timing of target gene expression. This is especially true when expressing foreign genes, including toxin genes in *E. coli*.

The most commonly used inducible promoter for target gene expression in *E. coli*, the *lac* promoter, is turned on when the nonhydrolyzable lactose analog isopropyl-β-D-1-thiogalactopyranoside (IPTG) is added to the growing culture *(35,36)*. This promoter is relatively weak and therefore is often suitable for expressing genes encoding toxic proteins *(35)*. Promoters dependent on IPTG induction are usually undesirable for large-scale production of therapeutic proteins because of the high cost of the inducer and potential toxicity *(37)*.

Alternative promoters have also been developed for this purpose. The arabinose promoter is induced upon addition of L-arabinose to the medium *(35,38)*. L-Arabinose is a good alternative to IPTG because it is a less-expensive compound and thus is cost-effective in large-scale fermentations. The cold-shock promoter based on the *cspA* gene of *E. coli* is induced upon temperature downshift *(39)*. However, this promoter becomes repressed within 2 hr after temperature downshift and is therefore unable to achieve high-level accumulation of desired proteins. This problem is overcome by using an *E. coli* strain carrying a null mutation in the *rbfA* gene and thereby allowing continuous expression of target genes *(40)*. The *phoA* and *trp* promoters are turned upon with phosphate or tryptophan depletion, respectively, in the medium *(35,36)*. These inducible expression systems are inexpensive to implement and therefore are worth consideration for industrial applications. Promoters controlled by pH, dissolved oxygen concentration, or osmolarity in the medium are possible attractive alternatives *(36,41)*.

2.4. Selectable Marker Genes

Selectable marker genes, which often encode proteins conferring resistance to antibiotics, are an important part of cloning vectors and are required for identification of transformed cells. Application of selection pressure is necessary because the number of transformed cells is often significantly less than the number of nontransformed cells. Transformed cells are identified using a toxic concentration of the selection agent to inhibit the growth of the nontransformed cells. Usually, high-level expression of a selectable marker gene is necessary to ensure complete elimination of nontransformed cells.

Antibiotic resistance marker genes, although routinely used, are not generally acceptable for the construction of recombinant organisms such as lactic acid bacteria and yeasts used for food fermentation *(34,42)*. For lactic acid bacteria, alternative selection systems based on plasmid-linked properties of the organism itself, including lactose metabolism, proteolytic activity, DNA synthesis, and bacteriocin resistance, have been developed and incorporated into cloning vectors *(11,42)*. One problem associated with these selection systems is that they tend to give more nontransformed background cells than the antibiotic resistance marker gene-based selection systems *(11)*.

For constructing recombinant yeast acceptable for food fermentations, a number of selection systems based on yeast genes instead of heterologous antibiotic resistance marker genes were developed *(43)*. One such system is based on the *YAP1* gene, which is responsible for stress adaptation in yeast *(34)*. Overexpression of this gene under the control of the constitutive yeast gene (*PGK*) promoter confers resistance in cells to the fatty acid synthesis inhibitor cerulenin and the protein synthesis inhibitor cycloheximide. An added advantage of this dual selection system is that it almost completely eliminates nontransformed background cells.

3. Strategies for Genetic Engineering of Microorganisms

Several strategies have been developed to create GMMs for desired traits. They include (1) disruption or complete removal of the target gene or pathway; (2) overexpression of the target gene in its native host or in a heterologous host; and (3) alteration of gene sequence, and thereby the amino acid sequence of the corresponding protein.

3.1. Disruption of Undesirable Gene Functions

Disruption of a gene function can be achieved by cloning a DNA fragment internal to the target gene into a suitable vector. Upon introducing the recombinant plasmid into the host organism, the internal fragment of the gene, along with the vector, is integrated into the host chromosome via single-crossover recombination. The integration results in the formation of two incomplete copies of the same gene separated by the inserted vector sequence, thereby disrupting the function of the target gene. However, such integration is unstable because of the presence of identical DNA sequences on either side of the vector. The recombinant strain often undergoes a second recombination that will "loop" out the recombinant plasmid from the chromosome, thus restoring normal function of the target gene.

To create a stable recombinant strain blocked in the unwanted gene function, a gene replacement plasmid carrying two selectable marker genes is required. The first selectable marker gene, originating from the cloning vector, is used to select the transformed cells, whereas the second selectable marker gene is inserted into the target gene. The recombinant plasmid is introduced into the host organism, followed by the selection of transformed cells based on the first selectable marker gene. Upon double-crossover recombination, the second selectable marker gene, now inserted into the target gene on the host chromosome, disrupts the sequence of the target gene and destroys gene function. The recombinant strain is selected based on its resistance to the second selectable marker gene product and its sensitivity to the first selectable marker gene product.

Another approach to disrupting gene functions relies on antisense technology. The technology is based on antisense ribonucleic acid (RNA) or DNA sequences that are complementary to the messenger RNAs (mRNAs) of the target genes *(44)*. The binding of an antisense molecule to its complementary mRNA results in the formation of a duplex RNA structure. The activity of the target gene is inhibited by the duplex RNA structure because of (1) an inaccessible ribosomal-binding site that prevents translation; (2) rapid degradation of mRNA; or (3) premature termination that prevents transcription *(45)*. Antisense technology has been used to downregulate target gene activities in bacteria *(45)*. The main advantages of this approach are rapid implementation and simultaneous downregulation of multiple target genes. In addition, this method is ideal for downregulation of primary metabolic gene activities without creation of lethal events.

3.2. Overexpression of Desired Genes

High-level expression of a target gene may be achieved by employing a high copy number vector. Eggeling et al. *(46)* constructed several *Corynebacterium glutamicum* recombinant strains containing increased copy numbers of *dapA*, a gene encoding dihydrodipicolinate synthase at the branch point of the lysine and methionine/threonine pathway. Lysine titer was higher in the recombinant strain containing one extra copy of *dapA* than the wild-type strain and was highest in the recombinant strain containing the highest copy number of the same gene.

However, gene expression systems based on high copy number vectors have a number of drawbacks. One is the segregational instability of recombinant plasmids, which results in the loss of recombinant plasmids and therefore loss of the desired traits. For example, expression of the *Bacillus thuringiensis* (*Bt*) toxin gene from a high copy number vector in *Pseudomonas fluorescens* was undetectable because of plasmid instability *(47)*. Segregational instability of plasmids is usually resolved by maintaining recombinant strains under selective pressures, usually by means of antibiotics. However, concerns about the use, release, and horizontal transfer of antibiotic resistance marker genes suggest that other means of maintaining plasmid stability need to be developed.

Baneyx *(35)* outlined a few options to achieve this goal. One method relies on creating a mutation in a critical chromosomal gene that is complemented with a functional copy of the same gene on the plasmid. As long as the plasmid housing the critical gene is present, the recombinant strain will survive. Major disadvantages of this method are the need to create a mutation in an essential gene of the host organism, the need to develop a specific growth medium, and the need to introduce an additional plasmid-encoded gene to complement the deficiency *(35)*.

Another concern about the use of high copy number vectors for high-level protein production in bacterial cells, especially in *E. coli*, is the formation of insoluble protein aggregates known as *inclusion bodies*. Inclusion bodies are biologically inactive because of protein misfolding, which is a consequence of rapid intracellular protein accumulation *(35,48)*. Although methods exist to isolate and renature inclusion bodies *(49–51)*, these systems are often inefficient and add steps in the purification of active proteins. Also, in the process of renaturation of proteins, a significant percentage of the proteins remains denatured and inactive *(52)*.

In most cases, the goal of protein production is to achieve acceptable levels of accumulation of desired proteins that retain biological activity. To achieve this goal, the rate of recombinant protein synthesis needs to be optimized *(36)*. An effective method to accomplish this is by lowering the growth temperature or by altering medium composition *(53)*. A molecular approach to maximizing active protein production is to coexpress the genes that facilitate protein folding and improve transportation of the recombinant protein out of the cell to decrease the intracellular concentration of the protein *(36,54,55)*. Another molecular approach is to carefully select a vector or promoter system that does not overwhelm the cell's capacity to produce active proteins *(56)*. However, both methods result in GMMs that harbor plasmids.

A different approach, which ensures expression of target genes at desired levels and avoids plasmid segregational instability and production of inclusion bodies, is to integrate target genes into the host's chromosome *(57)*. Although integration of a single copy of target genes may not be enough to achieve the desired level of protein production, integration of multiple copies of target genes has yielded very encouraging results *(58–60)*.

3.3. Improving Protein Properties

Site-directed mutagenesis and DNA shuffling are two powerful technologies that alter gene sequence in vitro to produce proteins that have improved characteristics. Site-directed mutagenesis is a technique used to change one or more specific nucleotides within a cloned gene to create an altered form of a protein via change in a specific amino acid *(61)*. This technique has been used successfully to identify catalytically important residues in new proteins. Two examples include the identification of catalytically essential residues in the *Aspergillus oryzae* Taka-amylase A *(62)* and the identification of the active site residue in the *Clostridium thermosulfurogenes* xylose isomerase *(63)*.

DNA shuffling, a technology introduced in 1994, is based on error-prone polymerase chain reaction and random recombination of DNA fragments *(64)*. DNA shuffling may involve a single gene or multiple genes of the same family. Family gene shuffling is more powerful than shuffling of single genes because it takes advantage of the natural diversity that already exists within homologous genes *(65,66)*.

Site-directed mutagenesis and DNA shuffling have been applied successfully for the improvement of numerous commercially important enzymes, notably the enzymes used in laundry detergents *(67,68)*. The goals sought commonly include altered substrate specificity, improved enzyme activity under broad washing conditions such as pH and temperature, enhanced resistance to detergent additives such as bleach, and longer shelf life. To improve enzyme characteristics using site-directed mutagenesis, prior knowledge regarding the enzyme, such as its active site and substrate-binding site, is required.

The advantage of site-directed mutagenesis is that only a limited number of recombinants will be screened. DNA shuffling, on the other hand, does not require specific knowledge about the enzymes of interest and can create new variants containing multiple beneficial mutations in the gene sequence for maximum benefit. High-throughput screening assays for identifying desired recombinants are necessary for using this method.

3.4. Approaches to Enhancing Product Yield

Usually, production levels of metabolites of commercial value, such as amino acids, vitamins, and antibiotics, by unaltered natural-producing microorganisms are quite low. Enhancing metabolite yield is therefore essential for meeting the product demands and for maintaining an economically viable process. Several approaches have proven successful in increasing the production of the desired products through manipulating the producing microorganisms. They include (1) overcoming rate-limiting steps; (2) eliminating feedback regulation; (3) manipulating regulatory genes; (4) perturbing central metabolism; (5) removing competing pathways; and (6) enhancing product transport.

3.4.1. Overcoming Rate-Limiting Steps

Rate-limiting steps refer to the steps in a biosynthetic pathway that restrain the flow of intermediates and thereby limit the overall production of the final product. The classical approach to identify rate-limiting steps is to feed pathway intermediates to the producing strain. If the intermediate is not converted to the final product, assuming it is transported into the cell, one or more steps between the intermediate and the final product are limiting. Once rate-limiting steps are identified, modification of the genes that encode the limiting pathway enzymes by either increasing gene dosage (amplification) or placing the gene under the control of a strong promoter often leads to relief of the bottleneck.

Ikeda et al. *(69)* successfully increased phenylalanine production in *C. glutamicum* by increasing the copy numbers of the genes encoding the rate-limiting enzymes in the phenylalanine pathway. Kennedy and Turner *(70)* increased penicillin production in *Aspergillus nidulans* by replacing the native promoter of the rate-limiting δ-(L-α-aminoadipyl)-L-cysteinyl-D-valine synthase gene with a strong inducible ethanol dehydrogenase promoter.

3.4.2. Eliminating Feedback Regulation

One mechanism by which microorganisms control the production of essential metabolites, such as amino acids, is feedback regulation *(71)*. Genetic engineering offers a promising solution to overcome feedback regulation and satisfy increasing demand. An excellent example is isoleucine, an amino acid produced by *C. glutamicum*. In *C. glutamicum*, threonine dehydratase, the first committed enzyme in the isoleucine biosynthetic pathway, is sensitive to inhibition by the end product isoleucine. Isoleucine production was increased by relieving the feedback inhibition either by amplifying the native gene (*ilvA*) encoding the enzyme or by expressing the *E. coli* gene *tcdB*, which is insensitive to isoleucine feedback inhibition, in *C. glutamicum (72)*.

3.4.3. Manipulating Transcription Regulatory Genes

Manipulation of transcription regulatory genes is another way of achieving increased production. There are two types of regulatory genes, categorized by the effect they have on the genes they control. Positive regulators turn on the expression of the genes they control; negative regulators repress the expression of the genes under their control. Regulatory genes control many biological processes, including antibiotic biosynthesis. Genes involved in antibiotic biosynthesis are often linked to form a cluster that usually contains pathway-specific regulatory genes *(73)*. Such regulatory genes encode proteins that directly bind to the promoter region of the biosynthetic gene for that antibiotic. This binding results in either an increase or decrease in the expression levels of the biosynthetic gene, which in turn either boost or hamper the production of the antibiotic. Manipulation of regulatory genes has proven to be rewarding in achieving increased production of the desired products *(73)*. Pathway-specific regulators are often controlled by higher-level regulators, called *global regulators*, that coordinate many metabolic activities. Manipulation of global regulators can be very fruitful, although their identification and the understanding of their functions is a time-consuming and labor-intensive task. Tatarko and Romeo *(74)* engineered a high phenylalanine-producing *E. coli* strain by disrupting the global regulatory gene *csrA*. Disruption of this gene enhanced gluconeogenesis and decreased glycolysis, which in turn resulted in increased accumulation of phosphoenoylpyruvate, one of the two starting molecules for phenylalanine biosynthesis.

3.4.4. Perturbing Central Metabolism

Central metabolism provides primary metabolites and energy to support the survival of microorganisms. In addition, central metabolism, under special growth conditions, contributes a small fraction of primary metabolites to pathways that produce secondary metabolites such as antibiotics. Perturbing central metabolism is complicated. Complications may arise from situations in which (1) disturbing the balance of metabolic activities is detrimental to the host or (2) the host is resistant to unnatural alternations imposed on it.

Despite these difficulties and complications, manipulation of central metabolism has proven to be rewarding. Butler et al. *(75)* engineered a superior actinorhodin-producing strain by deleting the genes responsible for either of the first two steps in the pentose phosphate pathway. Peters-Wendisch et al. *(76)* constructed a lysine-overproducing strain by expressing the *pyc* gene encoding pyruvate carboxylase at high levels. Apparently, overexpression of the gene increased the availability of oxaloacetate, the precursor for the starting material, aspartate, of the lysine pathway. Increased availability of oxaloacetate in turn enhanced lysine production.

3.4.5. Removing Competing Pathways

Competing pathways utilize the precursor or intermediate of the desired pathway. Two approaches may be used to remove competing pathways: (1) deletion of the entire competing pathway or (2) knocking out the function of the first gene of the competing pathway. Many antibiotic producers accumulate several compounds structurally related to the antibiotic of interest. Because these compounds do not usually possess useful biological activity, their synthesis is considered a waste of cellular energy, precursors, and intermediates. In addition, the presence of these compounds may complicate downstream purification. In this case, knocking out the pathways leading to the synthesis of the unwanted derivatives would simplify downstream processing and, more importantly, would redirect the precursor or intermediates toward the synthesis of the desired product. Backman et al. *(77)* engineered an elegant system in which tyrosine synthesis, competing for the common intermediate used for phenylalanine production, is interrupted during the phenylalanine production phase.

3.4.6. Enhancing Product Transport

Enhancing product transport across the cytoplasmic membrane is another commonly used approach. Most microorganisms producing metabolites of commercial interest have means of transporting the products across the membrane. Reinscheid et al. *(78)* showed that the intracellular concentration of threonine in *C. glutamicum* is proportional to the copy number of the deregulated threonine pathway genes and extracellular concentrations of the same amino acid remained unchanged, indicating the importance of optimizing export machinery. Existence of an active transport system in *C. glutamicum* involved in lysine secretion was identified through cloning and characterization of the *lysE* gene *(79)*. The *lysE* knockout mutant lost the ability to secrete lysine, and the *lysE*-overexpressed mutant was able to export the amino acid at an accelerated rate, thereby increasing extracellular lysine concentration.

3.5. Genomic Approaches

New technologies and novel approaches are revolutionizing the way microorganisms are engineered. Rapid advancements in sequencing technologies make it possible to unlock the genetic code of a target organism in a relatively short period of time. This wealth of information offers clues to the genes directly involved in the synthesis of the product of commercial interest. The information also provides insight into the metabolic potential of the organism, such as pathways supplying precursors or cofactors necessary for the production of the metabolite of interest, pathways competing for valuable precursors or intermediates, and global regulatory networks that may control metabolite production.

Genomic information is beginning to modernize the engineering of recombinant microorganisms. Ohnishi et al. *(80)* reported on the genetic engineering of a hyper-lysine-producing strain of *C. glutamicum*. The group identified the mutations carried by the lysine-producing strain derived from the conventional mutagenesis and random screening approach through comparative genomic analysis of the *C. glutamicum* wild-type strain and the lysine-producing strain. Introducing the mutations into the wild-type strain produced a mutant that was better than the classically derived lysine-producing strain in both lysine production and growth.

4. Applications of GMM-Derived Products

4.1. Human Health

4.1.1. Recombinant Therapeutic Proteins

Several proteins, such as insulin, interferons (IFNs), and interleukins, are now produced by GMMs for therapeutic use. The traditional method of supplying these proteins to patients requires purification of the proteins from cells, tissues, or organs of humans, cows, or pigs. Because it was impractical to treat diabetes with human insulin from cadaver sources, cow and pig insulin, which are somewhat different from human insulin, were substituted *(81)*. The problems with obtaining the proteins directly from animal sources included the limited supply and potential immunological responses *(81)*.

Limited supply translates into higher cost for the medication. Further concerns arose that therapeutic proteins of animal origin may be contaminated with viruses or other toxic substances *(81)*.

These problems can sometimes be avoided by producing the proteins in microorganisms. Human insulin, the first recombinant therapeutic protein approved by the Food and Drug Administration (FDA) in 1982, was produced by genetically engineered *E. coli* containing the human insulin genes *(81)*. Human growth hormone, approved by the FDA in 1985, was produced by a modified *E. coli* strain containing the native human growth hormone gene *(82)*. Recombinant IFN γ, under the trade name Actimmune® in North America or Imukin® in Europe, was jointly developed by Genentech and Boehringer Ingelheim *(83)*.

The gene encoding IFN γ was introduced into *E. coli* under the control of the tryptophan promoter and operator cassette. Therefore, IFN γ production is repressed in the presence of tryptophan in the medium during the first phase of fermentation and becomes derepressed when a tryptophan analog is added to the medium during the second phase of fermentation *(83)*. With this system, recombinant IFN γ production is regulated to ensure adequate accumulation of cell mass before production begins. Other examples of recombinant therapeutic proteins, such as IFN α-2a *(84)*, IFN β-1b *(85)*, and granulocyte-macrophage colony-stimulating factor *(86)*, are also produced using recombinant *E. coli* strains as production factories. Production of these therapeutic proteins in a fast-growing and easily manipulated organism ensures sufficient supply, free of contamination, reduced cost, and safe and consistent production.

4.1.2. Recombinant Vaccine

Hepatitis B is a serious disease caused by hepatitis B virus that attacks the liver. The first vaccine against hepatitis B was prepared with the purified hepatitis B surface antigen (HBsAg) extracted from blood samples of infected individuals *(87)*. This process is unsafe because of the risk in handling the infectious agent and expensive because of the required animal testing. In addition, the vaccine may be contaminated with other infectious agents. The second generation of hepatitis B vaccine was produced by expressing the gene coding for hepatitis B surface antigen in *Saccharomyces cerevisiae (87)*, common baker's yeast. The recombinant vaccine, under the trade name Engerix®-B, is identical to the first-generation hepatitis B vaccine and is produced safely, consistently, and economically.

4.2. Animal Health

4.2.1. Recombinant Proteins

Proteins benefiting animal health are also produced by recombinant microorganisms. Bovine somatotropin (bST), a natural protein hormone produced in the pituitary glands of cattle, regulates both animal growth and milk production in lactating dairy cows. Injection of pituitary extracts into lactating cows boosts milk production; however, pituitary glands from as many as 25 cows are needed to provide sufficient bST to supplement 1 cow for 1 day (http://www.monsantodairy.com/about/history).

To increase bST production for commercial use, the gene encoding bST was expressed in *E. coli*. The recombinant bST was approved by the FDA in 1994 and is marketed under the trade name Posilac™. The recombinant protein has the same chemical structure and biological activity as the native bST *(88)*. Studies showed that lactating dairy cows supplemented with recombinant bST produced 10–15% more milk. The same method has since been applied to produce a variety of animal growth hormones for other animals, including sheep, pig, buffalo, and goat for both meat and milk production *(89)*.

Another protein produced by a GMM and used to benefit animal health is phytase. Phytase catalyzes the release of phosphate from phytate, the primary storage form of phosphorus in plants *(90)*. The enzyme is present in ruminants, but is absent or nearly absent in nonruminants such as poultry and swine. Therefore, nonruminants are unable to obtain phosphate, an essential nutrient, from phytate present in the feed of plant sources. The lack of phosphate can be corrected by supplementing rock phosphate in animal feed. However, this method generates excessive amounts of released phosphate, both from the phosphate supplement and from the unused phytate, which is excreted in the animal's manure *(91)*. Excessive amounts of phosphate released to the environment contribute significantly to water pollution *(90,92)*.

One solution to the problem is to supplement phytase in animal feed if the enzyme can be produced economically. To achieve this goal, the gene encoding phytase was isolated from its natural host, *Aspergillus niger*, and placed under the control of the constitutively expressed glucoamylase gene promoter for high-level expression *(93)*. The recombinant gene cassette was introduced into an industrial strain of *A. niger* for phytase production and was found to integrate randomly at multiple locations in the host genome. With this system, phytase is produced in large quantities and is suitable for commercial application. Studies have shown that use of the recombinant phytase reduces the phosphate level in the feed by 20%. In addition, the level of phosphate in manure is reduced by 25–30% *(93)*.

4.2.2. Recombinant Vaccine to Eradicate Rabies

Rabies, a viral disease encountered by humans and other mammals, leads to more than 35,000 human deaths and several million animal deaths worldwide every year *(94,95)*. The rabies virus reservoir is primarily in wild animals, including fox, skunks, raccoons, wolves, mongooses, and raccoon-dogs. Humans normally become infected with the virus through bites from infected animals. Upon exposure to the virus, the current method to prevent development of rabies in humans and domestic animals is to inoculate with rabies vaccine prepared from an attenuated strain of rabies virus. This is normally in addition to treatment using antirabies γ-globulin. However, this method is impractical to eradicate rabies in wild animals.

A safe and cost-effective method to achieve vaccine production was through the development of a recombinant vaccinia virus expressing the glycoprotein G of rabies virus *(95)*. The recombinant virus is amplified in animal cells for preparation of vaccine suspensions, which are placed in animal baits for release into the wild. The recombinant vaccine is currently used in eradication programs in Europe and North America *(94)*.

4.3. Textile Industry

Microbial enzymes have been used in the textile industry since the early 1900s. To commercialize the enzymes, they must be produced at high levels. Conventional methods to enhance production include optimizing medium composition, growth conditions, and the fermentation process *(67)*. Random mutagenesis and screening commonly is used to achieve high yields. Genetic engineering offers a possibility in which high-level enzyme production is achieved in a heterologous host to overcome the limitations of the natural producing organism. Two examples are α-amylase of *Bacillus stearothermophilus (96)* and cellulase of the alkaliphilic *Bacillus* BCE103 *(97)*. Amylases have been used for many years to remove starch sizes from fabrics, known as *desizing*. Originally, amylases from plant or animal sources were used. Later, they were replaced by amylases of bacterial origin.

The first bacterial enzyme for desizing was α-amylase from *Bacillus subtilis*, which was commercialized in the early 1950s *(96)*. A novel α-amylase naturally produced by *B. stearothermophilus* is heat stable and is active over a broad pH range *(96)*. These characteristics make this new α-amylase an attractive alternative. However, the new α-amylase is produced at low levels by its natural producing organism. To increase the production of this new enzyme for commercial use, the gene encoding the enzyme was cloned into a heterologous host, *Bacillus licheniformis (96)*.

Cellulases prevent and remove fuzz and pills and provide color brightening of cellulose-based fabrics such as cotton. A novel cellulase, active under alkaline detergent conditions, is an attractive alternative *(97)*. It is also produced in low levels by its native producing organism, an extremophile, because of poor growth. The gene encoding the cellulase was cloned into a heterologous host, *B. subtilis*, for high-level expression and enzyme production. In both cases, the desired enzymes were produced efficiently from heterologous hosts during fermentation. In addition, the enzymes are secreted directly into the culture media, which simplifies the recovery of the respective enzymes.

4.4. Food Industry

Enzymes manufactured by GMMs have been used in the food industry for more than 15 years *(98)*. Well-known examples include the use of chymosin for cheese making and pectinases for fruit and beverage processing. Traditionally, cheese making requires chymosin-containing rennet from calf stomachs to provide the essential proteolytic activity for coagulation of milk proteins *(93)*. However, chymosin preparations could have animal sources of contaminants.

In the early 1980s, Gist-brocades began investigating the possibility of producing chymosin from a microorganism using the genetic engineering approach *(93)*. The gene encoding calf stomach chymosin was cloned and expressed in an industrial strain of *Kluyveromyces lactis*, a yeast that had been used for many years in the safe production of food ingredients. To facilitate prochymosin secretion into the culture medium for recovery, the yeast α-factor leader sequence was used. This system efficiently secreted prochymosin into the culture medium along with very few endogenous proteins. On fermentation, prochymosin is converted into active chymosin via a simple autolysis step, followed by recovery of the final product. Chymosin produced through this genetically engineered yeast strain has the same chemical and biological properties as that from calf rennet. The chymosin preparation, registered under the brand-name Maxiren®, has been commercially produced since 1988.

Production of pectinases via the genetic engineering approach focuses on economic enzyme production, enhanced enzyme purity, and environmentally friendly production processes *(98)*. Total pectin methyl esterase I (PME I) from *Aspergillus aculeatus* represents less than 1% of the total cellular protein *(99)*. In addition, this fungus accumulates a wide range of pectinolytic enzymes in the culture, making it difficult to acquire pure PME I *(99)*. These features make the natural producing organism less than ideal for commercial production of PME I.

The limitations were overcome by expressing the full-length cDNA (complementary DNA) encoding the enzyme in a heterologous host, *A. oryzae* *(99)*. The recombinant PME I represented 20–30% of the total cellular protein and was secreted directly into the culture medium for simplified purification. Complete pectin degradation was achieved by recombinant PME I in the presence of polygalacturonases *(99)*.

Complete degradation of pectin, a natural substance found in all fruits, is important for the beverage industry. The reason is that complete degradation of pectin increases juice extraction from fruits, enhances juice clarification, and helps the filtration step of the process *(98)*. Several enzyme preparations made by GMMs are currently used in the beverage industry. NovoShape™, containing a pure pectinesterase, helps retain the original shape and structure of individual fruit pieces during processing and thereby offers a finished product that is more appealing (http://www.novozymes.com). Pectinex® SMASH, containing a variety of different pectinases, is used for treating apple and pear mash for higher yield and capacity (http://www.novozymes.com).

4.5. Diagnostic Tools

Acquired immunodeficiency syndrome (AIDS) immunological tests are used for diagnosing the disease and for testing donated blood samples. The first generation of AIDS tests, commercialized in 1985, was based on inactivated human immunodeficiency virus (HIV) grown in tissue culture *(100)*. This production method is both expensive and, more importantly, hazardous because of the risk from handling the infectious agent. Further, this first-generation AIDS test was subject to false-positive reactions because of the cellular debris from virus-producing human cells. These problems were overcome by cloning the gene encoding the relevant antigenic coat protein of the virus into *E. coli* for large-scale production of the protein *(100)*.

Other diagnostic tests that have been developed using GMMs include one for diagnosing Alzheimer's disease. Noninvasive diagnosis of Alzheimer's disease was not possible until the development of an enzyme-linked immunosorbent assay kit in the mid-1990s *(101)*. The test kit, marketed under the trade name INNOTEST hTAU Antigen and used for the detection of tau proteins in human cerebrospinal fluid, is based on an Alzheimer's antigen produced by a modified *E. coli*.

4.6. Biodegradable Plastics

Conventionally, plastics polymers are made via petroleum-based processes. Because of the growing concerns over the environmental impact of petroleum-derived polymers, alternative methods to synthesize the polymers are under investigation. Many microorganisms naturally produce polyhydroxyalkanoates (PHAs) in the form of granules that the organisms use as an energy storage material *(102)*. PHAs are genuine polyester thermoplastics with properties similar to the petroleum-derived polymers. In addition, PHAs are degradable by depolymerase, an enzyme family widely distributed among bacteria and fungi *(103)*. These characteristics make PHAs an attractive replacement for the petroleum-based polymers. However, the microorganisms that naturally produce PHAs are not necessarily suitable for commercial PHA production, mostly because of slow growth and low yields.

PHA accumulation is achieved in *E. coli*, a microbe lacking PHA biosynthetic machinery, after receiving the PHA pathway genes through transformation. The advantages of producing PHAs in *E. coli* are (1) the organism is robust in growth; (2) the organism's metabolism is well characterized; and (3) the organism lacks PHA depolymerase, the enzyme that degrades PHAs. In 2002, Metabolix (Cambridge, MA; http://www.metabolix.com) demonstrated high-yield, commercial-scale manufacture of PHAs using *E. coli* as a host.

5. Applications of GMMs

5.1. Agriculture

5.1.1. Biological Control of Frost Injury in Plants

Frost damage is a major agricultural problem affecting many annual crops, deciduous fruit trees, and subtropical plants. In the United States alone, annual losses because of plant frost injury can reach over $1 billion *(104)*. In addition to the losses caused by frost injury, hundreds of millions of dollars are spent every year to reduce plant frost injury mechanically. These methods are both costly and ineffective *(104)*.

Frost damage is initiated by bacteria belonging to the genera *Pseudomonas*, *Xanthomonas*, and *Erwinia*, collectively called *ice-nucleating bacteria (105)*. The bacteria, living on the surface of the plants, possess a membrane protein that acts as an ice nucleus for initiation of ice crystal formation *(105)*. Ice crystals disrupt plant cell membranes, thus causing cell damage. The biological route of controlling the nucleating bacteria is through seed or foliar applications of non-ice-nucleating bacteria to outcompete ice-nucleating bacteria *(106)*. The non-ice-nucleating bacteria were isolated by treating the ice-nucleating bacteria with chemical mutagens *(107)*.

One disadvantage of the chemically induced mutants is that they often harbor multiple mutations that may adversely impact their genetic stability and ecological fitness. To avoid multiple mutations, ice-nucleation-deficient mutants of *Pseudomonas syringae* were constructed by deleting the genes conferring ice nucleation *(108)*. These genetically engineered mutants were able to compete successfully with ice-nucleating *P. syringae* for the colonization of plant leaf surfaces. Field tests showed that plants treated with ice-nucleation-deficient *P. syringae* suffered significantly less frost damage than the untreated control plants *(109)*.

5.1.2. Biological Control of Insect Pests

Bt, a naturally occurring soil-borne bacterium, produces unique crystal-like proteins that have larvicidal activities against different insect species and pose no harm to mammals, birds, or fish *(110)*. The crystal-like proteins bind to specific receptors on the intestinal lining of susceptible insects, causing the cells to rupture. Because of these unique features, *Bt* has been used as a safe alternative to chemical pesticides for several decades *(111)*. However, natural *Bt*-based products do possess some shortcomings, including instability in the natural environment, narrow host range, need for multiple applications, and difficulty in reaching the crop's internal regions where larvae feed *(111)*.

One way to overcome these problems is to use plant-associated bacteria as hosts for delivering the toxins. *Bt* toxin genes have been introduced successfully into several plant-associated bacteria, including *Clavibacter xyli* subsp *cynodontis (112)* and *Ancylobacter aquaticus (113)*. Genetically modified *C. xyli* subsp *cynodontis* containing the *Bt* toxin gene integrated into the chromosome showed moderate control of European corn borer *(112)*. The modified *A. aquaticus* strain expressing *Bt* toxin genes, introduced by electroporation, exhibited significant toxicity toward mosquito larvae, thus demonstrating its potential in mosquito control *(113)*.

5.1.3. Biological Control of Plant Disease

Plant pathogens, including fungi and bacteria, damage crops and thereby reduce crop yield. Plant diseases are conventionally fought with chemicals, a strategy that is expensive, inconvenient, potentially environmentally unfriendly, and sometimes ineffective. An alternative method is to develop biological control agents in which microorganisms are modified to deliver the desired chemicals.

Agrobacterium tumefaciens causes crown gall disease in a wide range of broad-leaved plants by transferring part of its DNA (T-DNA), located on a large tumor-inducing (Ti) plasmid, into the plant cell *(114)*. Upon integration of T-DNA into the plant host's chromosome, the genes on T-DNA are expressed, resulting in overproduction of plant growth hormones and opines *(115)*. Overproduction of plant growth hormones causes cancerous growth, whereas opines are believed to serve as nutrients for the bacterium. *Agrobacterium radiobacter* K84 produces a bacteriocin, agrocin 84, to which pathogenic *A. tumefaciens* strains are susceptible *(116)*. In addition, K84 competes for the nutrients on which the pathogen thrives. Therefore, *A. radiobacter* K84 became the first commercial biological control agent against crown gall disease.

One potential problem that threatens the continued success of this biological control agent is that the genes responsible for agrocin 84 production and resistance are located on a plasmid harbored by K84 *(115)*. Horizontal transfer of the plasmid from K84 to pathogenic *A. tumefaciens* allows the pathogen to acquire resistance against the toxic effect of agrocin 84 and therefore survive in the presence of K84. To prevent plasmid transfer, part of the transfer region was deleted from the plasmid. The genetically modified strain was as effective as the nonmodified K84 strain in preventing crown gall disease and was commercialized for use in Australia in 1989 *(115,117)*.

Other plant disease control methods using GMMs include (1) overproduction of oomycin A *(118)*; (2) synthesis of phenazine-1-carboxylic acid in a heterologous host *(119)*; and (3) heterologous expression of a lytic enzyme gene *(120)*.

5.1.4. Soil Improvements

Genetic modifications to improve soil fertility have also been developed. *Medicago sativa* (alfalfa), grown in soils with a high nitrogen concentration, has been shown to undergo better root nodulation when exposed to a genetically modified *Sinorhizobium (Rhizobium) meliloti* expressing the *Klebsiella pneumonia nif* A gene than plants in the same environment exposed to wild-type *S. meliloti (121)*. Another study showed that the recombinant *S. meliloti* significantly increased plant biomass when compared to the wild-type strain *(122)*.

5.2. Bioremediation

Bioremediation refers to the utilization of biological systems to detoxify environments contaminated with heavy metals such as mercury and lead, organic compounds such as petroleum hydrocarbons, radionuclides such as plutonium and uranium, and other compounds, including explosives, pesticides, and plastics *(123)*. The first field release of a GMM for bioremediation was *Pseudomonas fluorescens* HK44 for naphthalene degradation *(124)*. Strain HK44 was derived from *P. fluorescens* isolated from a site heavily contaminated with polyaromatic hydrocarbons. HK44 contains a plasmid capable of naphthalene catabolism. In addition, this genetically modified strain harbors a bioluminescence-producing reporter gene (*lux*) fused with the promoter that controls the naphthalene catabolic genes. Therefore, in the presence of naphthalene, the naphthalene genes are expressed, resulting in naphthalene degradation and emission of luminescence from the recombinant strain. The presence of the reporter system facilitates real-time monitoring of the bioremediation processes.

Despite the success, bioremediation based on GMMs is still limited to academic research. Commercial remediation currently relies on naturally occurring microbes identified at the contaminated sites. There are many issues surrounding the application of GMMs for use in bioremediation, including (1) their effectiveness compared with their counterparts present in nature; (2) their influence on indigenous microorganisms; (3) their fitness in nature; and (4) their containment. Until these issues are clarified, the future use of GMMs in bioremediation will remain uncertain.

6. Conclusion

GMMs have been developed to benefit human health, agriculture and the environment. Advances in functional genomics and bioinformatics tools, combined with existing recombinant DNA technologies, will help us better understand the physiology and metabolic potential of the organisms we study, and in turn, will lead to the development of GMMs best suited to our needs. In addition, functional genomics and bioinformatics can be applied for risk analysis of GMMs. A comprehensive safety assessment of GMMs is important both in addressing public concerns and in ensuring faster industrial applications of GMMs.

Acknowledgments

I wish to thank Arnold Demain, Douglas Pearson, Raghav Ram, and Donald Merlo for comments and suggestions.

References

1. Meyer, V., Mueller, D., Strowig, T., and Stahl, U. (2003) Comparison of different transformation methods for *Aspergillus giganteus*. *Curr. Genet.* **43**, 371–377.
2. Gietz, R. D. and Woods, R. A. (2001) Genetic transformation of yeast. *Biotechniques* **30**, 816–831.
3. Dubnau, D. (1991) Genetic competence in *Bacillus subtilis*. *Microbiol. Rev.* **55**, 395–424.
4. Bron, S., Meima, R., van Dijl, J. M., Wipat, A., and Harwood, C. R. (1999) Molecular biology and genetics of *Bacillus* spp. In *Manual of Industrial Microbiology and Biotechnology*, 2nd ed. (Demain, A. L. and Davies, J. E., eds.). ASM Press, Washington, DC, pp. 392–416.

5. Brigidi, P., De Rossi, E., Bertarini, M. L., Riccardi, G., and Matteuzzi, D. (1990) Genetic transformation of intact cells of *Bacillus subtilis* by electroporation. *FEMS Microbiol. Lett.* **55**, 135–138.
6. McDonald, I. R., Riley, P. W., Sharp, R. J., and McCarthy, A. J. (1995) Factors affecting the electroporation of *Bacillus subtilis. J. Appl. Bacteriol.* **79**, 213–218.
7. Thierbach, G., Schwarzer, A., and Pühler, A. (1988) Transformation of spheroplasts and protoplasts of *Corynebacterium glutamicum. Appl. Microbiol. Biotechnol.* **29**, 356–362.
8. Schäfer, A., Kalinowski, J., Simon, R., Seep-Feldhaus, A. H., and Pühler, A. (1990) High-frequency conjugal plasmid transfer from Gram-negative *Escherichia coli* to various Gram-positive coryneform bacteria. *J. Bacteriol.* **172**, 1663–1666.
9. Wolf, H., Pühler, A., and Neumann, E. (1989) Electrotransformation of intact and osmotically sensitive cells of *Corynebacterium glutamicum. Appl. Microbiol. Biotechnol.* **30**, 283–289.
10. Sambrook, J., Fritsch, E. F., and Maniatis, T. (eds.) (1989) *Molecular Cloning: A Laboratory Manual.* Cold Spring Harbor Press, Cold Spring Harbor, NY.
11. von Wright, A. and Sibakov, M. (1993) Genetic modification of lactic acid bacteria. In *Lactic Acid Bacteria* (Salminen, S. and von Wright, A., eds.). Marcel Dekker, New York, NY, pp. 161–198.
12. Itoh, N., Kouzai, T., and Koide, Y. (1994) Efficient transformation of *Pseudomonas* strains with pNI vectors by electroporation. *Biosci. Biotechnol. Biochem.* **58**, 1306–1308.
13. Iwasaki, K., Uchiyama, H., Yagi, O., Kurabayashi, T., Ishizuka, K., and Takamura, Y. (1994) Transformation of *Pseudomonas putida* by electroporation. *Biosci. Biotechnol. Biochem.* **58**, 851–854.
14. Sánchez-Romero, J. M. and de Lorenzo, V. (1999) Genetic engineering of nonpathogenic *Pseudomonas* strains as biocatalysts for industrial and environmental processes. In *Manual of Industrial Microbiology and Biotechnology*, 2nd ed. (Demain, A. L. and Davies, J. E., eds.). ASM Press, Washington, DC, pp. 460–474.
15. Kieser, T., Bibb, M. J., Buttner, M. J., Chater, K. F., and Hopwood, D. A. (eds.) (2000) *Practical Streptomyces Genetics*. John Innes Foundation, Norwich, UK.
16. Pigac, J. and Schrempf, H. (1995) A simple and rapid method of transformation of *Streptomyces rimosus* R6 and other streptomycetes by electroporation. *Appl. Environ. Microbiol.* **61**, 352–356.
17. Tyurin, M., Starodubtseva, L., Kudryavtseva, H., Voeykova, T., and Livshits, V. (1995) Electrotransformation of germinating spores of *Streptomyces* spp. *Biotech. Techniques.* **9**, 737–740.
18. Mazy-Servais, C., Baczkowski, D., and Dusart, J. (1997) Electroporation of intact cells of *Streptomyces parvulus* and *Streptomyces vinaceus. FEMS Microbiol. Lett.* **151**, 135–138.
19. Mazodier, P., Petter, R., and Thompson, C. J. (1989) Intergeneric conjugation between *Escherichia coli* and *Streptomyces species. J. Bacteriol.* **171**, 3583–3585.
20. Bierman, M., Logan, R., O'Brien, K., Seno, E. T., Rao, R. N., and Schoner, B. E. (1992) Plasmid cloning vectors for the conjugal transfer of DNA from *Escherichia coli* to *Streptomyces* spp. *Gene* **116**, 43–49.
21. Matsushima, P., Broughton, M. C., Turner, J. R., and Baltz, R. H. (1994) Conjugal transfer of cosmid DNA from *Escherichia coli* to *Saccharopolyspora spinosa*: effects of chromosomal insertions on macrolide A83543 production. *Gene* **146**, 39–45.
22. Sun, Y., Zhou, X., Liu, J., et al. (2002) "*Streptomyces nanchangensis*," a producer of the insecticidal polyether antibiotic nanchangmycin and the antiparasitic macrolide meilingmycin, contains multiple polyketide gene clusters. *Microbiology* **148**, 361–371.
23. Paranthaman, S. and Dharmalingam, K. (2003) Intergeneric conjugation in *Streptomyces peucetius* and *Streptomyces* sp. strain C5: chromosomal integration and expression of recombinant plasmids carrying the *chiC* gene. *J. Bacteriol.* **69**, 84–91.

24. Lorenz, M. G. and Wackernagel, W. (1994) Bacterial gene transfer by natural genetic transformation in the environment. *Microbiol. Rev.* **58**, 563–602.
25. Neumann, E., Schaefer-Ridder. M., Wang, Y., and Hofschneider, P. H. (1982) Gene transfer into mouse lyoma cells by electroporation in high electric fields. *EMBO J.* **1**, 841–845.
26. Kullen, M. J. and Klaenhammer, T. R. (1999) Genetic modification of intestinal lactobacilli and bifidobacteria. In *Probiotics: A Critical Review* (Tannock, G. ed.). Horizon Scientific Press, Wymondham, UK, pp. 63–83.
27. Lindgren, S. (1999) Biosafety aspects of genetically modified lactic acid bacteria in EU legislation. *Int. Dairy J.* **9**, 37–41.
28. Kilbane, J. J. and Bielaga, B. A. (1991) Instantaneous gene transfer from donor to recipient microorganisms via electroporation. *Biotechniques* **10**, 354–365.
29. Grohmann, E., Muth, G., and Espinosa, M. (2003) Conjugative plasmid transfer in Gram-positive bacteria. *Microbiol. Mol. Biol. Rev.* **67**, 277–301.
30. Rao, R. N., Richardson, M. A., and Kuhstoss, S. (1987) Cosmid shuttle vectors for cloning and analysis of *Streptomyces* DNA. *Methods Enzymol.* **153**, 166–198.
31. Sosio, M., Guisino, F., Cappellano, C., Bossi, E., Puglia, A. M., and Donadio, S. (2000) Artificial chromosomes for antibiotic-producing actinomycetes. *Nat. Biotechnol.* **18**, 343–345.
32. Verrips, C. T. and van den Berg, D. J. C. (1996) Barriers to application of genetically modified lactic acid bacteria. *Antonie Van Leeuwenhoek* **70**, 299–316.
33. Martin, M. C., Alonso, J. C., Suarez, J. E., and Alvarez, M. A. (2000) Generation of food-grade recombinant lactic acid bacterium strains by site-specific recombination. *Appl. Environ. Microbiol.* **66**, 2599–2604.
34. Akada, R., Shimizu, Y., Matsushita, Y., Kawahata, M., Hoshida, H., and Nishizawa, Y. (2002) Use of a *YAP1* overexpression cassette conferring specific resistance to cerulenin and cycloheximide as an efficient selectable marker in the yeast *Saccharomyces cerevisiae*. *Yeast* **19**, 17–28.
35. Baneyx, F. (1999) Recombinant protein expression in *Escherichia coli*. *Curr. Opin. Biotechnol.* **10**, 411–421.
36. Makrides, S. C. (1996) Strategies for achieving high-level expression of genes in *Escherichia coli*. *Microbiol. Rev.* **60**, 512–538.
37. Figge, J., Wright, C., Collins, C. J., Roberts, T. M., and Livingston, D. M. (1988) Stringent regulation of stably integrated chloramphenicol acetyl transferase genes by *E. coli lac* repressor in monkey cells. *Cell* **52**, 713–722.
38. Taylor, A., Brown, D. P., Kadam, S., et al. (1992) High-level expression and purification of mature HIV-1 protease in *Escherichia coli* under control of the *araBAD* promoter. *Appl. Microbiol. Biotechnol.* **37**, 205–210.
39. Phadtare, S., Alsina, J., and Inouye, M. (1999) Cold-shock response and cold-shock proteins. *Curr. Opin. Microbiol.* **2**, 175–180.
40. Vasina, J. A., Peterson, M. S., and Baneyx, F. (1998) Scale-up and optimization of the low-temperature inducible *cspA* promoter system. *Biotechnol. Prog.* **14**, 714–721.
41. Goldstein, M. A. and Doi, R. H. (1995) Prokaryotic promoters in biotechnology. *Biotechnol. Annu. Rev.* **1**, 105–128.
42. Renault, P. (2002) Genetically modified lactic acid bacteria: applications to food or health and risk assessment. *Biochimie* **84**, 1073–1087.
43. Akada, R. (2002) Genetically modified industrial yeast ready for application. *J. Biosci. Bioeng.* **94**, 536–544.
44. Brantl, S. (2002) Antisense RNAs in plasmids: control of replication and maintenance. *Plasmid* **48**, 165–173.

45. Desai, R. P. and Papoutsakis, E. T. (1999) Antisense RNA strategies for metabolic engineering of *Clostridium acetobutylicum*. *Appl. Environ. Microbiol.* **65**, 936–945.
46. Eggeling, L., Oberle, S., and Sahm, H. (1998) Improved L-lysine yield with *Corynebacterium glutamicum*: use of *dapA* resulting in increased flux combined with growth limitation. *Appl. Microbiol. Biotechnol.* **49**, 24–30.
47. Downing, K. J., Leslie, G., and Thomson, J.A. (2000) Biocontrol of the sugarcane borer *Eldana saccharina* by expression of the *Bacillus thuringiensis cry1Ac7* and *Serratia marcescens chiA* genes in sugarcane-associated bacteria. *Appl. Environ. Microbiol.* **66**, 2804–2810.
48. Le, H. V. and Trotta, P. P. (1991) Purification of secreted recombinant proteins from *Escherichia coli*. *Bioprocess Technol.* **12**, 163–181.
49. Lilie, H., Schwarz, E., and Rudolph, R. (1998) Advances in refolding of proteins produced in *E. coli*. *Curr. Opin. Biotechnol.* **9**, 497–501.
50. De Bernardez Clark, E., Schwarz, E., and Rudolph, R. (1999) Inhibition of aggregation side reactions during in vitro protein folding. *Methods Enzymol.* 309er, 217–236.
51. Altamirano, M. M., Garcia, C., Possani, L. D., and Fersht, A. R. (1999) Oxidative refolding chromatography: folding of the scorpion toxin Cn5. *Nat. Biotechnol.* **17,** 187–191.
52. Swartz, J. R. (2001) Advances in *Escherichia coli* production of therapeutic proteins. *Curr. Opin. Biotechnol.* **12**, 195–201.
53. Baneyx, F. (1999) In vivo folding of recombinant proteins in *Escherichia coli*. In *Manual of Industrial Microbiology and Biotechnology*, 2nd ed. (Demain, A. L. and Davies, J. E., eds.). ASM Press, Washington, DC, pp. 551–565.
54. Bergès, H., Joseph-Liauzun, E., and Fayet, O. (1996) Combined effects of the signal sequence and the major chaperone proteins on the export of human cytokines in *Escherichia coli*. *Appl. Environ. Microbiol.* **62**, 55–60.
55. Miksch, G., Neitzel, R., Friehs, K., and Flaschel, E. (1999) High-level expression of a recombinant protein in *Klebsiella planticola* owing to induced secretion into the culture medium. *Appl. Microbiol. Biotechnol.* **51**, 627–632.
56. Minas, W. and Bailey, J. E. (1995) Co-overexpression of *prlF* increases cell viability and enzyme yields in recombinant *Escherichia coli* expressing *Bacillus stearothermophilus* α-amylase. *Biotechnol. Prog.* **11**, 403–411.
57. Olson, P., Zhang, Y., Olsen, D., et al. (1998) High-level expression of eukaryotic polypeptides from bacterial chromosomes. *Protein Expr. Purif.* **14**, 160–166.
58. Kiel, J. A., ten Berge, A. M., Borger, P., and Venema, G. (1995) A general method for the consecutive integration of single copies of a heterologous gene at multiple locations in the *Bacillus subtilis* chromosome by replacement recombination. *Appl. Environ. Microbiol.* **61**, 4244–4250.
59. Peredelchuk, M. Y. and Bennett, G. N. (1997) A method for construction of *E. coli* strains with multiple DNA insertions in the chromosome. *Gene* **187,** 231–238.
60. Martinez-Morales, F., Borges, A. C., Martinez, A., Shanmugam, K. T., and Ingram, L. O. (1999) Chromosomal integration of heterologous DNA in *Escherichia coli* with precise removal of markers and replicons used during construction. *J. Bacteriol.* **181,** 7143–7148.
61. Glick, B. R. and Pasternak, J. J. (eds.) (1994) *Molecular Biotechnology: Principles and Applications of Recombinant DNA*. ASM Press, Washington, DC.
62. Nagashima, T., Tada, S., Kitamoto, K., Gomi, K., Kumagai, C., and Toda, H. (1992) Site-directed mutagenesis of catalytic active-site residues of Taka-amylase A. *Biosci. Biotechnol. Biochem.* **56**, 207–210.
63. Lee, C. Y., Bagdasarian, M., Meng, M. H., and Zeikus, J. G. (1990) Catalytic mechanism of xylose (glucose) isomerase from *Clostridium thermosulfurogenes*. Characterization of the structural gene and function of active site histidine. *J. Biol. Chem.* **265**, 19,082–19,090.

64. Stemmer, W. P. (1994) Rapid evolution of a protein in vitro by DNA shuffling. *Nature* **370**, 389–391.
65. Crameri, A., Raillard, S. A., Bermudez, E., and Stemmer, W. P. (1998) DNA shuffling of a family of genes from diverse species accelerates directed evolution. *Nature* **391**, 288–291.
66. Ness, J. E., Welch, M., Giver, L., et al. (1999) DNA shuffling of subgenomic sequences of subtilisin. *Nat. Biotechnol.* **17**, 893–896.
67. Gupta, R., Beg, Q. K., and Lorenz, P. (2002) Bacterial alkaline proteases: molecular approaches and industrial applications. *Appl. Microbiol. Biotechnol.* **59**, 15–32.
68. del Cardayre, S. and Powell, K. (2003) DNA shuffling for whole cell engineering. In *Handbook of Industrial Cell Culture: Mammalian, Microbial, and Plant Cells* (Vinci, V. A. and Parekh, S. R., eds.). Humana Press, Totowa, NJ, pp. 465–481.
69. Ikeda, M., Ozaki, A., and Katsumata, R. (1993) Phenylalanine production by metabolically engineered *Corynebacterium glutamicum* with the *pheA* gene of *Escherichia coli. Appl. Microbiol. Biotechnol.* **39**, 318–323.
70. Kennedy, J. and Turner, G. (1996) Delta-(L-alpha-aminoadipyl)-L-cysteinyl-D-valine synthetase is a rate limiting enzyme for penicillin production in *Aspergillus nidulans. Mol. Gen. Genet.* **253**, 189–197.
71. Wang, D. I. C., Cooney, C. L., Demain, A. L., Dunhill, P., Humphrey, A. E., and Lilly, M. D. (eds.) (1979) Biosynthesis of primary metabolites. In *Fermentation and Enzyme Technology.* John Wiley and Sons, New York, NY, pp. 14–25.
72. Guillouet, S., Rodal, A. A., An, G., Lessard, P. A., and Sinskey, A. J. (1999) Expression of the *Escherichia coli* catabolic threonine dehydratase in *Corynebacterium glutamicum* and its effect on isoleucine production. *Appl. Environ. Microbiol.* **65**, 3100–3107.
73. Champness, W. (1999) Cloning and analysis of regulatory genes involved in *Streptomyces* secondary metabolite biosynthesis. In *Manual of Industrial Microbiology and Biotechnology*, 2nd ed. (Demain, A. L. and Davies, J. E., eds.). ASM Press, Washington, DC, pp. 725–739.
74. Tatarko, M. and Romeo, T. (2001) Disruption of a global regulatory gene to enhance central carbon flux into phenylalanine biosynthesis in *Escherichia coli. Curr. Microbiol.* **43**, 26–32.
75. Butler, M. J., Bruheim, P., Jovetic, S., Marinelli, F., Postma, P. W., and Bibb, M. J. (2002) Engineering of primary carbon metabolism for improved antibiotic production in *Streptomyces lividans. Appl. Environ. Microbiol.* **68**, 4731–4739.
76. Peters-Wendisch, P. G., Schiel, B., Wendisch, V. F., et al. (2001) Pyruvate carboxylase is a major bottleneck for glutamate and lysine production by *Corynebacterium glutamicum. J. Mol. Microbiol. Biotechnol.* **3**, 295–300.
77. Backman, K., O'Conner, M. J., Maruya, A., et al. (1990) Genetic engineering of metabolic pathways applied to the production of phenylalanine. *Ann. N. Y. Acad. Sci.* **589**, 16–24.
78. Reinscheid, D. J., Kronemeyer, W., Eggeling, L. Eikmanns, B. J., and Sahm, H. (1994) Stable expression of *hom-1-thrB* in *Corynebacterium glutamicum* and its effect on the carbon flux to threonine and related amino acids. *Appl. Environ. Microbiol.* **60**, 126–132.
79. Vrljic, M., Sahm, H., and Eggeling, L. (1996) A new type of transporter with a new type of cellular function: L-lysine export from *Corynebacterium glutamicum. Mol Microbiol.* **22**, 815–826.
80. Ohnishi, J., Mitsuhashi, S., Hayashi, M., et al. (2002) A novel methodology employing *Corynebacterium glutamicum* genome information to generate a new L-lysine-producing mutant. *Appl. Microbiol. Biotechnol.* **58**, 217–223.
81. Johnson, I. S. (1983) Human insulin from recombinant DNA technology. *Science* **219**, 632–637.
82. Cronin, M. J. (1997) Pioneering recombinant growth hormone manufacturing: pounds produced per mile of height. *J. Pediatr.* **131**, S5–S7.

83. Falkner, E. and Maurer-Fogy, I. (1996) Recombinant interferon gamma for the therapy of immunological and oncological diseases. *J. Biotechnol.* **46**, 155–156.
84. Hjorth, R., Ryser, S., Pühler, A., and Diderichsen, B. (1992) Interferon-alfa—progress cancer therapy. *Biotech. Forum Europe* **9**, 641–642.
85. Petri, T. and Weber-Diehl, F. (1995) Interferon beta-1b for the treatment of multiple sclerosis. *J. Biotechnol.* **43**, 74–75.
86. Till, R. (1998) GM-CSF: more than a growth factor. *J. Biotechnol.* **61**, 158–160.
87. Crooy, P. (1991) Protection against hepatitis B, using a rec-DNA vaccine. *Biotech. Forum Europe* **8**, 247.
88. Tengerdy, R. P. and Szakács, G. (1998) Perspectives in agrobiotechnology. *J. Biotechnol.* **66**, 91–99.
89. Mukhopadhyay, U. K. and Sahni, G. (2002) Production of recombinant buffalo (*Bubalus bubalis*) and goat (*Capra hircus*) growth hormones from genetically modified *E. coli* strains. *J. Biotechnol.* **97**, 199–212.
90. Vohra, A. and Satyanarayana, T. (2003) Phytases: microbial sources, production, purification, and potential biotechnological applications. *Crit. Rev. Biotechnol.* **23**, 29–60.
91. Mullaney, E. J., Daly, C. B., and Ullah, A. H. (2000) Advances in phytase research. *Adv. Appl. Microbiol.* **47**, 157–199.
92. Abelson, P. H. (1999) A potential phosphate crisis. *Science* **283**, 2015.
93. van Dijck, P. W. M. (1999) Chymosin and phytase. *J. Biotechnol.* **67**, 77–80.
94. Rupprecht, C.E., Blass, L., Smith, K., et al. (2001) Human infection due to recombinant vaccinia-rabies glycoprotein virus. *N. Engl. J. Med.* **345**, 582–586.
95. Terré, J., Chappuis, G., Lombard, M., and Desmettre, P. (1996) Eradication of rabies, using a rec-DNA vaccine. *J. Biotechnol.* **46**, 156–157.
96. Diderichsen, B. (1995) A new desizing enzyme shows the changing nature of enzyme production. *J. Biotechnol.* **38**, 196.
97. Jones, B. and Quax, W. (1998) Brighter appearance to coloured textiles thanks to a new cellulase from an extremophilic bacterium. *J. Biotechnol.* **66**, 231–233.
98. Urlaub, R. (1999) Enzymes from genetically modified microorganisms and their use in the beverage industry. *Fruit Proc.* **9**, 158–163.
99. Christgau, S., Kofod, L. V., Halkier, T., et al. (1996) Pectin methyl esterase from *Aspergillus aculeatus*: expression cloning in yeast and characterization of the recombinant enzyme. *Biochem. J.* **319**, 705–712.
100. Baumann, E. (1995) AIDS diagnostics by use of recombinant antigens. *J. Biotechnol.* **38**, 195.
101. Vanmechelen, E. and Vanderstichele, H. (1998) Towards an earlier diagnosis of Alzheimer's disease. *J. Biotechnol.* **66**, 229–231.
102. Anderson, A. J. and Dawes, E. A. (1990) Occurrence, metabolism, metabolic role, and industrial uses of bacterial polyhydroxyalkanoates. *Microbiol. Rev.* **54**, 450–472.
103. Jendrossek, D. and Handrick, R. (2002) Microbial degradation of polyhydroxyalkanoates. *Annu. Rev. Microbiol.* **56**, 403–432.
104. Fuller, M., Hamed, F., Wisniewski, M. E., and Glenn, D. M. (2001) *Protection of Crops From Frost Using a Hydrophobic Particle Film and an Acrylic Polymer*. Agricultural Research Service, US Department of Agriculture. Available at http://warp.nal.usda.gov/ttic/tektran/data/000012/39/0000123937.html.
105. Gurian-Sherman, D. and Lindow, S. E. (1993) Bacterial ice nucleation: significance and molecular basis. *FASEB J.* **7**, 1338–1343.
106. Lindow, S. E. (1983) Methods of preventing frost injury caused by epiphytic ice nucleation active bacteria. *Plant Dis.* **67**, 327–333.
107. Lindow, S. E. (1987) Competitive exclusion of epiphytic bacteria by ice *Pseudomonas syringae* mutants. *Appl. Environ. Microbiol.* **53**, 2520–2527.

108. Lindow, S. E. (1985) Ecology of *Pseudomonas syringae* relevant to the field use of ice-deletion mutants constructed in vitro for plant frost control. In *Engineered Organisms in the Environment Scientific Issues*. (Halvorson, H. O., Pramer, D., and Rogul, M., eds.). ASM, Washington, DC, pp. 23–35.
109. Lindow, S. E. and Panopoulos, N. J. (1988) Field tests of recombinant ice$^-$ *Pseudomonas syringae* for biological frost control in potato. In *Release of Genetically Engineered Microorganisms* (Sussman, M., Collins, C. H., and Skinner, F. A., eds.). Academic Press, London, UK, pp. 121–138.
110. Höfte, H. and Whiteley, H. R. (1989) Insecticidal crystal proteins of *Bacillus thuringiensis*. *Microbiol. Rev.* **53**, 242–255.
111. Amarger, N. (2002) Genetically modified bacteria in agriculture. *Biochimie* **84**, 1061–1072.
112. Lampel, J. S., Canter, G. L., Dimock, M. B., et al. (1994) Integrative cloning, expression, and stability of the *cryIA(c)* gene from *Bacillus thuringiensis* subsp. *kurstaki* in a recombinant strain of *Clavibacter xyli* subsp. *cynodontis*. *Appl. Environ. Microbiol.* **60**, 501–508.
113. Yap, W. H., Thanabalu, T., and Porter, A. G. (1994) Expression of mosquitocidal toxin genes in a gas-vacuolated strain of *Ancylobacter aquaticus*. *Appl. Environ. Microbiol.* **60**, 4199–4202.
114. Chilton, M. (2001) Agrobacterium. A memoir. *Plant Physiol.* **125**, 9–14.
115. Jones, D. A. and Kerr, A. (1989) *Agrobacterium radiobacter* strain K1026, a genetically engineered derivative of strain K84, for biological control of crown gall. *Plant Dis.* **73**, 15–18.
116. Kerr, A. (1980) Biological control of crown gall through production of agrocin 84. *Plant Dis.* **64**, 25–30.
117. Ryder, M. (1994) Key issues in the deliberate release of genetically-manipulated bacteria. *FEMS Microbiol. Ecol.* **15**, 139–146.
118. Gutterson, N., Howie, W., and Suslow, T. (1990) Enhancing efficacies of biocontrol agents by the use of biotechnology. In *New Directions in Biocontrol* (Baker, R. and Dunn, P., eds.). A. R. Liss, New York, pp. 749–765.
119. Glandorf, D. C. M., Verheggen, P., Jansen, T., et al. (2001) Effect of genetically modified *Pseudomonas putida* WCS358r on the fungal rhizosphere microflora of field-grown wheat. *Appl. Environ. Microbiol.* **67**, 3371–3378.
120. Sundheim, L., Poplawski, A. R., and Ellingboe, A. H. (1988) Molecular cloning of two chitinase genes from *Serratia marcescens* and their expression in *Pseudomonas* species. *Physiol. Mol. Plant Pathol.* **33**, 483–491.
121. Vázquez, M.M., Barea, J.M., and Azcón, R. (2002) Influence of arbuscular mycorrhizae and a genetically modified strain of *Sinorhizobium* on growth, nitrate reductase activity and protein content in shoots and roots of *Medicago sativa* as affected by nitrogen concentrations. *Soil Biol. Biotechnol.* **34**, 899–905.
122. Galleguillos, C., Aguirre, C., Barea, J. M., and Azcón, R (2000) Growth promoting effect of two *Sinorhizobium meliloti* strains (a wild type and its genetically modified derivative) on a non-legume plant species in specific interaction with two arbuscular mycorrhizal fungi. *Plant Sci.* **159**, 57–63.
123. Plaza, G., Ulfig, K., Hazen, T. C., and Brigmon, R. L. (2001) Use of molecular techniques in bioremediation. *Acta Microbiol. Pol.* **50**, 205–218.
124. Sayler, G. S. and Ripp, S. (2001) Field applications of genetically engineered microorganisms for bioremediation processes. *Curr. Opin. Biotechnol.* **11**, 286–289.

3

Producing Proteins Derived From Genetically Modified Organisms for Toxicology and Environmental Fate Assessment of Biopesticides

Steven L. Evans

1. Introduction

Adoption of genetically modified crops in agriculture occurred very rapidly. The first US regulatory approval for an insect-resistant crop came in 1995 for potatoes modified to express a resistance gene targeting the Colorado potato beetle, followed in 1996 by approvals for corn modified to resist European corn borer and cotton modified to resist budworms and bollworms *(1)*. Soybeans modified to tolerate the herbicide glyphosate likewise made their first domestic impact in 1996 *(2)*. Thereafter, a 25-fold increase in hectares planted to genetically modified crops occurred globally between 1996 and 2000.

James suggested that the adoption rate for these genetically modified crops is unprecedented, exceeding that of any prior technology innovation in agriculture *(2)*. According to figures tabulated for 2000, there were 15 countries that participated in this agricultural revolution, with plantings in the United States, Argentina, Canada, and China accounting for most of the acreage. Dominant crops were soybean, corn, cotton, and canola, with an estimated 58% of the global soybean crop planted to genetically modified seed, 23% of corn, and 12 and 7% of cotton and canola, respectively. Crops modified to exhibit herbicide tolerance and insect resistance led the way during this initial wave of truly commercially viable genetically enhanced crops. The estimated global value of all genetically enhanced crops in 2000 exceeded $3 billion *(2)*.

These crops contain genes inserted with tools of modern molecular biology. This unprecedented acceptance of a technology signified the successful convergence of several disciplines of basic science with the practice of large-scale agriculture, governmental regulation, and public perception. The focus of this chapter is on protein test substances necessary for safety evaluation, so it might be expected to focus on protein standards for crops expressing herbicide tolerance (HT). This is because the largest single application for genetically modified plant technology is herbicide-tolerant soybeans, which control 59% of all hectares planted to genetically enhanced crops *(2)*. However, the history of this successful convergence in plant biotechnology was crafted

From: *The GMO Handbook: Genetically Modified Animals, Microbes, and Plants in Biotechnology*
Edited by: S. R. Parekh © Humana Press Inc., Totowa, NJ

by decades of prior science, regulation, and commercial-scale practice of biological pest control, particularly with respect to bioinsecticides derived from the soil bacterium *Bacillus thuringiensis* (*Bt*). Genes derived from *Bt* represent the dominant insect control transgene in genetically enhanced crops today *(3)*, and decades of safe use of this class of proteins in agriculture provided a unique opportunity to examine health, safety, and environmental attributes. Moreover, the regulatory framework created to enable safe and successful use of native and genetically modified microbial *Bt* products served to guide the process developed to evaluate current genetically enhanced crops *(4)*.

This chapter covers methodologies to produce protein test substances used to evaluate the safety of genetically modified products in agriculture. At the onset, it should be appreciated that the actual production methodologies employed on registered products usually are difficult to obtain. Much of the primary literature resides in documents submitted to regulatory agencies; only some of this literature is readily accessible to the public. One notable exception is the methodology behind production of cry3 protein at the 1000-L scale *(5)*. Summaries of the methods used to produce test substances for domestic products are available in the supplemental registration material published on the biopesticide Web site of the US Environmental Protection Agency (EPA) (http://www.epa.gov/pesticides/biopesticides/ingredients/index.htm). This chapter provides a synthesis of the historical rationale behind the types of protein test substances used by registrants to date and provides illustrations of applicable methodologies for commonly encountered proteins expressed in genetically modified microbial and plant products.

The EPA refers to pesticidal genes and proteins in genetically enhanced crops as plant-incorporated protectants (PIPs). It is important to understand the genesis of current PIP regulations because history plays a significant role in the nature of test substances used to examine toxicology and environmental effects of PIP proteins. The current state of regulating PIPs began under the framework first initiated by the EPA to enable a regulatory path for *Bt* insecticides derived from bacterial strains.

In the 1960s, these products were registered on a case-by-case review involving the US Department of Agriculture and the Food and Drug Administration (FDA) *(6)*. Beginning in 1974, the EPA began sponsoring scientific interchanges seeking input on safety concerns for microbial pesticides *(7)*. This pathway was formalized in the 1980s by publication of guidelines for registrants seeking to develop microbial or biochemically based pesticides (Subdivision M, initially published in 1983 *[8]*). The regulatory guidelines were aimed at natural strains of microorganisms and indicated a case-by-case evaluation of genetically altered microorganisms.

Products based on heterologous expression of *Bt* toxin genes in non-*Bacillus* bacterial strains continued to be developed throughout the 1980s, and expanded microbial guidelines were subsequently published in 1989. The first proposed rules for plant pesticides were published by the EPA in 1994 and were based partly on principles developed during the 30 years of experience with the microbial products *(4,9)*. The proposed rules for plant pesticides were subsequently finalized in 2001 (*Federal Register* 40 CFR Parts 152 and 174).

Thus, current requirements for test substances used in toxicology and environmental fate testing of genetically modified crops have their origins in this historical pathway, beginning with sprayable *Bt* biopesticides. In fact, technology enabling efficient large-scale production of *Bt* toxins by fermentation continues to play a significant role in successful registrations of *Bt*-expressing plants because of the concept of equivalency (*see* Section 4.1).

This chapter outlines the historical framework of *Bt* regulation, first to explain the practices used to create test substances for proteins expressed in genetically modified microbial biopesticides, then to bridge those practices to protein test substances for genetically modified crops. Next, it examines practices and options for large-scale protein production for insect control proteins and herbicide tolerance proteins. Last, it examines potential opportunities and future challenges in production of supporting test substances for the next generation of commercial plants enhanced by genetic techniques (output traits, industrial products, therapeutics).

2. Overview of Microbial Pesticides and Plant-Incorporated Protectants

The natural antagonistic propensity of one organism toward another has no doubt shaped the course of human history. Population density was able to increase above that which could be sustained by the native flora and fauna because humankind domesticated animal and plant species. In the process of selecting beneficial domesticated crop traits, restriction of genetic diversity could tip the balance of pathogen vs crop strongly in favor of the pathogen. Localized epidemics, such as the Irish potato blight of 1845, not only can reshape local populations, but also can lead to a dramatic transoceanic emigration. Such disruptions to the balance of nature illustrate the devastating impact natural pathogenic relationships can have at their extreme. However, the norm between a pathogen and its target tends more toward a standoff, without either gaining definitive advantage. Scientists have long wished to understand these host–pathogen relationships in hopes of harnessing for beneficial purposes the power and specificity often exemplified in the disease cycle.

2.1. Historical Overview of Biological Pest Control

Debach and Rosen produced a thorough review of the history of biological control *(10)*. They documented practices of biological control spanning centuries and provided a detailed view of developments over the past 400 years that resulted in elaboration of the scientific foundations of biological control. They defined biological control as a specialized part of the more encompassing concept of natural control, by which natural control constitutes the regulation of populations within some upper and lower limits over extended periods of time. Natural control is effected by both biotic and abiotic factors. Biological control focuses on the use of a natural enemy to regulate the population density of another organism at a level lower than it would otherwise tend to exist. This often is accomplished by focused importation of natural enemies from vastly different geographies.

The successful deployment of natural agents to protect or benefit humankind stands as a tribute to the incredible observational powers of early naturalists and scientists. Illustrations from history can be found in protection of livestock, plants, and even humans *(10)*. For instance, the use of spiders for pest control can be documented in China over 2000 years ago, and there are even earlier allusions to analogous practices in several civilizations. Diseases of the silkworm were studied in Europe, Japan, and China prior to and during the Renaissance. Work by Bassi, Pasteur, Metchnikoff, and LeConte in the 1800s is considered the foundational footing of modern biocontrol. Their work established clear cause-and-effect relationships in pathogenesis and began the documentation of principles of disease transmission. LeConte is credited with initially proposing the concept of microbially based biocontrol in August 1873 at a meeting of the American Association for the Advancement of Science.

Only a few natural enemies can meet both the challenging biological constraints necessary for effective pest management and the practical constraints necessary for sustaining a biocontrol agent as a viable commercial product (for example, the ability to produce consistent product at an appropriate scale and price). One such agent with relevant attributes is the bacterium *Bt*. In fact, *Bt* is the most widely utilized biocontrol agent worldwide *(1)*. The history of development of this organism spans a century *(11)*, and it played a central role in several significant technological and regulatory advances, first in the field of industrial microbiology and ultimately in the field of plant biotechnology. As such, *Bt* is a good case study of the intersection of science, industry, regulation, and society.

2.2. Biological Control Based on **Bacillus thuringiensis**

Bacillus thuringiensis is a Gram-positive, spore-forming bacterium found in soil and other environmental habitats. Members of the species often display an entomopathogenic phenotype *(12–14)*, but the genus *Bacillus* also is used widely as a source of industrial enzymes and other fermentation-derived biochemicals *(15)*. Initial observations of the insecticidal properties of *Bacillus* date to 1901 and 1911. An insect pathogenic bacillus was isolated by the German scientist Ernst Berliner in 1911. The isolate was found in Thuringia, Germany, so the species was named *B. thuringiensis* Berliner. However, in 1901, the Japanese scientist Sigetane Ishiwata had isolated a similar *Bacillus* infecting silkworm. He named it using a Japanese term describing the rapid death observed in the larvae, "sotto," yielding the binomial *Bacillus sotto*. Only much later was this earlier discovery recognized, largely because of language barriers. A fascinating review of over 100 years of history in *Bt* research and development was published by Yamamoto *(11)*.

Strains of *Bt* are characterized by production of spores and of insoluble parasporal proteins that condense in the cell as a crystalline protein complex *(12,13)*. These crystals can account for as much as 25% of the dry weight of a sporulated culture *(15)*. The genes responsible for production of the crystalline proteins are referred to as *cry* genes (= *cry*stalline). To date, there are 40 distinct classes of *cry* genes (reviewed in ref. *16* and updated at the *Bt* nomenclature Web site http://www.biols.susx.ac.uk/home/Neil_Crickmore/Bt/).

The pathogenic armory of *Bt* extends well beyond the spore and cry proteins. In 1967, Heimpel *(12)* grouped the known *Bt* toxins according to their location in relation to the cell (endotoxins or exotoxins) and afforded groups a sequential Greek letter designation (α to δ). Exotoxins were secreted into the culture medium. α-Exotoxin was identified as lecithinase C or phospholipase. β-Exotoxin was identified as a thermostable small molecule with extremely high potency toward flies. γ-Exotoxin was an unidentified protein component with debatable insect toxic potential. δ-Endotoxins were the parasporal crystalline proteins, otherwise known as insecticidal crystal proteins (ICPs). The δ-endotoxins and β-exotoxins ultimately emerged to receive the majority of attention. Subsequently, *Bt* has been shown to produce two other classes of insecticidal proteins that can be isolated from early growth stage supernatant *(17,18)*. The landmark activity setting the stage for the ultimate exploitation of *Bt* occurred in 1981 when Schnepf and Whiteley first cloned and functionally expressed a *Bt* endotoxin gene in a heterologus system *(19)*.

The entomopathogenic bacilli have played prominent roles in industry and in the regulatory aspects of biological pesticides. Projects in at least six countries attempted to exploit *Bacillus* strains for control of insects prior to the 1960s. *Bt* was the first biocontrol organism produced by fermentation on a large scale (1950s) and was the earliest biological pesticide with the potential for broad utility to receive US regulatory approval (1960–1961) *(6,11)*. In 1977, a strain with a novel, nonlepidopteran host range was first isolated. *B. thuringiensis* subsp *israelensis* remains one of the more effective and potent agents for the control of mosquitoes and blackflies *(14)* and represents the point in history at which the potential for broad-spectrum insect control by *Bt* was first realized.

In the 1980s, *Bt* crystalline protein genes were first cloned and functionally expressed *(19)*, thus allowing *Bt* to become a central player in the emerging field of plant biotechnology *(20,21)*. In the 1990s and beyond, genetic engineering with *Bt* genes resulted in the first US EPA approval for a genetically engineered microbial pesticide under the Federal Insecticide, Fungicide, and Rodenticide Act *(22)* and the first US EPA approval for a genetically engineered plant expressing a pesticidal gene *(1)*.

The last half of the decade of the 1990s witnessed broad adoption of *Bt*-expressing commercial crops. In 1999 and 2000, *Bt* became, respectively, the subject of intense scientific and public debate on nontarget effects of cry1Ab on the Monarch butterfly and on prediction of allergenic potential and risk assessment with the development of corn transformed with a novel *cry9* gene *(3)*.

Each of these steps has been intimately tied to characterization of the toxicological dimensions of *Bt* and to assessing worker exposure, impact on nontarget organisms, and environmental impact. Reliable test substances have been crucial to this examination for 50 years *(23)*.

B. thuringiensis was first introduced into large-scale domestic agriculture in 1961 by the commercialization of Thuricide®. This action pre-dated the existence of the US EPA, which was not created until 1970 *(6)*. At that time, the active ingredient of *Bt* was considered to be primarily the spores, and the product was thus examined for safety because it could be an infective agent. The predominant questions addressed by test substances were (1) whether the organism was an acute infective agent and (2) whether the organism was distinguishable from a taxonomic relative, *Bacillus anthracis (6)*. Concentrated spores were used for both tests. In the next two decades, the significance of the β-exotoxin and δ-endotoxins to both biological potency and mammalian safety became clearer, and new test substances were created to document the effects of those components.

The critical thing to note here is that biopesticide commercialization and the regulatory framework around *Bt* developed in parallel. New information on mechanisms of pathogenicity obtained in the 1960s clarified the nature of the protein endotoxins. They were confirmed to be protoxins, thus requiring a proteolytic activation step to convert them to the form toxic to the insect *(24)*. They also were shown to be degraded by the mammalian gut protease pepsin *(6)*. This mechanistic work not only incited the need to develop protein test substances for toxicology, but also resulted in the earliest promulgation of a testing strategy—endotoxin susceptibility to pepsin—that has resurfaced today as central to registration of genetically modified plants producing *Bt* toxins *(4)*.

This pattern of parallel development of technology and regulation continued in the 1970s with the discovery of the mode of action of β-exotoxin. This purine derivative was found to inhibit deoxyribonucleic acid (DNA)-dependent ribonucleic acid (RNA) polymerase and was shown to be mutagenic to flies (discussed more fully in Section 3.3). This initiated extensive testing of commercial *Bt* production strains for the ability to produce β-exotoxin and created new regulatory guidelines and toxicology tests. It also ended domestic exploratory work on β-exotoxin as a commercial insecticide *(6)*.

The 1980s were the decade of parallel development of regulations and products involving numerous genetically modified microbial production strains. The 1990s was the decade of parallel development of regulations for products based on genetically modified plants expressing the *Bt* toxins. It was this close relationship of development and regulation in the microbial product arena over a 30-year period that set the stage for regulatory pathways and testing strategies relevant to genetically modified crops. As the example with pepsin digestion shows, concepts elaborated during development of *Bt* as a microbial pesticide often resurface as testing requirements for genetically modified crops. It is with this concept in mind that such attention is given to the history and practice of *Bt* biopesticide development in this chapter. Because of the confidential nature of many test substances, describing why it is made tends to be easier than describing how it is made. However, by understanding the former, one should have a better chance of producing appropriate test materials for their subject proteins.

2.2.1. Sprayable Native Bt *Insecticides*

The 1940s witnessed the emergence of synthetic chemical agents that were economical and more effective than the naturally derived agents in use. Synthetic chemistry came to fruition in areas such as polymers (synthetic rubbers, early plastics, nylon) and agricultural chemistry (DDT). Synthetic pesticides were thus primed to replace products based on nicotine extracts, rotenone, and various inorganic toxicants. Thus, the 1940s represent the start of the boom in agricultural chemicals. At the same time, World War II ushered in a tremendous need for fermentation capability to make antibiotics *(25)*. Efforts at strain optimization, advances in large-scale aseptic practices, and mechanical improvements in fermentors led to phenomenal gains in productivity for antibiotics.

These advances in industrial microbiology yielded benefits in areas outside that of antibiotic production. One beneficiary was the insecticidal bacterium *Bt*. The development of new fermentation technologies was central to commercialization of microbial biopesticides. Although the ultimate trajectories were different, it is interesting to observe that fundamental technologies related to large-scale practice of both traditional agrochemicals and microbial pesticides could be traced to a coincident decade of origin *(26)*.

Early efforts at commercial use of strains of *Bacillus* as an insect control agent took place in France (1930; Sporeine *[3]*) and the United States (1948, Doom®, *Bacillus popilliae*; *10*). However, these efforts did not make a significant impact on the market. Pioneering work in the 1950s by Edward Steinhaus at the University of California, Berkeley, is credited with launching the domestic effort that became what is known today as the sprayable *Bt* product concept *(6,11,26)*. Solid-state production in a laboratory setting was used to produce test material that successfully demonstrated the field performance of *Bt*. Furthermore, Steinhaus successfully convinced a fermentation company (Pacific Yeast, Wasco, CA) to produce *Bt* in stirred tanks. This early boutique fermentation became the basis of Thuricide biological insecticide, the first truly large-scale *Bt* product. It was registered in 1961 by Bioferm Corporation and subsequently ushered in a decade of active pursuit of biological pest control in the form of fermentation-derived products based on *Bt*.

During the 1960s, broader commercial interest in biocontrol emerged as it became clear that *Bt* had the potential to offer competitive products in certain areas of use. For instance, *Bt* products came to enjoy wide usage to control cabbage looper in cole crops in California *(26)* and as a significant tool to control gypsy moth in forestry applications *(14)*. The species was of industrial interest because it was not only potent enough for commercial application, but also extremely specific, in contrast to some of the insecticidal chemistries in use during the 1960s. In fact, prior to 1977, the specificity of *Bt* was believed to be limited solely to lepidopteran pests, although now examples are available of *Bt* toxins active on a broad variety of insects, nematodes, flatworms, and protozoa *(27)*. This specificity was important because it also was manifested by a very favorable toxicity profile with respect to mammals *(28)*.

The *Bt* formulations based on native bacterial strains produced in deep-tank fermentation were the first biopesticide products taken to registration. *B. thuringiensis* subsp *kurstaki*, one of the earliest production organisms, remains even today one of the most widely utilized production hosts for native products *(1,26,29)*. The desire to add new genes to a strain to improve the spectrum led to the need to modify production strains genetically. Conjugal transfer of *cry* genes residing on megaplasmids was demonstrated *(30)* and became an important path to products for Ecogen Corporation *(26,29)*.

Bacterial conjugation as a means of genetic modification did not prompt significant new review hurdles for a production strain. However, it suffered several disadvantages, some of which were addressable by advances in electroporation of *Bt* strains *(11)*. These techniques allowed genetic advances to take place in production strains, but did not significantly raise the level of scrutiny for registration because the techniques were regarded simply to overcome barriers to natural gene transfer *(4)*. In contrast, as technologies enabled intergenic transfer by recombinant methods, the hurdle for regulatory review increased dramatically. The 5-year delay in field testing of live recombinant Ice$^-$ strains because of regulatory and public concerns *(31)* dampened, but did not end, recombinant work in *Bt* directed at commercial products. These intergenic recombinant efforts did not yield significant fruit until the late 1980s.

2.2.2. Recombinant Sprayable Bt Insecticides

Currently, there are over 200 registered microbial pest control products in the United States (http://www.epa.gov/pesticides/biopesticides/product_lists/microbial_prods_by_ai.pdf). *Bt* still dominates product usage, accounting for 95% of all microbial pesticide products *(32)*. A subset of these products results from use of recombinant DNA technology to express *Bt* genes in a non-*Bacillus* microbial host.

In the United States, seven *Bt*-based products are produced by expression of a *Bt* insecticidal gene in *Pseudomonas fluorescens* in the CellCap® bioencapsulation system *(33,34)*. The *P. fluorescens* strain was originally selected to compete successfully as an epiphyte and was envisioned for release as a viable, heterologous recombinant organism that would propagate on treated leaf surfaces. In light of the regulatory climate in the mid-1980s regarding release of viable recombinant organisms, focus shifted to development of a novel means of killing the bacterium at the end of the fermentation cycle *(33)*. This was successfully accomplished and documented to the satisfaction of EPA, resulting in the first registration of a genetically engineered biopesticide *(22)*.

Work continued on *Bt* strains modified by recombinant procedures. Such efforts included techniques to enable expression of multiple *cry* genes *(29)*, modification of fundamental strain properties such as sporulation dependence of toxin production *(35)*, protease production *(36)*, and elimination of β-exotoxin production *(37)* or diarrheal toxin production *(38)*. During the 1990s, the stage was set for exploitation of *Bt* in a variety of heterologous production systems beyond recombinant *Bacillus* and *Pseudomonas*.

For example, *Bacillus* mosquitocidal genes were expressed in cyanobacteria (*Synechococcus, Anabena, Agmenellum*) *(39,40)* or the aquatic bacterium *Caulobacter crescentus (41)*, organisms on which mosquito larvae feed. This was an attempt to place the toxin in a more normal component of the larval diet and to overcome the rapid settling of the toxin that plagues standard *Bt* formulations in an aqueous setting *(26,39)*. In addition, Crop Genetics Corporation introduced *Bt* genes into the monocot endophyte *Clavibacter xyli (42)*, an organism limited to reproduction in the xylem. The product concept involved inoculation of corn seed with the recombinant bacterial product (InCide®), which would grow in the xylem of treated plants, express a *Bt* gene, and protect the crop against the European corn borer.

The late 1980s signified the crossroads in *Bt* biotechnology. Native *Bt* products were clearly performing in market areas. Robust heterologous expression in microbial systems was clearly possible on the commercial scale and would enter markets in the 1990s. Novel expression systems were under development to provide unique delivery systems for recombinant biopesticides; by 1987, *Bt* genes were in early transgenic plants *(43)*. At this juncture, the concluding comment in Heimpel's seminal review of *Bt* 20 years earlier *(12)* had proven quite prophetic: "The future looks bright indeed for insect control with *Bacillus thuringiensis*!"

2.2.3. Other Microbial Biocontrol Agents

Other bacilli have been developed into commercial products. *B. popilliae* (Doom, toxic against Japanese beetle) has been in use since 1948, but reportedly is no longer registered *(26)*. Several *Bacillus* products are active against dipteran pests, with an emphasis on mosquito. These include products based on *B. sphaericus* (Vectolex®, toxic against mosquito larvae) and *B. thuringiensis* subsp *israelenesis* (for instance, Vectobac®, Teknar®, and Bactimos®, active against mosquitoes and blackflies) *(10,26)*. However, these agents do not have a spectrum or use pattern consistent with transgenic crops. In New Zealand, a strain of *Serratia entomophila*, the causal agent of amber disease, is registered to control a grub in pasture (Invade®) *(26)*. One new recombinant biopesticide product, Messenger® (Eden Bioscience, Bothell, WA), is based on the harpin protein derived from *Erwinia amylovora (44)* and expressed in *Escherichia coli (45)*.

3. Test Materials for Microbial Pesticides

Current guidelines for registration of microbial pesticides are found in the *Federal Register* 40 CFR Parts 158 (for conventional biopesticides) and 172 (for genetically altered organisms). A set of testing and protocol guidelines known as Subdivision M was published as a companion to 40 CFR Part 158 *(8)*. These regulations and their application have been reviewed in detail *(9,46–48)*. These guidelines are quite complex and have to be coordinated domestically with the Animal and Plant Health Inspection Service division of the US Department of Agriculture and with the FDA. The guidelines are further harmonized with testing requirements of the European Union and the Organization for Economic Cooperation and Development *(48)*.

To understand test material requirements better, it is useful to examine briefly the evolution of these guidelines from their first promulgation in the early 1980s. In 1984, the EPA published guidelines to facilitate registration of products based on microbes or that had "biochemical" modes of action (pheromones, etc.). Central to their framework, these guidelines codified the toxicological tests needed for registration in a tiered manner. A base set of tests was to be conducted on all candidate products (Tier I tests), and a second and third tier of tests was defined in the event any results from Tier I required further amplification. With the toxicology tests defined, characteristics of the test substances could then be derived *(7,47)*.

The underlying framework for decision making on test substances follows a traditional scale-up metaphor *(47)*. Because the microbial pesticide is a fermentation product, it has a well-defined process to proceed from initial seed culture to full production volumes, then through the steps of recovery, concentration, and formulation. The expected range of concentration of active ingredient present in each of these process steps could be ascertained through analysis of representative samples. Also, one could capture representative samples of the material at these process steps to serve as test materials for toxicology studies. Viewed in this light, microbial biopesticides are analogous to most any other manufactured product, including chemical pesticides.

Microbial pesticides are examined in Tier I tests for mammalian toxicity, worker safety, and environmental fate. Tier I test substances emphasize two to three main components of this overall scale-up process. First is the actual end-use product. Second is the active ingredient in its purest or highest concentration form, also known as the technical grade active ingredient. Third is the fermentation product at the end of the run, which on occasion is utilized as a test substance. These represent the predominant materials to which a production worker, end user, or the environment would be exposed.

Although built on a framework of chemical safety assessment, these Tier I tests take into account a novel property of microbial pesticides compared to chemicals—the ability to propagate. In Tier I, animal models are used to determine the presence of acutely toxic or pathogenic properties. The test substance also is applied to animal models by various routes of entry (oral, dermal, ocular, or pulmonary and by injection) to assess irritation or pathogenicity. These tests also provide data for assessing the impact of potential exposure of manufacturing employees during production of the final product. Tier I nontarget tests encompass exposure of suitable aquatic, terrestrial, and avian species to the test materials *(47)*.

3.1. Bt *Spore Concentrates*

First-generation sprayable biopesticides were based on natural strains of *Bt* produced in stirred tank fermentation. The *Bt* fermentation life cycle includes a period of vegetative cell growth that subsequently shifts to production of endospores and the insecticidal proteins, finally terminating in cell lysis *(7,35)*. Thus, these early products were a complex mixture of vegetative and lysed cells, spores, insecticidal proteins, and fermentation by-products. The earliest domestic product, Thuricide, was manufactured on the basis of spore counts *(49)*. As discussed by Rogoff, the primary concern in the late 1950s with this "new" class of pesticide was pathogenicity testing "to eliminate the question of frank infective potential" *(6)*. This was accomplished by performing a series of tests using spores as the material administered to the animals. The seminal review on mammalian toxicity of pesticides based on *Bt*, published in 1995, focused intensively on spore toxicology and pathogenicity *(28)*.

Generalized production schemes for *Bt* at the pilot and larger scale have been published (*7,50–52*, and references cited therein), although actual commercial production routes often remain confidential *(52)*. In general, a spore preparation used for toxicology would be derived from an actual commercial pilot or production run using the relevant strain and a proven production medium. For instance, commercial powders or suspensions provided by the manufacturer were used when Siegel et al. examined the mammalian toxicity of products based on *B. thuringiensis* var *israelensis (53)*.

The fermentation run would proceed through sporulation and insecticidal protein production and terminate in lysis. According to Couch, typical fermentation broths might achieve final concentrations of 6–8% solids, with 1–3% representing a mix of spores and crystals *(52)*. Lopez-y-Lopez et al. indicated the typical commercial broth might attain spore concentrations of up to 10^9 ml^{-1} *(51)*.

The next step involves concentration of the fermentation broth, usually by some form of centrifugation or filtration. Analysis and quality control of *Bt* spores relies primarily on determination of their concentration by spore-counting procedures involving microbial plate count techniques *(4)*. Following the administration of the spore concentrates to the test animal, tests often employed viable microbial counts on tissue or organ samples to assess growth rate or clearance of the spores by the mammalian defense systems *(23,28,53)*. Despite their historical use, spore counts do not provide a reliable estimate of product potency or concentration of the biopesticide *(49)*.

Purity of spore preparations is a complicated issue, but is not normally dealt with in these regulatory studies. If a pure preparation of spores is desired, physical separation of the spores and crystalline proteins would be required. Spores and the protein endotoxins can be separated by centrifugation on sucrose *(54)*, sodium bromide *(55)* or Renografin® gradients *(56)*. The aspect of purity that has been discussed is potential contamination of test preparations with β-exotoxin (*see* Section 3.3). β-Exotoxin has an acute oral median lethal dose (LD_{50}) of 170 mg kg^{-1} in rats, making it highly potent to mammalian test subjects. The presence of this compound as a contaminant in test substances has been implicated in unexpected toxicity of certain spore preparations *(28,53)*.

An interesting historic test has taken on new meaning in today's focus on bioterrorism. *Bt* resides taxonomically in a triad along with *B. cereus* and *B. anthracis (7)* (*see* Section 3.4.1). The organisms are difficult to distinguish based on biochemical tests, and in 1960, there were no molecular tests. Because the commercial objective is to produce literally kilograms of spores, confirmation that a batch is not unexpectedly contaminated with anthrax is crucial. Scientists in the FDA Bureau of Biologics unit developed a test to distinguish *B. anthracis* from the other bacilli based on its markedly toxic effect to mice on intraperitoneal injection *(6)*. This intraperitoneal injection test of spores is still performed today on all fermentation batches of bacillus-based *Bt (4)*.

Although the discussion of spore preparations for toxicology might seem distantly removed from the practice of genetically modified crops, it is helpful to recognize the convergence of historical precedence leading to the registration of today's genetically modified crops. In conducting the assessment to exempt certain cry proteins from the requirement of tolerance, the EPA utilized data collected from just such mixed spore–protein preparations in making its final opinion *(57)*. Data from spore–protein concentrates were cited as supporting data indicating the lack of toxicity or other ill effects of *Bt* proteins used for transgenic plants. This long history of use of *Bt*, coupled with the extensive database of spore–protein toxicology, facilitated registration of *Bt* endotoxin-expressing plants *(1,4)*.

3.2. Bt *Endotoxin Preparations*

Early efforts to categorize *Bt* strains employed such techniques as flagellar serotyping and crystal morphology determination *(7)*. In the late 1970s, multiple efforts yielded antibodies capable of immunological distinction between crystal protein components *(58)*. However, the real advances in toxin classification occurred once molecular techniques were applied when the genes corresponding to the endotoxins became known *(59)*. Strains could possess multiple *Bt* genes, but differential expression during fermentation meant the actual toxin composition of a *Bt* product was not definitively known.

In the early 1990s, analytical techniques that resolved the crystalline toxins were developed and applied to characterization of commercial products *(60)*. Not surprisingly, most commercial products were found to contain a mixture of genes and expressed endotoxins *(61)*. However, purified protein extracts from commercial biopesticides have been used only sparingly in *Bt* toxicology. The primary case involved characterization of the alkaline soluble cytotoxic/hemolytic toxins from *B. thuringiensis* var *israelensis (53)*.

The primary impetus to initiate a more focused examination of *Bt* protein toxicology came as a result of development of genetically modified *Bt* strains. Initially, novel *Bt* strains were constructed by conjugation to add specific new toxins to existing strains *(59)*. This was considered a nonrecombinant strategy and was a primary commercial strategy of Ecogen Corporation. A *Bt* endotoxin gene was first cloned and heterologously expressed by Schnepf and Whiteley in 1981 *(19)*, thus placing *Bt* toxins squarely in the recombinant DNA revolution. Heterologous expression, first in microbes and ultimately in plants, drove the need for protein test substances to characterize *Bt* toxin effects *(4)*.

A hallmark of *Bt* crystal proteins is their extreme insolubility. Endotoxins typically are protoxins, which need a proteolytic processing event to aid solubility and activate the toxic component *(24)*. These properties are used to advantage during large-scale production of purified endotoxin. Expression and purification of various *Bt* toxins in heterologous systems at the laboratory scale are well documented in the scientific and patent literature. These are discussed in Section 4.1 as they directly relate to protein test substances for genetically modified crops.

3.3. Exotoxins

Part of the insecticidal arsenal of *Bt* includes an adenine nucleotide derivative known as β-exotoxin *(12)*. This molecule is produced by certain strains of *Bt* and is secreted into the culture broth. The common name for the best-studied form of the toxin is thuringiensin, although Levinson reported the existence of another structurally uncharacterized molecule referred to as type II β-exotoxin *(37)*. The mechanism of toxicity is inhibition of DNA-dependent RNA polymerase. β-Exotoxin is markedly potent against a variety of insects, and the molecule was under commercial development both domestically and abroad. Domestic development ceased once the mutagenic properties were discovered *(6)*. Nonetheless, limited commercial activity is ongoing in Taiwan *(62)*.

Production of exotoxin is strain dependent *(37,63)*. Commercial producers began shifting production strains to those documented not to produce the metabolite upon the discovery of nontarget effects of β-exotoxin *(4,6,28,37)*. Federal regulations require commercial *Bt* formulations to be free of β-exotoxin, so batch surveillance is conducted by the fly bioassay method *(64)*. Libraries of *Bt* strains have been screened for β-exotoxin production to identify new production strains *(37,63)*. One attraction of heterologous production of δ-endotoxins is the elimination of the possibility of β-exotoxin production in commercial products. *Bt* β-exotoxin is of no concern in transgenic *Bt* crops.

β-Exotoxin is only 701 Da, but it can be isolated from broth by micellar-enhanced ultrafiltration. Cetylpyridinium chloride forms micelles, which have high affinity for β-exotoxin *(62)*, facilitating recovery of 90% of the product by 10,000 molecular weight cutoff membranes. Final purification is accomplished by high-performance liquid chromatography. Analytical methods for β-exotoxin include high-performance liquid chromatography *(37,62)*, capillary electrophoresis *(62)*, enzyme-linked immunosorbent assay *(65)*, and the fly bioassay *(64)*. The bioassay is based on the heat stability of β-exotoxin and its exquisite potency on the housefly.

3.4. Other Bt Fermentation Test Substances

3.4.1. Anthrax

The phylogeny of *Bt* places it in a triad of closely related Gram-positive spore-forming species: *B. thuringiensis, B. cereus*, and *B. anthracis (7,66)*. These species are sufficiently similar that biochemical typing does not adequately distinguish them. The presence of plasmid-borne toxins is the most reliable molecular approach because genome similarity is very high *(66)*. Nonetheless, the mammalian pathogenicity of *B. anthracis* allows ready and unambiguous differentiation to occur. A test was devised based on intraperitoneal injection of spores for detection of *B. anthracis*, and it is used to eliminate the possibility of accidental contamination of spore-based *Bt* products by *B. anthracis (6)*. Therefore, pathogenicity testing to rule out the presence of anthrax was a fundamental requirement of *Bt*-based pesticides *(4)*. The typical test sample was derived from a relevant in-process spore concentrate from the batch. Analytical procedures generally involved determining the spore concentration by spore-counting procedures.

3.4.2. Diarrheal Enterotoxins

Bt is closely related to the opportunistic food pathogen, *B. cereus*. A primary pathogenic determinant of *B. cereus* is a mammalian-active diarrheal toxin *(67)*. Existence of the gene in a variety of *Bt* isolates has been demonstrated by polymerase chain reaction *(68)*, and production of this toxin has been shown by immunoassay in commercial formulations *(69)*. Of note, heterologous production of δ-endotoxins in non-*Bacillus* strains eliminates concern for the presence of this agent. The recombinant product MVP™ bioinsecticide, which contains a *Bt* endotoxin in killed *P. fluorescens* cells, was the only commercial formulation tested that did not show the presence of diarrheal toxin by immunoassay *(69)*.

Isolation of the *B. cereus* toxin has been reported *(70)*. The procedure employed ammonium sulfate precipitation, ion exchange, chromatofocusing, and size exclusion chromatography. Testing for this agent is not a current requirement for commercial biopesticides, although the increased attention it is receiving in the literature may cause increased scrutiny of new products in strains without a long history of commercial use *(4)*. Some groups are developing nonenterotoxigenic *Bt* strains for endotoxin production *(38)*. This component is not a constituent of *Bt*-expressing crops, so no test substances focused on this toxin would be required in development of a genetically modified crop.

4. Test Materials for Plant-Incorporated Protectants

The first field trials involving a genetically modified crop occurred in 1987; the cry1Ab protein was expressed in tomato to control *Heliothis zea (20,71)*. The EPA registered the first PIP in 1995. The cry3A insecticidal endotoxin was expressed in potato for the control of Colorado potato beetle, and the product was introduced under the name NewLeaf® Potato *(3)*.

This introduction of genetically modified crops for insect control crystallized two significant new challenges with respect to generation of test substances in the regulatory and risk assessment pathway. First, the classical industrial scale-up metaphor, which effectively served the fermentation-based microbial biopesticide registration process, was no longer relevant. Second, protein expression levels in genetically modified plants can be extremely low, thus limiting availability of the plant-derived protein for specialized testing.

The scale-up metaphor was central to generation of test substances for fermentation-derived biopesticides *(4)*. With a typical sprayable biopesticide, the producer uses a fermentation process to make large quantities of toxin, then manipulates the fermentation broth to provide the bioinsecticide in a suitable formulation to a user. Thus, toxin is present throughout the later stages of scale-up and obviously is present in the actual article in commerce. In a typical fermentation-and-recovery process, the protein toxin is present at high concentration in a certain step or steps. This might correspond to a concentrated liquid or dried intermediate, depending on the manufacturing process, and this concentrate is referred to as the technical grade active ingredient. Although it is central to the manufacturing process, it also provides a convenient starting point for the isolation of large quantities of protein to enable specialized testing for toxicology. Because the pesticidal protein is manufactured, tests for regulatory or toxicology purposes can be conducted on an authentic substance isolated directly from the production stream.

With a genetically modified plant, the scale-up metaphor could no longer be sustained. The product actually "manufactured" and sold is a seed or other propagative tissue, not the insecticidal toxin itself. Although the gene is present throughout the plant's growth cycle and tissues, the protein might or might not be, depending on the nature of the regulatory elements used in assembling the molecular construct (tissue-specific or inducible promoters) *(72)*. Furthermore, there is no point during the normal plant production process at which the protein obligately would be found at sufficient concentration to enable isolation of gram quantities of protein.

To frame the magnitude of the issue, it is helpful to note the protein expression levels in some early plant-incorporated pesticides. Fuchs et al. observed that proteins in engineered plants could be present at levels as much as 10,000 times lower than those obtained from the same gene in fermentation *(73)*. Perlak et al. described the levels of cry1Ac in Bollgard® cotton as ranging from 0.0005 to 0.0020% by fresh weight of tissue *(20)*. Finally, in the EPA's *Pesticide Fact Sheet* for BT11 corn, a computation was produced describing the amount of toxin and biomass produced per acre (http://www.epa.gov/fedrgstr/EPA-PEST/1998/August/Day-17/attri.htm). In this corn line, an acre of crop would produce a total of 259 g of cry1Ac toxin, and this would be present in 89,300 lb of biomass. Quantities of protein for even a minimum series of toxicology tests are on the order of 10 g (*see* Section 4.1). Based on the expression range for cotton mentioned by Perlak et al., 10 g of cry1Ac would be contained in 500–2000 kg fresh plant material; for corn, this would be about 1570 kg fresh plant material.

The EPA recognized this issue and responded by elaboration of the concept of equivalency *(4)*. This concept permits protein production via fermentation to be used as a surrogate protein in toxicology tests to support registration of the genetically modified plant.

4.1. The Concept of Equivalency

As the EPA considered the earliest genetically enhanced insect-resistant crops, it was clear that the plants themselves could not provide the quantity of material necessary for even preliminary toxicology assessment. A significant component of this assessment is determination of the acute oral toxicity of the protein to mice. With biorational pesticides, it is ideal to seek a no observable effect level of 2000–5000 mg kg^{-1} in the experimental animal, the EPA's lowest level of toxicity. This corresponds to a dosing level of 100–125 mg test substance per experimental subject when conducting the analysis with appropriately staged mice. Test populations can reach 40–50 animals when subjects from both genders are included and a reasonable number of replicates are performed. Thus, the required test material for this one study alone easily surpasses 6 g of protein.

The EPA allows applicants to produce large quantities of test protein in fermentation provided small quantities of authentic material are extracted from the plants for comparison. The fermentation-derived material is compared to the authentic protein in biological and biochemical tests to document equivalency *(4,20,74)*. These materials are tested for potency and spectrum in relevant insect bioassays, examined by sodium dodecylsulfate-polyacrylamide gel electrophoresis (SDS-PAGE) for similarity in apparent molecular weight under denatured conditions, tested for immunoreactivity in Western blots and enzyme-linked immunosorbent assay, subjected to tests for post-translational modification such as glycosylation, and have the primary sequence of their N-terminus determined *(74)*. The point of these tests is not to determine that the respective protein samples are identical, but rather to use the combination of results to "indicate a high probability that these two sources produce proteins that are *essentially* identical" (italics added; *75*).

The concept of equivalency is central to the creation of suitable protein test materials for safety and environmental testing of genetically modified crops. Interestingly, it also links progress in plant biotechnology with the historic predecessor it supplanted, fermentation-derived protein production. The multigram quantities of protein needed for registration studies generally are out of the range of laboratory-scale production. For instance, the yield MacIntosh et al. reported for cry3A in a 10-L fermentation in recombinant *E. coli* was 300 mg *(76)*, and large-scale preparation of the same protein ultimately resulted from a 1000-L fermentation *(5)*.

This points out an anachronism of modern plant biotechnology. For many reasons, the technology shifted away from fermentation toward plant-incorporated production of proteins, yet the ability to suitably test and successfully register a genetically modified plant currently requires access to protein production via classical fermentation scale-up. In general, a suitable microbial expression strain must be developed, and process development to the 100- to 1000-L scale must be performed. Time spent optimizing microbial expression can result in the benefit of processing smaller volumes, but that establishes the need for an expression optimization program in a bacterial system. The protein that results from this bacterial scale-up needs to be timed to support the application for the Environmental Use Permit for field release. This illustrates how microbial expression and scale-up are inserted into the critical path of development of a genetically modified crop.

4.2. The Scope of Test Substances Required for Genetically Modified Plants

Guidelines for the required testing of genetically modified plants somewhat followed those set for biopesticides in Subdivision M *(4,48)* with a few key differences. A key issue with biopesticides was their potential for infection and pathogenicity, but that issue is moot in plants expressing only the insecticidal toxin gene. Likewise, certain tests aimed at worker exposure (intraperitoneal, dermal irritation) or process-specific by-products (β-exotoxin) likewise no longer were relevant. However, new issues arose as a consequence of the protein being expressed continuously in the plant and because it could be present in a potential food source. Thus, tests aimed at digestive and environmental fate became more prominent in testing and assessment of genetically modified plants *(4)*.

Unlike microbial pesticides, plant-incorporated pesticide registrations tend to proliferate toxicology and environmental fate test substances. Often, this is because of the nature of the exposure system; for instance, bees preferably are tested with pollen expressing the protein rather than with simply protein itself. However, to dose the test subjects adequately, amendment of toxin to the pollen might be required. Depending on the nature of the question under investigation, the test substance can be (1) purified microbial protein; (2) purified or enriched protein from the modified plant; (3) control protein isolated from an isogenic plant lacking the transgene, but extracted in a way identical to the material in item 2; (4) pollen from modified and isogenic plants; (5) dried plant tissue (often leaf or root) from modified and isogenic plants, (6) various animal feeds made from plant material from modified and isogenic plants; and (7) any of the above amended with purified microbial protein to increase the dose. As mentioned in Section 4.1, the equivalency determination established the need for biochemical tests employing substances 1, 2, and 3 to demonstrate interchangeable use.

Insect resistance management (IRM) programs also require access to protein test substances. This element of the registration process usually has experimental and field components. The experimental part involves biochemical testing to estimate likelihood of cross-resistance, often by ligand blotting or other protein–receptor techniques *(77)*. It can include bioassays on resistant insects if available and performance of selected tests from a defined menu to establish that the plant expresses the toxin at a level sufficient to exceed the 50% lethal concentration by a factor of 25 *(78)*. In any of these tests, purified microbial proteins or dried plant material are typical test substances, and required quantities are small.

Field studies can entail longitudinal monitoring of target organism susceptibility using field-collected insects in a bioassay. This can also be expanded to include other target and nontarget species. For these longitudinal studies, microbial protein usually is the material of choice because a robust IRM program can consume multigram quantities annually for the life of the product.

In summary, the testing scheme for a genetically modified plant requires a robust supply of test substances. The predominant material used is derived from microbial expression of the relevant gene, followed by purification. Critical uses for the purified microbial protein include that for the acute oral toxicity determination, as an amending agent to other test substances to elevate the dose for better estimation of hazard of exposure, for biochemical tests to demonstrate equivalency to plant-produced protein, and for IRM studies. The need is for multigram quantities and, depending on the IRM program, can recur for the life of the product. Additional test substances are enriched or further purified from plant tissue and, because of their complexities, frequently require a control batch extracted in parallel from an isogenic line differing only in the presence of the relevant gene. Finally, animal feeds derived from appropriate tissues of transgenic and isogenic lines round out the feeding studies required of the products for registration.

4.3. Protein Derived From Bt *Toxin Genes*

Bt genes representing the cry1Ab, cry1Ac, cry1F, cry2Ab2, cry3A, cry3Bb1, cry9c, cry 34Ab1/cry35Ab1, and vip3A proteins have been presented to the EPA for registration in transgenic crops. In each case, test substances comprising proteins produced in microbial systems would be a part of the submission package. The toxicology conclusions obtained from tests involving some of these substances are documented in the open scientific literature and are summarized in EPA reports; however, the methodology of preparation of these substances is more difficult to collect. The material best documented in the scientific literature is cry3A protein, used in potato for the control of Colorado potato beetle *(5)*. Because of a large amount of regulatory activity surrounding Starlink™ corn and assessment of its potential for allergenicity, there also is quite an extensive documentation of protein test substances available on the EPA Web site (http://www.epa.gov/pesticides/biopesticides/pips/starlink_corn_archive.htm).

Laboratory-scale purification of the cry3A toxin from *B. thuringiensis* var *tenebrionis* expressed in *E. coli* has been described. The procedure exploited a propensity of the toxin to solubilize at high pH and precipitate at near-neutral pH. Protein purity was 96%, as assessed by densitometry following SDS-PAGE, and the yield from 10 L of fermentation was 300 mg *(76)*. Large-scale purification was described 7 years later at a scale of 1000 L *(5)*.

The contrasts between the two methods illustrated components critical to successful scale-up. First, the ultimate production strain was modified to eliminate one ribosome-binding site, thus yielding a single 68-kDa protein rather than a mix of that protein and the full-length, 72-kDa protein. Second, the size of inclusion bodies obtained on scale-up was smaller than previously experienced, so the large-scale procedure required increased centrifugation time, but fewer washing steps. The method of solubilizing used at the smaller scale (pH 10.0) was only partly effective. At the larger scale, a higher pH solubilization process, coupled with a lower temperature precipitation methodology, was needed. The reported yield was 52 g of lyophilized cry3A, which is about 1.7 times the productivity reported at the 10-L scale.

The analytical details are instructive regarding protein test substances. Based on protein analysis by SDS-PAGE, the protein purity was estimated as 100%. That is, no other proteins were detectable by SDS-PAGE and Coomassie staining. However, the product purity was determined as 90% by weight because of the presence of water and sodium ion. Appropriate analytical tools were applied to confirm identity of the protein (amino acid composition, N-terminal sequencing, bioassay), although immunochemical analysis was absent in that report.

Noteborn et al. *(79)* briefly described production of cry1Ab protein for toxicology assessment of genetically modified tomato. The production strain was identified as *E. coli* strain K514, and the laboratory methods of Hofmann et al. *(77)* and Höfte et al. *(80)* were used. Cells of *E. coli* containing the *Bt* inclusions were disrupted with lysozyme and sonication. Cell debris was washed with detergent and salt. Inclusions were solubilized in carbonate buffer containing 0.2*M* sodium thioglycolate at pH 9.5. Solubilized protein was dissolved in a buffer containing 1*M* urea and 2-mercaptoethanol and applied to a diethylaminoethylcellulose (DEAE)–cellulose column. Purified protein was precipitated at pH 4.0, then solubilized in an alkaline buffer and subjected to size exclusion chromatography. The protoxin was subjected to trypsin or chymotrypsin digestion to generate the 60-kDa active fragment.

This material was subsequently determined to contain copurified DNA *(77)* (*see* Section 4.5.1), so the following additional steps were taken to remove it. The toxin was first precipitated with ammonium sulfate, then solubilized in a Tris buffer with moderate salt content. DNA was precipitated by addition of streptomycin sulfate. Following centrifugation and dialysis, the material was subjected to anion exchange chromatography. Quantitative yield or analysis of the proteins was not provided.

Chaturvedi et al. described a procedure for isolation of cry1Ac that claimed to eliminate components responsible for *Bt* toxin oligomerization *(81)*. An *E. coli* strain expressing the full-length *cry1Ac* gene was cultured under batch conditions and harvested. Cells were lysed by sonication, and the inclusions were collected by centrifugation, then washed in 0.5*M* salt with detergent. Inclusions were solubilized at pH 10.5 overnight, then activated with trypsin overnight. The trypsin treatment removed the bulk of contaminating proteins from the cell lysate. The activated toxin was applied to a phenyl sepharose column equilibrated with 2*M* KCl at 4°C and eluted with an inverse salt gradient. The authors noted that cry1Ac was irreversibly bound to the column if chromatography was conducted at room temperature in the presence of 2*M* salt. The toxin resisted elution with organic modifiers or even 6*M* urea. Contaminating DNA eluted from the column during the high-salt wash, and the *Bt* toxin eluted at the end of the inverse gradient. Their purification yield was calculated as 43%, and the protein purity was >95%. The method was performed at the laboratory scale (12-mL column, approx 100 mg starting material).

Other cry *Bt* proteins in the 72- to 130-kDa range can be purified by procedures like those just described. A binary cry toxin active on corn rootworm and comprised of a 14-kDa component (cry34Ab1) and a 44-kDa component (cry35Ab1) *(82)* has been submitted to the EPA for registration in corn. These proteins were heterologously expressed in a *P. fluorescens* strain and purified from inclusions. Proteins were extracted from inclusions using citrate buffer at pH 3.3, and the 14-kDa protein was subjected to further purification via ion exchange chromatography. A noncrystalline *Bt* toxin, vip3A, has been submitted for registration in cotton. This class of protein typically is purified from *E. coli* lysates by pH-induced precipitation at pH 4.5. Following resolubilization at alkaline pH, the material is subjected to anion exchange chromatography *(17,18,83)*.

4.4. Selectable Markers and Herbicide Tolerance Proteins

Transgenic *Bt* crops contain additional genes with different functions in the product. The EPA requires that test substances be created for all genes producing expressed proteins in the genetically modified plant. Selectable markers can be included to enable identification of transformants during the early portion of the transformation process. These typically are genes coding for resistance to antibiotics or herbicides. The genes widely used for this purpose are the *nptII* gene (neomycin phosphotransferase II, kanamycin resistance) *(84)* and the *bar* or *pat* genes (bialaphos or L-phosphinothricin resistance) *(85)*. Other genes, such as *hyg* (hygromycin resistance), have been used in plant transformation, but have not been widely used in domestic crops *(86)*.

Crops also can be modified for field tolerance to commercial herbicides. By far the most prevalent example is resistance to the herbicide glyphosate (RoundUp®). The most common mechanism of resistance is via mutant 5-enolpyruvylshikimate-3-phosphate synthase (EPSPS) *(87,88)*. Some genes discussed not only are suitable for selection of transgenic events (*bar* and *pat*) *(89,90)*, but also are usable in the field as a herbicide tolerance trait (Liberty® herbicide, glufosinate sodium).

The *nptII* gene has been used extensively in agricultural biotechnology. Fuchs et al. isolated nptII protein from recombinant *E. coli* and from genetically modified potato, cotton seed and tomato fruit *(74)*. Two fermentation batches derived from 150-L tanks were pooled to yield 7.5 kg cell paste. The purification procedure involved extensive washing of the inclusions, followed by solubilization in urea and dithiothreitol. The protein was diluted to lower urea concentration, then contaminants were precipitated with ammonium sulfate. Approximately 40% of the nptII also precipitated but, sufficient material was left in solution to obtain the desired quantity of product. The final yield of product on lyophilization was 30 g. Amino acid analysis indicated an overall purity by weight of 74%. Bacterial endotoxin (*see* Section 4.5.2) was present, but the level was not disclosed.

Isolation of nptII protein from potato and tomato plant tissues was described in ref. *74*. Isolation from potato tuber was accomplished by homogenizing 1.5 kg tubers in extraction buffer. The material was filtered and concentrated by ultrafiltration, then purified on a series of ion exchange and size exclusion columns. The yield was not disclosed, but the ultimate product was described as 20% pure by SDS-PAGE. The protein was isolated from tomato by a combination of hydrophobic interaction chromatography and affinity chromatography using the antibiotic amikacin as a ligand. This sample was described as 55% pure by SDS-PAGE. These proteins were used in equivalency studies with the microbial protein.

Extraction from cottonseed was described in ref. *84*. Cottonseed (500 g) was powdered in a blender and extracted with acetone to remove interfering lipids. The acetone was air-dried, and the powder was used as starting material for further purification. The material was fractionated by precipitation and ion exchange chromatography. It was finally purified on an amikacin affinity column. Prior to affinity purification, the yield was estimated as 0.26 mg nptII. Extraction buffers contained components to minimize interferences because of organic compounds in the plant tissue. Borate was employed to minimize effects of gossypol, and ascorbic acid was added as an antioxidant.

The CP4 EPSPS was purified from 100 g soybean by powdering the seed in liquid nitrogen, then extracting with acetone. Following homogenization and filtration, the protein was precipitated with ammonium sulfate, then purified by anion exchange chromatography and hydrophobic interaction chromatography. The authors indicated that the general method was applicable to canola and cotton expressing EPSPS and mentioned but did not describe purification of CP4 EPSPS from *E. coli (91)*. The microbial EPSPS (product of *aroA* gene) was purified following overexpression in *E. coli (92)*. A 60-L fermentation in minimal medium was concentrated by tangential flow filtration and centrifugation. The cell pellet was lysed by freeze–thaw, lysozyme, and DNase treatment. The clarified lysate was taken to 50% saturation with ammonium sulfate, and the precipitate was discarded. The resulting supernatant was further treated to 70% saturation, and the resulting precipitate was taken to hydrophobic interaction chromatography. The material was polished by anion exchange chromatography. The resulting yield was 260 mg.

Purification of pat protein from *E. coli* was performed after fermentation at the 50-L scale. Following lysis, the protein was precipitated with ammonium sulfate and applied to ion exchange, hydrophobic interaction, hydroxylapatite, and size exclusion chromatography. The protein was characterized as at least 90% pure by SDS-PAGE. Total yield was less than 1 mg *(93)*.

4.5. Well-Known Contaminants of Purified Heterologous Protein

Recombinant expression undeniably altered the course of scientific and commercial exploitation of proteins by providing a ready source of the material. However, the proteins produced can harbor new or different contaminants compared to proteins from their native source. For obvious reasons, the presence of these potential contaminants has been the source of much work in pharmaceutical protein expression and purification. Several classes of contaminants appear common to proteins produced via recombinant expression, and their presence in test materials for PIPs should be evaluated.

4.5.1. DNA

Cellular DNA is a well-known contaminant of heterologous protein expression. As a charged polymer, it interacts with proteins and can be copurified with the target protein. Procedures aimed at removal of DNA include use of DNase, use of precipitants such as streptomycin sulfate *(77)*, or chromatography *(81)*. Newer technologies include high-performance tangential flow filtration *(94)*. There are several examples of tight association between fragments of DNA and *Bt* toxins. Specific association of cry1A toxins with cellular DNA has been shown by several groups *(95,96)*.

Bietlot et al. examined trypsin-activated toxin isolated from a native *Bt* strain and found it to chromatograph in two different fractions during anion exchange separation *(95)*. The fractions appeared indistinguishable on SDS-PAGE, but the later-eluting fraction was shown to contain DNA by absorbance at 260 nm, DNase susceptibility, and staining.

Clairmont et al. extended these observations to demonstrate the presence of a 20-kbp DNA fragment in cry1A isolated from six different *Bt* subspecies and from inclusions of three *cry1A* genes in *E. coli (96)*. Evidence suggested that the interaction is specific to the N-terminus and is involved in the stepwise proteolytic activation of this class of toxins.

Chaturvedi et al. observed that cry1Ac expressed in *E. coli* accumulated as an insoluble inclusion in specific association with approx 25-kb DNA fragments *(81)*. The DNA was removed during hydrophobic interaction chromatography by a 2*M* KCl wash step.

B. thuringiensis subsp *israelensis* produces a 27-kDa endotoxin called cytA. Unlike the classical *Bt* endotoxins, this protein is highly cytotoxic to mammalian cells. During purification of cytA on anion exchange, Yokoyama et al. observed a fraction appearing to be a high molecular weight aggregate, and the ultraviolet spectrum of the material was consistent with the presence of DNA *(97)*. The presence of DNA in the inclusions was confirmed with the DNA stain 4',6 diamino-2-phenylindole dihydrochloride and with ethidium bromide staining before and after treatment with DNase. Ability of purified cytA to associate with DNA was demonstrated in gel shift assays. It was concluded that little sequence specificity was manifested because cytA bound equally well to *Bacillus*, *Escherichia*, and phage ΦX174 DNA. The N-terminal arginine appeared to play a role in the toxin–DNA interaction.

In addition to the procedures mentioned here, sensitive analytical detection of *E. coli* genomic DNA to 1 pg mL^{-1} in protein samples relevant to pharmaceuticals has been reported by polymerase chain reaction-based amplification of genes encoding 5S ribosomal RNA *(98)*.

4.5.2. Bacterial Lipopolysaccharide

The vast majority of work in bacterial expression takes place in Gram-negative hosts such as *E. coli* or *Pseudomonas* species. These organisms possess a complex double membrane structure that contains a specific lipopolysaccharide (LPS) known as bacterial endotoxin. This endotoxin is the agent responsible for endotoxic shock in Gram-negative septicemia *(99)*. LPS is amphiphilic, comprised of a hydrophobic core lipid called lipid-A (the toxic component), 2-keto-3-deoxyoctosonic acid substituted with phosphate and ethanolamine, and various species of hydrophilic polysaccharides *(100)*.

Unfortunately the term *endotoxin* is used to describe both this lipopolysaccharide toxin from Gram-negative bacteria and the insecticidal protein toxins from *Bt*. These are substantially different agents and should not be confused. For the sake of clarity, some authors choose not to use the term endotoxin in reference to *Bt* toxins given the larger frame of reference for the customary medical use of the term *(7)*. However, within the agricultural biotechnology field, endotoxin is used heavily in reference to *Bt* crystalline toxins. Therefore, we refer to Gram-negative bacterial endotoxin by its general abbreviation, LPS.

Like DNA, LPS often copurifies with the target protein *(99–103)* and even with bacterially produced polymers such as poly(3-hydroxybutyrate) *(104)*. It can interact via its charged or hydrophobic groups. LPS forms macromolecular aggregates *(105)* and can be retained by standard ultrafiltration procedures *(101)*. Common procedures for removal of LPS from proteins employ chromatographic processing. Specific resins bind the endotoxin and can be used to purify the protein *(100)*. Anion exchange at alkaline pH also tends to remove LPS from proteins *(106)*. It is also reportedly unstable to high pH *(104)*.

Analytically, the LPS content of a protein can be determined via the *Limulus* amebocyte lysate assay (BioWhittaker, Walkersville, MD). This is a functional assay for the presence of LPS. It can be shown qualitatively in samples by polyacrylamide gel electrophoresis followed by a modified silver stain *(107)*, ethidium bromide *(108)*, or techniques to detect 2-keto-3-deoxyoctosonic acid *(100)*. Because this is a well-known toxicological agent, care should be taken to ensure that toxicology test samples destined for injection studies are free of LPS, as would be done for a protein-based injectable therapeutic agent. Irritation of the respiratory tract by LPS from proteins administered for oral toxicology has been observed *(103)*.

4.5.3. δ-Endotoxin Fragments

The analytical chemistry of *Bt* endotoxins is quite fascinating. In general, the procedures to clean up used chromatography resins or otherwise destroy proteins are the exact procedures used to handle *Bt* endotoxins properly. This includes solubilization at pH in excess of 11, addition of copious quantities of sulfhydral reagents and chaotropes, and liberal use of proteases. One of the hallmark articles in *Bt* structure demonstrated the retention of the secondary and tertiary structure of *Bt*s in 6–8*M* urea *(109)*. In fact, these authors found that *Bt* toxin could be incubated for hours in 8*M* urea and yet remain resistant to various proteases. These extreme conditions are often part of the purification process for cry proteins, particularly the use of enzyme digestion to convert the full-length protoxins to their truncated, activated forms. This activation proceeds through a complex proteolysis involving removal of both N- and C-terminal portions of the protoxin protein *(80)*.

In the case of cry1C, 10- and 25-kDa components derived from the protoxin primary sequence were shown to adhere tightly to the truncated toxin *(110)*. These peptide contaminants were pursued because they blocked the ability to iodinate truncated cry1C for ligand-binding assays. Neither reducing agents nor 6*M* urea allowed resolution of the peptides from the 62-kDa truncated toxin, but chromatography in the presence of 1% SDS did. Purification and sequencing of the resolved peptides showed that they originated from the N-terminus and reflected products of overdigestion beyond the anticipated toxin N-terminus of residue 29. The SDS treatment reduced the expected potency by approximately 70%, but activity was restored using guanidine to remove the detergent. Luo and Adang cited additional examples of this phenomenon in other cry1 toxins *(110)*.

4.5.4. Glycosylation and Carbohydrates

Prokaryotes generally have been thought not to glycosylate proteins because they lack the complex machinery used by eukaryotic organisms to effect protein glycosylation. Nonetheless, the primary sequences of prokaryotic proteins do contain potential eukaryotic glycosylation signals, and expression of bacterial proteins in eukaryotic hosts can result in glycosylated products *(111)*. At least one report exists of a *Bt* toxin glycosylated during heterologous expression in insect cells with a baculovirus vector *(112)*. However, according to information in the product fact sheets on the EPA's biopesticide Web site (http://www.epa.gov/pesticides/biopesticides/ingredients/index.htm), none of the *Bt* proteins registered in plants to date has been shown to be glycosylated *in planta*.

The determination of glycosylation status is part of the equivalency study performed for each registration. Thus, expression of proteins in eukaryotic hosts for the purpose of generating test substances should proceed with caution and an eye to detection of glycosylation so that a surrogate test substance is not produced that is glycosylated when the plant-derived material is not.

Specific instances of prokaryotic glycosylation are now known to occur *(113,114)*, and the existence of specific glucosyltransferases has also been implicated as an important virulence determinant in some Gram-negative pathogens. Historically, *Bt* toxins were part of the landscape of prokaryotic glycochemistry. Early *Bt* toxins were referred to as glycoproteins based on colorimetric sugar analysis *(115)*. Nonenzymatic glycosylation, also sometimes referred to as glycation, occurs to proteins as a consequence of a condensation reaction between reducing sugars and primary amine groups during fermentation or formulation *(116,117)*. Basically, a Maillard-like reaction occurs that results in a protein with covalently attached sugar entities. The reaction is pH dependent. This type of reaction has been suggested to account for the earlier reports of glycosylation of *Bt* toxins, especially because the reaction is accelerated under alkaline conditions *(117)*. The current state of the art indicates that *Bt* toxins are not true glycoproteins as expressed in their native strains and in bacterial and plant expression systems. Thus, toxicology test substances should reflect this lack of glycosylation.

Recombinant hosts can produce carbohydrate components that copurify with the protein of interest. The problem in detection is that these tend not to manifest themselves in standard protein assays. Use of *Pichia pastoris* to express recombinant human serum albumin resulted in copurification of *P. pastoris* mannans, which have undesirable immunological implications in an injectable therapeutic *(118)*. Two detection methodologies were developed to quantify mannans in the presence of high levels of protein. The first was an immunochemistry system and the second an analytical methodology based on anion exchange chromatography and pulsed amperometric detection of the sugars. Although these carbohydrates are critical in injectable therapeutics, the impact of such molecules in a toxicology lot destined for oral toxicology might be minimal given that numerous carbohydrates are present in most diets. An understanding of the range of contaminants elaborated by the production host is important in selection of a scale-up organism.

4.5.5. Proteins Derived From the Expression Host

In pharmaceutical expression systems, verification of the absence of even trace levels of proteins derived from the expression host is critical. Such requirements to do likewise have not been suggested for PIP surrogate proteins, and generally protein purity as adjudged by SDS-PAGE has been adequate to date. A methodology relying on total cellular immunochemistry frequently is employed on therapeutic proteins to detect host cell proteins (*119* and references cited therein). These are strain-dependent, process-dependent test systems created by generating antibodies to cellular proteins in mock fermentation runs. The production strain containing the production plasmid but lacking the gene of interest is used in the production fermentation process to create the expressed protein repertoire from which antibodies are developed. These antibodies can then be used to probe production lots for the presence of host cell proteins during processing or to allow product release.

5. Beyond Today

The history of widespread commercial use of genetically modified organisms cannot be viewed properly without attention to *Bt*. The use of natural and engineered strains of this organism since the 1960s enabled a relatively straightforward transition to the registration of genetically modified plants in the 1990s. The *Bt* legacy will no doubt continue to be important as new classes of pesticidal proteins are brought forward for consideration. Protein test substances required for such registrations should be reliably predicted from history. As proteins are expressed in plants for purposes other than insect protection, it would seem that two issues would drive the nature of test substances required: (1) the history of previous recombinant production with the protein and (2) the intended uses for, and consequences of exposure to, the new proteins *in planta*.

The existence of history of recombinant production of the class of proteins considered for plant expression would lend important information to safety and risk assessment. Just as the regulation of plant-incorporated pesticides relied on the historical framework of fermentation-based pesticidal proteins, it is reasonable to conclude that such parallels will be sought if plants are used to produce, for instance, recombinant enzymes for food production. The pathway for evaluating food-related recombinant enzymes in microbes is established *(120,121)*, so adapting relevant parts of this process to plant-produced enzymes could be envisioned. For instance, if a history of safe use of the enzyme as a recombinant microbial product exists, that should weigh into the safety evaluation of the same enzyme expressed in genetically modified plants.

One conceptual difference between today's use of PIPs and use of plants as factories is that the PIP-producing plants are not necessarily a good source of large quantities of pesticidal protein because the potency of the PIPs does not often require high-level plant expression. Therefore, equivalency processes are used to allow microbial production for toxicology studies. However, using plants as factories would imply that such plants would be an adequate source for extracting gram or kilogram quantities of proteins. The concept of equivalency might still have a role, however, if the registrant needs to perform initial toxicology tests to establish mammalian safety prior to attaining regulatory permission to scale the field production to acreage sufficient to produce grams of the product.

The intended use of the protein or the potential for exposure should also play a role in the nature of toxicology test substances required for evaluation. The consequences of exposure to proteins made in plant "factories" could have an impact on the nature of required test substances. For instance, therapeutic proteins could be produced in one scenario where the plant functions simply as a process intermediate from which protein would be purified, and in another, with the plant as part of a delivery system, as in edible oral vaccines.

The current framework for toxicology test substances to enable safety evaluation relies on principles that should be transferable to new areas. When the number of inserted genes is small, as it is today, the pathway seems clearly elucidated. However, as more complicated traits are developed involving heterologous expression of numerous genes, for instance, with expression of entire biosynthetic pathways involving dozens of genes, then the ability to do testing on multigram quantities of protein from all translated open reading frames might become technically unfeasible. Creation of suitable test substances to support the evaluation of such genetically enhanced plants of the future will bring exciting challenges in the area of expression and large-scale purification.

References

1. Betz, F. S., Hammond, B. G., and Fuchs, R. L. (2000) Safety and advantages of *Bacillus thuringiensis*-protected plants to control insect pests. *Reg. Toxicol. Pharmacol.* **32**, 156–173.
2. James, C. (2000) *Global Status of Commercialized Transgenic Crops: 2000.* ISAAA Briefs No. 23. ISAAA, Ithaca, NY.
3. Shelton, A. M., Zhao, J.-Z., and Roush, R. T. (2002) Economic, ecological, food safety, and social consequences of the deployment of *Bt* transgenic plants. *Annu. Rev. Entomol.* **47**, 845–881.
4. McClintock, J. T., van Beek, N. A. M., Kough, J. L., Mendelsohn, M. L., and Hutton, P. O. (2000) Regulatory aspects of biological control agents and products derived by biotechnology. In *Biological and Biotechnological Control of Insect Pests* (Rechcigl, J. E. and Rechcigl, N. A., eds.). Lewis Publishers, Boca Raton, FL, pp. 305–357.
5. Gustafson, M. E., Clayton, R. A., Lavrik, P. B., et al. (1997) Large-scale production and characterization of *Bacillus thuringiensis* subsp *tenebrionis* insecticidal protein from *Escherichia coli*. *Appl. Microbiol. Biotechnol.* **47**, 255–261.
6. Rogoff, M. H. (1982) Regulatory safety data requirements for registration of microbial pesticides. In *Microbial and Viral Pesticides* (Kurstak, E., ed.). Marcel Dekker., New York, pp. 645–679.

7. Andrews, R. E., Jr., Faust, R. M., Wabiko, H., Raymond, K. C., and Bulla, L. A., Jr. (1987) The biotechnology of *Bacillus thuringiensis*. *Crit. Rev. Biotechnol.* **6**, 163–232.
8. US Environmental Protection Agency, Office of Pesticides and Toxic Substances. (1983) *Pesticide Assessment Guidelines Subdivision M Biorational Pesticides*. EPA 540/9-82-028. Docket No. PB83153965. National Technical Information Service, US Department of Commerce, Springfield, VA.
9. Scientific Advisory Panel. (2000) *FIFRA Scientific Advisory Panel Meeting, Session II, A Set of Scientific Issues Considered by the Environmental Protection Agency Regarding: Mammalian Toxicity Assessment Guidelines for Protein Plant Pesticides*. SAP Report No. 2000-03B. US Environmental Protection Agency, Washington, DC.
10. Debach, P. and Rosen, D. (1991) *Biological Control by Natural Enemies*, 2nd ed. Cambridge University Press, Cambridge, UK.
11. Yamamoto, T. (2001) One hundred years of *Bacillus thuringiensis* research and development: discovery to transgenic crops. *J. Insect Biotechnol. Sericol.* **70**, 1–23.
12. Heimpel, A. M. (1967) A critical review of *Bacillus thuringiensis* var *thuringiensis* Berliner and other crystalliferous bacteria. *Annu. Rev. Entomol.* **12**, 287–322.
13. Höfte, H. and Whiteley, H. R. (1989) Insecticidal crystal proteins of *Bacillus thuringiensis*. *Microbiol Rev.* **53**, 242–255.
14. Schnepf, E., Crickmore, N., Van Rie, J., et al. (1998) *Bacillus thuringiensis* and its pesticidal crystal proteins. *Microbiol. Mol. Biol. Rev.* **62**, 775–806.
15. Agaisse, H. and Lereclus, D. (1995) How does *Bacillus thuringiensis* produce so much insecticidal crystal protein? *J. Bacteriol.* **177**, 6027–6032.
16. Crickmore, N., Zeigler, D. R., Feitelson, J., et al. (1998) Revision of the nomenclature of the *Bacillus thuringiensis* pesticidal crystal proteins. *Microbiol. Mol. Biol. Rev.* **62**, 807–813.
17. Yu, C. G., Mullins, M. A., Warren, G. W., Koziel, M. G., and Estruch, J. J. (1997) The *Bacillus thuringiensis* vegetative insecticidal protein Vip3A lyses midgut epithelium cells of susceptible insects. *Appl. Environ. Microbiol.* **63**, 532–536.
18. Warren, G. W., Koziel, M. G., Mullins, M. A., et al. (1996) Novel pesticidal proteins and strains. Patent WO 96/10083. World Intellectual Property Organization.
19. Schnepf, H. E. and Whiteley, H. R. (1981) Cloning and expression of the *Bacillus thuringiensis* crystal protein gene in *Escherichia coli*. *Proc. Natl. Acad. Sci. USA* **78**, 2893–2897.
20. Perlak, F. J., Oppenhuizen, M., Gustafson, K., et al. (2001) Development and commercial use of Bollgard® cotton in the USA—early promises vs today's reality. *Plant J.* **27**, 489–501.
21. Gatehouse, J. A. and Gatehouse, A. M. R. (2000) Genetic engineering of plants for insect resistance. In *Biological and Biotechnological Control of Insect Pests* (Rechcigl, J. E. and Rechcigl, N. A., eds.). Lewis Publishers, Boca Raton, FL, pp. 211–241.
22. Panetta, J. D. (1993) Engineered microbes: the CellCap® system. In *Advanced Engineered Pesticides* (Kim, L., Ed.). Marcel Dekker, New York, pp. 379–392.
23. Siegel, J. P. (2001) The mammalian safety of *Bacillus thuringiensis*-based insecticides. *J. Invertebr. Pathol.* **77**, 13–21.
24. Andrews, R. E., Jr., Bibilos, M. M., and Bulla, L. A., Jr. (1985) Protease activation of the entomocidal protoxin of *Bacillus thuringiensis* subsp *kurstaki*. *Appl. Environ. Microbiol.* **50**, 737–742.
25. Crueger, W. and Crueger, A. (1984) *Biotechnology: A Textbook of Industrial Microbiology* (Brock, T. D., ed.). Sinauer, Sunderland, MA, pp. 1–4.
26. Flexner, J. L. and Belnavis, D. L. (2000) Microbial insecticides. In *Biological and Biotechnological Control of Insect Pests* (Rechcigl, J. E. and Rechcigl, N. A., eds.). Lewis Publishers, Boca Raton, FL, pp. 35–62.
27. Feitelson, J. S. (1993) The *Bacillus thuringiensis* family tree. In *Advanced Engineered Pesticides* (Kim, L., ed.). Marcel Dekker, New York, pp. 63–71.

28. McClintock, J. T., Schaffer, C. R., and Sjoblad, R. D. (1995) A comparative review of the mammalian toxicity of *Bacillus thuringiensis*-based pesticides. *Pesticide Sci.* **45**, 95–105.
29. Baum, J. A., Johnson, T. B., and Carlton, B. C. (1999) *Bacillus thuringiensis* natural and recombinant bioinsecticide products. In *Biopesticides Use and Delivery* (Hall, F. R. and Menn, J. J., eds.). Humana, Totowa, NJ, pp. 189–209.
30. González, J. M., Jr., Brown, B. J., and Carlton, B. C. (1982) Transfer of *Bacillus thuringiensis* plasmids coding for δ-endotoxin among strains of *B. thuringiensis* and *B. cereus*. *Proc. Natl. Acad. Sci. USA* **79**, 6951–6955.
31. Lindow, S. E. (1993) Biological control of plant frost injury: the Ice$^-$ story. In *Advanced Engineered Pesticides* (Kim, L., ed.). Marcel Dekker, New York, pp. 113–128.
32. Sanchis, V. (2000) Biotechnological improvement of *Bacillus thuringiensis* for agricultural control of insect pests: benefits and ecological implications. In *Entomopathogenic Bacteria: From Laboratory to Field Application* (Charles, J.-F., Delécluse, A., and Nielson-Le Roux, C., eds.). Kluwer, Dordrecht, The Netherlands, pp. 441–459.
33. Gaertner, F. H., Quick, T. C., and Thompson, M. A. (1993) CellCap®: an encapsulation system for insecticidal biotoxin proteins. In *Advanced Engineered Pesticides* (Kim, L., ed.). Marcel Dekker, New York, pp. 73–83.
34. Panetta, J. D. (1999) Environmental and regulatory aspects, industry view and approach. In *Biopesticides Use and Delivery* (Hall, F. R. and Menn, J. J., eds.). Humana, Totowa, NJ, pp. 473–484.
35. Sanchis, V., Agaisse, H., Chaufaux, J, and Lereclus, D. (1996) Construction of new insecticidal *Bacillus thuringiensis* recombinant strains by using the sporulation non-dependent expression system of *cryIIIA* and a site specific recombination vector. *J. Biotechnol.* **48**, 81–96.
36. Donovan, W. P., Tan, Y., and Slaney, A. C. (1997) Cloning of the *nprA* gene for neutral protease A of *Bacillus thuringiensis* and effect of in vivo deletion of *nprA* on insecticidal crystal protein. *Appl. Environ. Microbiol.* **63**, 2311–2317.
37. Levinson, B. L. (1990) High-performance liquid chromatography analysis of two β-exotoxins produced by some *Bacillus thuringiensis* strains. In *Analytical Chemistry of* Bacillus thuringiensis (Hickle, L. A. and Fitch, W. L., eds.). ACS Symposium Series 432. American Chemical Society, Washington, DC, pp. 114–136.
38. Yang, C.-Y., Pang, J.-C., Kao, S.-S., and Tsen, H.-Y. (2003) Enterotoxigenicity and cytotoxicity of *Bacillus thuringiensis* strains and development of a process for cry1Ac production. *J. Agric. Food Chem.* **51**, 100–105.
39. Soltes-Rac, E., Kushner, D. J., Williams, D. D., and Coleman, J. R. (1993) Effect of promoter modification on mosquitocidal *cryIVB* gene expression in *Synechococcus* sp. strain PCC 7942. *Appl. Environ. Microbiol.* **59**, 2404–2410.
40. Stevens, S. E., Jr., Murphy, R. C., Lamoreaux, W. J., and Coons, L. B. (1994) A genetically engineered mosquitocidal cyanobacterium. *J. Appl. Phycol.* **6**, 187–197.
41. Thanabalu, T., Hindley, J., Brenner, S., Oei, C., and Berry, C. (1992) Expression of the mosquitocidal toxins of *Bacillus sphaericus* and *Bacillus thuringiensis* subsp *israelensis* by recombinant *Caulobacter crescentus*, a vehicle for biological control of aquatic insect larvae. *Appl. Environ. Microbiol.* **58**, 905–910.
42. Dimock, M., Turner, J., and Lampel, J. (1993) Endophytic microorganisms for deliver of genetically engineered microbial pesticides in plants. In *Advanced Engineered Pesticides* (Kim, L., ed.). Marcel Dekker, New York, pp. 85–97.
43. Barton, K. A., Whiteley, H. R., and Yang. N.-S. (1987) *Bacillus thuringiensis* δ-endotoxin expressed in transgenic *Nicotiana tabacum* provides resistance to lepidopteran insects. *Plant Physiol.* **85**, 1103–1109.
44. Wei, Z.-M., Laby, R. J., Zumoff, C. H., et al. (1992) Harpin, elicitor of the hypersensitive response produced by the plant pathogen *Erwinia amylovora*. *Science* **257**, 85–88.

45. US Environmental Protection Agency. (2002) *Harpin Protein, Biopesticide Regulatory Action Document*. PC Code 006477. US Environmental Protection Agency, Washington, DC.
46. Matten, S. R., Milewski, E. A., Schneider, W. R., and Slutsky, B. I. (1993) Biological pesticides and the US Environmental Protection Agency. In *Advanced Engineered Pesticides* (Kim, L., ed.). Marcel Dekker, New York, pp. 321–335.
47. McClintock, J. T. (1999) The federal registration process and requirements for the United States. In *Biopesticides Use and Delivery* (Hall, F. R. and Menn, J. J., eds.). Humana, Totowa, NJ, pp. 415–441.
48. Libman, G. N. and MacIntosh, S. C. (2000) Registration of biopesticides. In *Entomopathogenic Bacteria: From Laboratory to Field Application* (Charles, J.-F., Delécluse, A., and Nielson-Le Roux, C., eds.). Kluwer, Dordrecht, The Netherlands, pp. 333–353.
49. Beegle, C. (1990) Bioassay methods for quantification of *Bacillus thuringiensis* δ-endotoxin. In *Analytical Chemistry of* Bacillus thuringiensis (Hickle, L. A. and Fitch, W. L., eds.). ACS Symposium Series 432. American Chemical Society, Washington, DC, pp. 14–21.
50. Yang, X.-M. and Wang, S. S. (1998) Development of *Bacillus thuringiensis* fermentation and process control from a practical perspective. *Biotechnol. Appl. Biochem.* **28**, 95–98.
51. Lopez-y-Lopez, E. V., Chavarria-Hernandez, N., Fernandez-Sumano, P., and de la Torre, M. (2000) Fermentation processes for biopesticide production. An overview. *Recent Res. Dev. Biotechnol. Bioeng.* **3**, 1–20.
52. Couch, T. L. (2000) Industrial fermentation and formulation of entomopathogenic bacteria. In *Entomopathogenic Bacteria: From Laboratory to Field Applications* (Charles, J.-F., Delécluse, A., and Nielson-Le Roux, C., eds.). Kluwer, Dordrecht, The Netherlands, pp. 297–316.
53. Siegel, J. P., Shadduck, J. A., and Szabo, J. (1987) Safety of the entomopathogen *Bacillus thuringiensis* var *israelensis* for mammals. *J. Econ. Entomol.* **80**, 717–723.
54. Chow, E., Singh, G. J. P., and Gill, S. S. (1989) Binding and aggregation of the 25-kilodalton toxin of *Bacillus thuringiensis* subsp *israelensis* to cell membranes and alteration by monoclonal antibodies and amino acid modifiers. *Appl. Environ. Microbiol.* **55**, 2779–2788.
55. Ang, B. J. and Nickerson, K. W. (1978) Purification of the protein crystal from *Bacillus thuringiensis* by zonal gradient centrifugation. *Appl. Environ. Microbiol.* **36**, 625–626.
56. Sharpe, E. S., Nickerson, K. W., Bulla, L. A., Jr., and Aronson, J. N. (1975) Separation of spores and parasporal crystals of *Bacillus thuringiensis* in gradients of certain X-ray contrasting agents. *Appl. Microbiol.* **30**, 1052–1053.
57. US Environmental Protection Agency. (1997) Notice of filing of pesticide petitions. *Fed. Register* **62**, 52,988–53,001.
58. Krywienczyk, J., Dulmage, H. T., and Fast, P.G. (1978) Occurrence of two serologically distinct groups within *Bacillus thuringiensis* serotype 3ab var *kurstaki*. *J. Inverterbr. Pathol.* **31**, 372–375.
59. Carlton, B. C. and Gawron-Burke, C. (1993) Genetic improvement of *Bacillus thuringiensis* for bioinsecticide development. In *Advanced Engineered Pesticides* (Kim, L., ed.). Marcel Dekker, New York, pp. 43–61.
60. Pusztai-Carey, M., Carey, P. R., Lessard, T., and Yaguchi, M. (1994) Isolation, quantitation and purification of insecticidal proteins from *Bacillus thuringiensis*. Patent No. 5,356,788. US Patent Office.
61. Skovmand, O., Thiéry, I., and Benzon, G. (2000) Is *Bacillus thuringiensis* standardisation still possible? In *Entomopathogenic Bacteria: From Laboratory to Field Applications* (Charles, J.-F., Delécluse, A., and Nielson-Le Roux, C., eds.). Kluwer, Dordrecht, The Netherlands, pp. 275–295.
62. Tsun, H.-Y., Lui, C.-M., and Tzeng, Y.-M. (1999) Recovery and purification of thuringiensin from the fermentation broth of *Bacillus thuringiensis*. *Bioseparation* **7**, 309–316.

63. Hernandez, C. S., Martinez, C., Porcar, M., Caballero, P., and Ferre, J. (2003) Correlation between serovars of *Bacillus thuringiensis* and type I beta-exotoxin production. *J. Invertebr. Pathol.* **82**, 57–62.
64. Tompkins, G., Engler, R., Mendelsohn, M., and Hutton, P. (1990) Historical aspects of the quantification of the active ingredient percentage for *Bacillus thuringiensis* products. In *Analytical chemistry of* Bacillus thuringiensis (Hickle, L. A. and Fitch, W. L., eds.). ACS Symposium Series 432. American Chemical Society, Washington, DC, pp. 9–13.
65. Bekheit, H. K. M., Lucas, A. D., Gee, S. J., Harrison, R. O., and Hammock, B. D. (1993) Development of an enzyme-linked immunosorbent assay for the β-exotoxin of *Bacillus thuringiensis. J. Agric. Food Chem.* **41**, 1530–1536.
66. Ivanova, N., Sorokin, A., Anderson, I., et al. (2003) Genome sequence of *Bacillus cereus* and comparative analysis with *Bacillus anthracis. Nature* **423**, 87–91.
67. Rowan, N. J., Caldow, G., Gemmell, C. G., and Hunter, I. S. (2003) Production of diarrheal enterotoxins and other potential virulence factors by veterinary isolates of *Bacillus* species associated with nongastrointestinal infections. *Appl. Environ. Microbiol.* **69**, 2372–2376.
68. Perani, M., Bishop, A. H., and Vaid, A. (1998) Prevalence of β-exotoxin, diarrhoeal toxin and specific δ-endotoxin in natural isolates of *Bacillus thuringiensis. FEMS Microbiol. Lett.* **160**, 55–60.
69. Damgaard, P. H. (1995) Diarrhoeal enterotoxin produced by strains of *Bacillus thuringiensis* isolated from commercial *Bacillus thuringiensis*-based insecticides. *FEMS Immunol. Med. Microbiol.* **12**, 245–250.
70. Shinagawa, K., Sugiyama, J., Terada, T., Matsusaka, N., and Sugii, S. (1991) Improved methods for purification of an enterotoxin produced by *Bacillus cereus. FEMS Microbiol. Lett.* **64**, 1–5.
71. Delannay, X., LaValle, B. J., Proksch, R. K., et al. (1989) Field performance of transgenic tomato plants expressing the *Bacillus thuringiensis* var *kurstaki* insect control protein. *Biotechnology (NY)* **7**, 1265–1269.
72. Koziel, M. G., Beland, G. L., Bowman, C., et al. (1993) Field performance of elite transgenic maize plants expressing an insecticidal protein derived from *Bacillus thuringiensis. Biotechnology (NY)* **11**, 194–200.
73. Fuchs, R. L., MacIntosh, S. C., Dean, D. A., et al. (1990) Quantification of *Bacillus thuringiensis* insect control protein as expressed in transgenic plants. In *Analytical chemistry of* Bacillus thuringiensis (Hickle, L. A. and Fitch, W. L., eds.). ACS Symposium Series 432. American Chemical Society, Washington, DC, pp. 105–113.
74. Fuchs, R. L., Heere, R. A., Gustafson, M. E., et al. (1993) Purification and characterization of microbially expressed neomycin phosphotransferase II (NPTII) protein and its equivalence to the plant expressed protein. *Biotechnology (NY)* **11**, 1537–1542.
75. US Environmental Protection Agency, Office of Pesticides Programs, Biopesticides and Pollution Prevention Division. (2001) *Bacillus thuringiensis (Bt) Plant Incorporated Protectants. Biopesticides Registration Action Document.* US Environmental Protection Agency, Washington, DC.
76. MacIntosh, S. C., McPherson, S. L., Perlak, F. J., Marrone, P. G., and Fuchs, R. L. (1990) Purification and characterization of *Bacillus thuringiensis* var *tenebrionis* insecticidal proteins produced in *E. coli. Biochem. Biophys. Res. Commun.* **170**, 665–672.
77. Hofmann, C., Vanderbruggen, H., Höfte, H., Van Rie, J., Jansens, S. and Van Mellaert, H. (1988) Specificity of *Bacillus thuringiensis* δ-endotoxins is correlated with the presence of high-affinity binding sites in the brush boarder membranes of target insect midguts. *Proc. Natl. Acad. Sci. USA* **85**, 7844–7848.
78. US Environmental Protection Agency, Office of Prevention, Pesticides and Toxic Substances. (1998) *The Environmental Protection Agency's White Paper on* Bacillus thuringiensis *Plant-*

Pesticide Resistance Management. EPA 739-S-98-001. US Environmental Protection Agency, Washington, DC.

79. Noteborn, H. P. J. M., Bienemann-Ploum, M. E., Alink, G. M., et al. (1996) Safety assessment of the *Bacillus thuringiensis* insecticidal crystal protein cryIA(b) expressed in transgenic tomatoes. In *Agri-Food Quality. An Interdisciplinary Approach*. (Fenwick, G. R., Hedley, C., Richards, R. L., and Shokhar, S., eds.). Special Publication - Royal Society of Chemistry, **179**, 23–26.
80. Höfte, H., de Greve, H., Seurinck, J., et al. (1986) Structural and functional analysis of a cloned delta endotoxin of *Bacillus thuringiensis berliner* 1715. *Eur. J. Biochem.* **161**, 273–280.
81. Chaturvedi, R., Bhakuni, V., and Tuli, R. (2000) The δ-endotoxin proteins accumulate in *Escherichia coli* as a protein–DNA complex that can be dissociated by hydrophobic interaction chromatography. *Protein Expr. Purif.* **20**, 21–26.
82. Ellis, R. T., Stockhoff, B. A., Stamp, L. et al. (2002) Novel *Bacillus thuringiensis* binary insecticidal crystal proteins active on western corn rootworm, *Diabrotica virgifera virgifera* LeConte. *Appl. Environ. Microbiol.* **68**, 1137–1145.
83. Doss, V. A., Kumar, K. A., Jayakumar, R., and Sekar, V. (2002) Cloning and expression of the vegetative insecticidal protein (*vip3V*) [sic] gene of *Bacillus thuringiensis* in *Escherichia coli*. *Protein Expr. Purif.* **26**, 82–88.
84. Wood. D. C., Vu, L. V., Kimack, N. M., Rogan, G. J., Ream, J. E., and Nickson, T. E. (1995) Purification and characterization of neomycin phosphotransferase II from genetically modified cottonseed (*Gossypum hirsutum*). *J. Agric. Food Chem.* **43**, 1105–1109.
85. Wohlleben, W., Arnold, W., Broer, I., Hillemann, D., Strauch, E., and Pühler, A. (1988) Nucleotide sequence of the phosphinothricin *N*-acetyltransferase gene from *Streptomyces viridochromogenes* Tü494 and its expression in *Nicotiana tabacum*. *Gene* **70**, 25–37.
86. Datta, S. K., Datta, K., Soltanifar, N., Donn, G. and Potrykus, I. (1992) Herbicide-resistant Indica rice plants from IRI breeding line IR72 after PEG-mediated transformation of protoplasts. *Plant Mol. Biol.* **20**, 619–629.
87. Shah, D. M., Horsch, R. B., Klee, H. J., et al. (1986) Engineering herbicide tolerance in transgenic plants. *Science* **233**, 478–481.
88. Saroha, M. K., Sridhar, P., and Malik, V. S. (1998) Glyphosate-tolerant crops: genes and enzymes. *J. Plant Biochem. Biotechnol.* **7**, 65–72.
89. De Block, M., Botterman, J., Vandewiele, M., et al. (1987) Engineering herbicide resistance in plants by expression of a detoxifying enzyme. *EMBO J.* **6**, 2513–2518.
90. Wehrmann, A., Van Vliet, A., Opsomer, C., Botterman, J., Schulz, A. (1996) The similarities of *bar* and *pat* gene products make them equally applicable for plant engineers. *Nat. Biotechnol.* **14**, 1274–1278.
91. Harrison, L. A., Bailey, M. R., Naylor, M. W., et al. (1996) The expressed protein in glyphosate-tolerant soybean, 5-enoylpyruvylshikimate-3-phosphate synthase from *Agrobacterium* sp. strain CP4, is rapidly digested in vitro and is not toxic to acutely gavaged mice. *J. Nutr.* **126**, 728–740.
92. Shuttleworth, W. A., Hough, C. D., Bertrand, K. P., and Evans, J. N. S. (1992) Over-production of 5-enolpyruvylshikimate-3-phosphate synthase in *Escherichia coli*: use of the T7 promoter. *Protein Eng.* **5**, 461–466.
93. Vinnemeier, J., Dröge-Laser, W., Pistorius, E. K., and Broer, I. (1995) Purification and partial characterization of the *Streptomyces viridochromogenes* Tü494 phosphinothricin-*N*-acetyltransferase mediating resistance to the herbicide phosphinothricin in transgenic plants. *Z. Naturforsch. [C]* **50**, 796–805.
94. Knudsen, H. L., Fahrner, R. L., Xu, Y., Norling, L. A., and Blank, G. S. (2001) Membrane ion-exchange chromatography for process-scale antibody purification. *J. Chromatogr. A* **907**, 145–154.

95. Bietlot, H. P., Schernthaner, J. P., Milne, R. E., Clairmont, F. R., Bhella, R. S. and Kaplan, H. (1993) Evidence that the cryIA crystal protein from *Bacillus thuringiensis* is associated with DNA. *J. Biol. Chem.* **268**, 8240–8245.
96. Clairmont, F. R., Milne, R. E., Pham, V. T., Carrière, M. B., and Kaplan, H. (1998) Role of DNA in the activation of the Cry1A insecticidal crystal protein from *Bacillus thuringiensis. J. Biol. Chem.* **273**, 9292–9296.
97. Yokoyama, Y., Kohda, K., and Okamoto, M. (1998) CytA protein, a δ-endotoxin of *Bacillus thuringiensis* subsp *israelensis* is associated with DNA. *Biol. Pharm. Bull.* **21**, 1263–1266.
98. Gregory, C. A., Rigg, G. P., Illidge, C. M., and Matthews, R. C. (2001) Quantification of *Escherichia coli* genomic DNA contamination in recombinant protein preparations by polymerase chain reaction and affinity-based collection. *Anal. Biochem.* **296**, 114–121.
99. Zhang, G.-H., Mann, D. M., and Tsai, C. M. (1999) Neutralization of endotoxin in vitro and in vivo by a human lactoferrin-derived peptide. *Infect. Immun.* **67**, 1353–1358.
100. Li, J. and Clinkenbeard, K. D. (1999) Lipopolysaccharide complexes with *Pasteurella haemolytica* leukotoxin. *Infect. Immun.* **67**, 2920–2927.
101. Stuer, W., Jaeger, K. E., and Winkler, U. K. (1986) Purification of extracellular lipase from *Pseudomonas aeruginosa. J. Bacteriol.* **168**, 1070–1074.
102. Bohach, G. A. and Snyder, I. S. (1986) Composition of affinity purified α-hemolysin of *Escherichia coli. Infect. Immun.* **53**, 435–437.
103. Landry, T. D., Chew, L., Davis, J. W., et al. (2003) Safety evaluation of an α-amylase enzyme preparation derived from the archaeal order *Thermococcales* as expressed in *Pseudomonas fluorescens* biovar I. *Reg. Toxicol. Pharmacol.* **37**, 149–168.
104. Lee, S. Y., Choi, J.-I., Han, K., and Song, J. Y. (1999) Removal of endotoxin during purification of poly(3-hydroxybutyrate) from Gram-negative bacteria. *Appl. Environ. Microbiol.* **65**, 2762–2764.
105. Wang, Y. and Hollingsworth, R. I. (1996) An NMR spectroscopy and molecular mechanics study of the molecular basis for the supramolecular structure of lipopolysaccharides. *Biochemistry* **35**, 5647–5654.
106. Colangeli, R., Heijbel, A., Williams, A. M., Manca, C., Chan, J., Lyashchenko, K., and Gennaro, M. L. (1998) Three-step purification of lipopolysaccharide-free, polyhistidine-tagged recombinant antigens of *Mycobacterium tuberculosis. J. Chromatogr. B Biomed. Sci. Appl.* **714**, 223–235.
107. Tsai, C.-M. and Frasch, C. E. (1982) A sensitive silver stain for detection of lipopolysaccharides in polyacrylamide gels. *Anal. Biochem.* **119**, 115–119.
108. Kido, N., Ohta, M., and Kato, N. (1990) Detection of lipopolysaccharides by ethidium bromide staining after sodium dodecyl sulfate-polyacrylamide gel electrophoresis. *J. Bacteriol.* **172**, 1145–1147.
109. Choma, C. T. and Kaplan, H. (1990) Folding and unfolding of the protoxin from *Bacillus thuringiensis*: evidence that the toxic moiety is present in an active confirmation. *Biochemistry* **29**, 10971–10977.
110. Luo, K. and Adang, M. J. (1994) Removal of adsorbed toxin fragments that modify *Bacillus thuringiensis* cryIC δ-endotoxin iodination and binding by sodium dodecyl sulfate treatment and renaturation. *Appl. Environ. Microbiol.* **60**, 2905–2910.
111. Tull, D., Gottschalk, T. E., Svendsen, I., et al. (2001) Extensive *N*-glycosylation reduces the thermal stability of a recombinant alkalophilic bacillus alpha-amylase produced in *Pichia pastoris. Protein Expr. Pur.* **21**, 13–23.
112. Martens, J. W., Knoester, M., Weijts, F., et al. (1995) Characterization of baculovirus insecticides expressing tailored *Bacillus thuringiensis* cryIA(b) crystal proteins. *J. Invertebr. Pathol.* **66**, 249–257.

113. Benz, I., and Schmidt, M. A. (2002) Never say never again: protein glycosylation in pathogenic bacteria. *Mol. Microbiol.* **45**, 267–276.
114. Upreti, R. K., Kumar, M., and Shankar, V. (2003) Bacterial glycoproteins: functions, biosynthesis and applications. *Proteomics* **3**, 363–379.
115. Bulla, L. A., Jr., Kramer, K. J., and Davidson, L. I. (1977) Characterization of the entomocidial parasporal crystal of *Bacillus thuringiensis*. *J. Bacteriol.* **130**, 375–383.
116. Mironova, R., Niwa, T., Hayashi, H., Dimitrova, R., and Ivanov, I. (2001) Evidence for nonenzymatic glycosylation in *Escherichia coli*. *Mol. Microbiol.* **39**, 1061–1068.
117. Bhattacharya, M., Plantz, B. A., Swanson-Kobler, J. D., and Nickerson, K. W. (1993) Nonenzymatic glycosylation of lepidopteran-active *Bacillus thuringiensis* protein crystals. *Appl. Environ. Microbiol.* **59**, 2666–2672.
118. Ohtani, W., Ohda, T., Sumi, A., Kobayashi, K., and Ohmura, T. (1998) Analysis of *Pichia pastoris* components in recombinant human serum albumin by immunological assays and by HPLC with pulsed amperometric detection. *Anal. Chem.* **70**, 425–429.
119. Wan, M., Wang, Y., Rabideau, S., Moreadith, R., Schrimsher, J., and Conn, G. (2002) An enzyme-linked immunosorbent assay for host cell protein contaminants in recombinant PEGylated staphylokinase mutant SY161. *J. Pharm. Biomed. Anal.* **28**, 953–963.
120. Pariza, M. W. and Johnson, E. A. (2001) Evaluating the safety of microbial enzyme preparations used in food processing: update for a new century. *Reg. Toxicol. Pharmacol.* **33**, 173–186.
121. Harbak, L. and Thygesen, H. V. (2002) Safety evaluation of a xylanase expressed in *Bacillus subtilis*. *Food Chem. Toxicol.* **40**, 1–8.

4

Genetically Modified Microorganisms

Biosafety and Ethical Issues

Douglas J. Stemke

1. Introduction

Over the last 30 years, the ability to modify specific genes in microorganisms has revolutionized numerous fields of the biosciences, including medicine, agriculture, and basic research into life processes. However, this capability raises concerns about the potential hazards posed by the technology. In response to these concerns, specific protocols have been developed to safely monitor the use of genetically modified microorganisms (GMMs). It is the scope of this chapter to review safety issues that have arisen and address bioethical issues that have become apparent through GMM use.

GMMs are defined as bacteria, fungi, or viruses in which the genetic material has been altered principally through recombinant DNA technology, in other words, by means that do not occur naturally. The first section of this chapter addresses GMM safety through risk assessment, identification of hazards, and the methods to use GMMs safely. Subtopics include safety issues of GMM foods and food products, environmental release of GMMs, and concerns arising from horizontal transfer of GMM deoxyribonucleic acid (DNA) to other organisms. Next, a brief review of protocols and recommendations developed by regulatory agencies for the safe use of GMMs is given. A final section on safety reviews strategies used to engineer suicide GMMs.

The second part of this work is devoted to looking at issues of bioethics and GMMs. Specific attention is devoted to issues of patents and GMMs, labeling of GMM foods, concerns over releasing *GMMs* perception of GMMs by the general public, biological warfare using GMMs, and the consequences of not using this technology.

2. Historical Developments in GMM Risk Analysis

The issues surrounding GMMs have been controversial from their inception. In 1972, shortly after Drs. Herbert Boyer and Stanley Cohen first published their breakthrough research in DNA recombination, a self-imposed moratorium on certain types of cloning deemed hazardous was enacted by many who pioneered the field. A year later, potential hazards from the release of genetically modified organisms (GMOs) were raised at a Gordon Conference on Nucleic Acids. In an open letter to the National Academy of Sciences, attendees of the conference agreed to halt progress in the area

From: *The GMO Handbook: Genetically Modified Animals, Microbes, and Plants in Biotechnology*
Edited by: S. R. Parekh © Humana Press Inc., Totowa, NJ

until an international panel could review the subject. Those initial concerns in led to the formation of the Recombinant DNA Advisory Committee (RAC) in the United States in 1974 and, internationally, to the formation of the Asilomar Conference in 1975. Both were charged with addressing these issues. The findings from the Asilomar Conference recommended replacing the moratorium with a set of guiding rules for some types of recombinant research that were identified as posing minimal risk and prohibiting other research deemed too hazardous, such as the cloning of DNA from "highly pathogenic organisms." These recommendations were used by the RAC in developing guidelines, in 1976, for recombinant work, the basis of which were adopted internationally by other government agencies *(1–3)*.

The strict guidelines laid out through these initial regulations were relaxed by RAC and international agencies after mounting evidence demonstrated that the technology itself was safe. In the United States, regulations of GMMs were moved from the RAC into the Food and Drug Administration (FDA), the Environmental Protection Agency (EPA), and the US Department of Agriculture (USDA); whereas RAC regulated recombinant issues relevant to human therapeutic uses *(1)*. International safety regulations for GMMs have been developed through several national and international agencies, including the World Health Organization (WHO), the Food and Agriculture Organization (FAO) of the United Nations, the European Union (EU), and the Organization for Economic Cooperation and Development (OECD), as well as other national and international regional agencies. The specific regulatory roles of these agencies regarding GMM safety regulations are noted elsewhere (*see* Sections 5.1 and 5.2) in this chapter.

Although it has been up to key government agencies to enforce regulation, it has primarily been the responsibility of the scientific community and the developers of specific technologies to accurately identify and define specific GMM safety issues. For example, to address issues regarding the release of *GMMs* and their potential impact on the environment, an international symposium, sponsored by the American Society for Microbiology, was held in June 1985 *(4)*. The seminars at this conference addressed several ethical issues pertaining to GMMs, including an analysis on the impact of Frost Free Ice^{-} *Pseudomonas* (the first intentionally released GMM) on leaf surfaces *(5)*, the potential use of recombinant vaccinia virus *(6)*, methods to monitor modified *Pseudomonas* released into the environment *(7)*, previously identifiable exchanges of DNA between different bacterial genera as a prelude to potential exchanges of DNA between GMMs and microbial communities *(2,8)*, and model systems to apply established chemical environmental risk analysis to assess GMM environmental impact *(9)*. Subsequently, numerous publications *(2,10–20)* and international symposia *(21–32)* have attempted to address these and other GMM safety and ethical issues.

As GMM technologies have become more refined and developed in increasing applications, society's questions about the technology have become more widespread and vocal. Increasingly, concerns have been expressed not only by researchers and regulators, but also by large segments of the public, who deem recombinant technologies unnatural, dangerous, or unnecessary. This vocal opposition has developed into political discussions that produced mandates that, in many cases, are no longer based purely on scientific arguments. Therefore, the future development and use of GMM technologies lies not only in their proven safety and success record, but also in how safe GMMs are perceived by the public *(33)*.

3. Biosafety and Risk Analysis

The major misconception in risk analysis is that the term *safety* does not imply a 0% chance that a given hazardous event will occur *(11)*. Rather, safety is about identification of risk factors and minimization of the likelihood that a given adverse event will occur. The problem would therefore seem to be to identify accurately the risk factors associated with GMMs and their probability of occurring. However, the potential risk factors are often difficult to define, especially considering the evolutionary nature of organisms *(34)*.

In some cases, risk factors are fairly obvious. For example, in a medical research study, it is conceivable that a cloned virulence gene in a host such as *Escherichia coli* could produce a novel GM pathogen. Production of such GMMs is not uncommon when fundamental questions of pathogens are under investigation. However, if the GMM is properly contained and the laboratory closely follows safety guidelines, the research can generally be carried out safely.

In other cases, risk factors are less obvious, such is for Ice^- *Pseudomonas*. The strain is produced through a knockout of a surface protein that serves as nuclei for ice crystal formation. In principle, the GM pseudomonads will no longer serve as ice nuclei; therefore, fruits containing Ice^- *Pseudomonas* will not be damaged by light frosts *(35)*. On initial analysis, it would seem difficult to imagine a reasonable scenario wherein an Ice^- GM strain would pose a significant risk to humans or the environment. However, it was argued that such bacterial surface proteins may be needed to initiate water droplet formation in the atmosphere, and that bacteria normally blown up into the atmosphere might serve as the initial nuclei to produce rain. The concern that ice^- bacteria would disrupt weather patterns has been widely reported *(36,37)*. However, because ice^- bacteria are a natural part of plant microflora and classically induced ice^- mutants failed to show an impact on weather patterns, it is generally accepted that the impact on rainfall by Ice^- GMM would be extremely small or nonexistent *(35)*.

Herein characterizes the complexity of identifying GMM risk factors. Which risk factors raise reasonable concerns? There are some general principles involving GMMs and their genes that must be taken into account when assessing risk *(38)*. Consider, for example, GMMs released into the environment. GMMs or their recombinant genes have the potential to interfere with indigenous organisms by disrupting complex biological interactions *(39)*. That is not to infer that there is an inherent risk posed by these GMMs, but merely that the potential is there for such an outcome. If this risk potential is coupled to the realization that genes are transferred between different members of microbial communities and that the comprehension of the maintenance of genes in a population is still incomplete *(40,41)*, then there is some understanding of the difficulty identifying the true risks that GMMs pose to the environment.

The reality that science cannot provide absolute assurance about the safe use of GMMs has left policymakers looking for methods of regulation that address political realities. The 1992 United Nations Conference on Environmental and Development meeting in Rio de Janeiro, Brazil, formulated the *precautionary principle*. In essence, the precautionary principle states that politicians, when faced with uncertainty and potential risk from recombinant technologies, may act to prohibit the technology in the absence of scientific proof of the true nature of the hazards *(42)*. Therefore, based on the precautionary principle, contamination and persistence of GMMs and their genes are regarded as a potential risk and are acceptable reasons to prohibit GMM releases into the environment or their presence in foods.

This interpretation of the precautionary principle has been used more extensively to restrict industrial use of GMMs in European nations than in the United States *(43)*. In the European Union, the precautionary principle has been interpreted as meaning that, despite current limited evidence of hazards from a given GMM, new risks may become evident in the future, and the prudent action is to ban a given practice. In the United States, the precautionary principle is interpreted more conservatively and is not based exclusively on a risk-free policy. Uncertainty in the US approach is addressed through regulatory agencies, which produce policy directives to maintain safe uses of GMMs *(44)*.

If there is a perceived risk with GMMs and their DNA, the question pondered is, why should GMMs be used in the first place? The obvious answer is the enormous potential that recombinant technologies bring to bioremediation, medicine production, food production, and a wide variety of other industrial processes. However, despite the promise of the technology, there has to be a point at which a given risk is unacceptable. At the Second International Symposium on the Biosafety Results of Genetically Modified Plants and Microorganisms, Hull proposed the following formula to assess risk *(29)*:

$$\text{Acceptable risk} = \frac{\text{Probability of hazard } A \times \text{Magnitude of the hazard } B}{\text{Benefit from this product } C}$$

Risk assessment is also, and perhaps more frequently, determined without the denominator benefit variable of this equation *(11,14,19,45)*.

Whereas the probability of hazard *A* has been estimated experimentally from the stability and transmittability of the DNA or product of the GMMs *(46–49)*, the magnitude of the hazard *B* seems the more difficult variable to quantify. Typically, biotechnology watch groups concentrate their efforts primarily on the magnitude variable and project grave consequences for GMM applications *(50,51)*, whereas industries using GMMs tend to concentrate their safety efforts on reducing the probability variable *(52)*.

One way that biotechnology firms accomplish their measure of safety is by testing and defining the probability variable as safe through the *substantial equivalence principle*. This principle defines the safe use of a GMM product as minimal risk if, for all practical purposes, it has the same impact on the environment as the non-GMM form. Finally, when risk is calculated using the Hull equation, the benefit of the product *C* is normally defined by direct comparisons of the result or value derived from the GMMs vs those technologies that do not use GMM technologies *(53)*. For example, if bioremediation is to be carried out using a GMM compared to a non-GMM, comparisons might include how completely the compound is mineralized in both systems and a cost–benefit analysis of both approaches.

4. GMMs and Safety Issues

4.1. Human Risk

Humans may encounter GMMs or their associated DNA in a variety of ways, including in food products, in GMM vaccines, or as interactions with released GMMs in the environment or laboratory. In the future, exposure to GMMs may also include whole GMM foods and probiotics (viable organisms that have medical beneficial effects when ingested). Currently, human risks are generally placed in the following categories: increased exposure to antibiotic resistance genes; which may result in transfer of antibiotic resistance genes to indigenous flora, transfer of genes accidentally or intentionally that might produce human pathogens; production of GMM toxins; and activation of human immune allergies *(54,55)*.

4.1.1. Risk From Antibiotic Resistance Genes

Antibiotic resistance genes are the most widely used selectable markers for general cloning. However, there are credible concerns that antibiotic resistance genes might be transferred into other microorganisms, including known pathogens or opportunistic pathogens. In one obvious example, the most widely used microorganism in genetic research, *E. coli*, can readily exchange DNA with a host of known enteric pathogens via plasmids or transposons through methods such as conjugation, transduction, or transformation *(56,57)*. If a pathogen does successfully pick up an antibiotic resistance gene, it is in effect picking up a potentially novel virulence factor *(58)*. For this reason, regulatory agencies worldwide generally ban outright use of viable GMMs containing antibiotic resistance in foods and are attempting to minimize antibiotic genes in GMMs used to produce processed foods *(59)*. Certainly, widespread use and misuse of antibiotics is a major contributing factor to the worldwide epidemic of antibiotic resistance in a host of microbes; however, introducing antibiotic resistance genes may exacerbate the problem.

4.1.2. Risks of Human Toxicity

To determine the potential human toxicity of a specific GMM or GMM product, a series of defined toxicity tests must be completed *(20)*. To evaluate the toxicity levels of the product, analysis is used to determine the dosage at which no adverse effects are measured. This value is defined as the *no observed adverse effect level* (NOAEL) *(20)*. The NOAEL value is then used to develop a safe level for human consumption.

Toxic metabolites are typically analyzed through in vitro analysis, including Ames tests and cell line cytotoxicity assays and through in vivo animal testing analyses, including acute oral, subcutaneous, interperitoneal, and inhalation toxicity tests *(14,20)*. Typically, a large margin of safety is added to generate a safe level for human consumption. The NOAEL of whole food is calculated because whole foods are too complex to identify all effects the GMM products might have on the food. Foods are considered safe once it is established that the GM food is as safe as the traditional, non-GM food, thus complying with the concept of substantial equivalence. Hazard and risk analysis flow diagrams have been developed to use this paradigm *(20)*.

4.1.3. Risk of Allergies

Allergies are caused by the specific activation of an inflammatory process resulting from allergens interacting with immune effector mechanisms *(60)*. Specific allergies show geographic distribution primarily because of dietary considerations. Some of the best-documented examples of food allergies include peanuts, milk, hen's eggs, Brazil nuts, hazelnuts, walnuts, shellfish, celery, kiwi fruit, and rice *(54,61–64)*. Although typical symptoms of food allergens are not life threatening, severe anaphylactic reactions may be fatal *(60,61)*.

The concern derived from GMM foods is whether a novel protein expressed in a GMM will produce an allergic reaction. To test for the likelihood of an allergic response, the GMM protein can be tested for serological cross-reactivity with known allergens *(65)*, its amino acid sequence can be compared to that of known allergens *(66)*, or the GMM protein's stability in simulated gastric fluid can be determined *(67)*. Direct immunoglobulin E responses can be analyzed in animal models *(60,67–69)* or through skin prick tests *(70)*. To date, only one genetically modified (GM) food was ever found to have transferred an allergenic protein. The allergen was identified before the product was marketed, and its founder, Hybrid International, never marketed the GM soy *(64)*.

4.1.4. Risk From Unknown Pathogenicity

The lessons learned from the agricultural use of *Burkholderia cepacia* might serve as a model to identify the risks from released GMMs. Strains of this organism have been developed and used for their diverse metabolic properties, including their use as a biofungicide and their ability to biodegrade herbicides *(71–74)*. It is becoming increasingly evident that *B. cepacia* is also a pathogen in cystic fibrosis patients, causing serious pulmonary deterioration and associated fatal bacteremia *(74–79)*. These organisms were developed by classical selection methods; hence, they are not technically GMMs. However, even as the connection between the widespread use of *B. cepacia* and its frequency in cystic fibrosis patients is still under investigation, the connection is considered a warning by some against the widespread introduction of a novel microorganism into the environment *(80)*.

A second example perhaps better illustrates the potential harm that a specific GMM might inflict. During the course of a mouse sterility research program, a gene encoding interleukin-4 was inserted into mouse pox. Inadvertently, a GM virus that was highly lethal to mice was engineered *(81)*. Because the research group followed the proper safeguards, the GM virus was properly contained, and the experiment was terminated without further incident. However, the potential release of an unintended pathogen constructed through recombinant manipulation is an unacceptable consequence of this technology.

To assess the potential risk of a given ingested GMM as a human pathogen, animal models have been suggested and developed *(82)*. The first of these systems was developed using streptomycin-treated mice to develop complete human gut microflora. The findings from these model systems paralleled results between human and mouse responses to diet and the colonization by enteric flora. This suggests that these mouse gut models should be an excellent system to analyze the impact of GMMs in the human gut. Potentially, the flora of other parts of the human body could likewise be mimicked in such models *(82)*.

Another area in which unknown pathogenicity may arise is in the development of GM viruses used in gene therapy. This technology has shown remarkable promise in the treatment or therapy of many human diseases *(83–85)*. Unfortunately, there have been well-documented tragic failures with this technology, including the death of Jesse Gelsinger, who was under treatment for an ornithine transcarbamylase deficiency *(86)*. Also, the onset of leukemia in patients under treatment for the X-linked form of severe combined immune deficiency disease further raised concerns about the technology *(87)*.

Gene therapy, with its inherent risks, must be examined on a case-by-case basis by institutional and federal medical review boards before any trials are conducted *(86)*.

4.1.5. Risk From Known GMM Pathogens

There are relatively few scenarios by which a known GMM pathogen might infect a person. Interaction with known GM bacteria, fungi, or viruses containing pathogenic genes or pathogens genetically altered to analyze the organism could occur as a result of a laboratory accident. For example, a known pathogen with a gene knockout may inadvertently infect a laboratory researcher during the course of an experiment. The likelihood of such accidents is small if such experiments are conducted in the proper biosafety containment facility (*see* Sections 6.2 and 6.3). However, such accidents do happen.

In reviewing the cases of known incidences, many would have been prevented if proper safety measures were followed. In an extensive survey made in 1979, over 4070 cases of laboratory-acquired infections were identified *(88)*. These infections were collectively caused by 38 different species of bacteria, 84 different types of viruses, 16 species of parasites, 9 species of rickettsia, 9 species of fungi, and 3 species of chlamydia. Although it is not completely clear what percentage of these infections were actually obtained in the laboratory, there is an evident association with the type of work conducted in the laboratory and the acquired infection. In 20% of these cases, an identifiable incident resulted in the infection; in many of the other cases, workers contracted the infectious agent through an unknown incident *(89,90)*. Although these laboratory-acquired infections were presumably caused by non-GMMs, these incidents do indicate an area of concern regarding GMM pathogens.

Humans also might logically encounter pathogenic GMMs as the result of biological warfare or a terrorist action with GMMs used as the weapon. Obviously, the only way to prevent human exposure to these weaponized GMMs is to prevent their use. The 1972 international Biological Weapons Convention (BWC) with 150 signatories unfortunately has not been the last word on the development of weaponized GMMs *(91)*. The threat from GM weapons will need to be met by vigilance of the international community to prevent such weapons from development and deployment.

4.2. Environmental Impact

Although the number of GMMs released directly into the environment is currently relatively low, the large number of proposals to release organisms in the environment to remediate contaminated soils, improve soil fertility, manage pest control, and vaccinate livestock and wildlife has prompted active research and regulation into the safety of released GMMs *(92)*. Introduced here are some of the general concerns from different types of environmental releases, including determination of how to assess risk from released GMMs, methods to control the dissemination of GMMs, methods to monitor the impact of a GMM on microbial flora, and methods to monitor GMM activity.

4.2.1. Risk Analysis

As noted in Section 3, risk analysis of a GMM minimally involves the magnitude of a risk multiplied by the likelihood the risk will occur. To address these two variables, risk assessment for GMMs released into the environment needs to answer several questions *(93)*:

1. Are there potential hazards the GMM might impose on the environment?
2. How likely is it that the potential hazard will actually happen?
3. What are the consequences if the hazards are realized?
4. What management procedures, if any, are needed to control the risk?
5. What level of monitoring is necessary to confirm the risk assessment and determine whether control measures are efficacious?

A flowchart for events that might lead to an environmental catastrophe was developed by the US Office of Technology *(94)*. In this scenario, (1) a hazardous gene is inserted into a microorganism; (2) the GMM escapes into the environment; (3) the GMM multiplies and establishes a niche in the environment; (4) the GMM produces some "factor" that causes disease or damage; and (5) the hazardous effect is manifested in humans or other hosts. Each step can be assigned a probability for actually occurring.

In considering GMMs for release into the environment, it is imperative that the GMM demonstrate no adverse human health effects, be nonharmful to agricultural interests, and produce no irreversible damage to the ecosystem into which it is released *(95)*. Therefore, before the organism is released, an environmental impact study must be conducted in a controlled environment, such as in growth chambers or in greenhouses *(14)*. Data from these studies are used to identify the variables necessary to make a risk analysis of the release. In addition to any toxicological analysis of the GMM, a profile of the natural microbial population should be determined prior to the GMM release.

There are several issues that need to be addressed when considering how a given GMM will behave in a complex environment. For the industry using GMMs, an important issue is determining if the organism actually completes its task in its complex environment and how long the organism remains active *(96)*. Because of this need to maintain a stable form of the organism in the environment, safety concerns arise as to how that GMM will behave in an uncontrolled environment. First, an evaluation must be made to determine if conditions are favorable to permit cell growth beyond the release area. Once growth patterns are determined, an analysis must be undertaken to identify gene expression pattern changes when the GMM is released into the environment *(47)*. This might be the result of mutations that the GMM acquires as a result of stress factors from the environment. Such mutations could fundamentally change the niche of the organism *(97,98)*.

After the organism is evaluated, the impact the organism has on the local environment must also be evaluated. For example, chemical and physical impacts on the environment need to be determined *(99)*. Also, the impact generated by the GMM on indigenous microflora needs to be analyzed and evaluated *(28,93,96,100)*. The GMM itself needs to be monitored both inside and outside the release area to assess local impact. To aid in this analysis, evaluation trees have been devised to determine the potential hazards of GMM releases *(101)*.

In comparison to GM plants, there have been relatively few GMMs released directly into the environment that can be used directly as case studies for safety assessment. An analysis completed in late 1998 of the OECD databases identified that only about 1% of intentional GMOs released were bacterial, 0.3% were viral, and 0.2% were fungal.

Released GMMs included a variety of different organisms. *Sinorhizobium (102,103)* and *Bradyrhizobium* have been used to improve soil nutrition. *Pseudomonas (104)* has been used in a variety of plant or microbe model studies and to model GMMs released into soils *(10,105)*. *Clavibacter xyli* modified with the δ-endotoxin has been used in environmental releases in pest control *(106,107)*. GMM release studies have also been carried out in animals, such as in the use of GM *Lactococcus lactis* in pigs *(108)*. Viral GM releases to date include the use of recombinant baculovirus for pest control *(109)*, of bacteriophages to trace and monitor pollution *(110)*, and in recombinant vaccines for rabies control *(19)*. Although not a comprehensive listing of all GMMs developed for environmental release, these examples give an overall indication of the types of organisms used for GMM releases.

To provide a framework for the safe use of GMMs to be released into the environment, the Safety in Biotechnology Working Party of the European Federation of Biotechnology developed a risk assessment classification system for the environmental release of GMMs *(19)*.

- Risk Class 1: GMM generally considered safe and highly unlikely to produce adverse effects.
- Risk Class 2: GMM that may produce local adverse effects, but are generally unlikely to produce a serious hazard. This includes localized transient displacement of indigenous species and temporary minor biological or chemical changes to soils. GMM can be treated or controlled. The organism is unlikely to disseminate beyond the treated site.
- Risk Class 3: GMM likely to produce serious adverse effects that act locally at the release site. GMM or application area can be treated or controlled. Impact includes a localized displacement of indigenous organisms and major changes in soil biochemistry that are restorable. The organism is unlikely to disseminate beyond the treated site.
- Risk Class 4: GMM will likely produce serious adverse effects likely to disseminate beyond the treated site. GMM and its DNA persist at the site and may result in serious damage to the soil fertility and vegetation. No treatment or control of the GMM is available.

When using GMMs *in situ*, care should be taken to ensure that the GMM does not become disseminated beyond its trial site. However, it may actually be impossible to restrict a GMM completely to its inoculation site *(38)*. Wind *(38)*, water *(38,111,112)*, and insects such as grasshoppers *(113)*, caterpillars *(114)*, Southern corn rootworm *(115)*, and earthworms *(116)* have all been shown to disseminate recombinant microorganisms from the initial inoculation site. It is reasonable to assume other physical processes that displace microorganisms will also actively transport GMMs to distal sites.

It is also important to have a fundamental understanding of the biology of the organism used to safely construct GMMs. Does the organism produce spores (i.e., *Bacillus, Clostridium*, etc.) or enter a dormant state (i.e., *Arthrobacter*)? What are the physical growth requirements for the organism? How does the organism behave physiologically once it is stressed? What is its affected host range? All of these are essential questions that require answers before the organism is released *(17,117)*.

4.2.2. Environmental Releases of Recombinant Viruses Used as Vaccines

In the United States, evaluations for the safe vaccination of organisms are reported through the "Summary Information Format for Environmental Release" and are carried out through the Center for Veterinary Biologists, which evaluates experimental vaccines. Because such vaccinations may include recombinant GMMs, their assessment is noted here. Several criteria are evaluated, including: (1) the location of the study; (2) the physical characteristics of the test site; (3) the qualifications of test personnel; (4) the design of the experiment; (5) the potential for dissemination of the organism; (6) the potential that the organism may become established in the environment, and (7) contingency plans if problems arise in the study *(58)*. Active programs have been developed to release GM vaccinia virus to eradicate rabies in fox, raccoon, and coyote populations in Europe and North America *(118)*. An episode in which a woman in Ohio apparently became infected with GM vaccinia virus after removing bait laced with the virus from her pet dog has raised concerns about this practice *(119)*.

4.2.3. Pest Control Measures Using Released GMMs

Several GMMs are under development for the use of pest management. These include mammal, insect, fungal, and plant pests *(120–124)*. It is essential in the development of a risk assessment for pest control that the GMM is proven safe for humans and nontarget organisms before the release is conducted *(124)*. Particular care should be taken to identify susceptible species indigenous to the release area. Microbes with especially broad host ranges should not be considered for pest control.

When a pathogen is released into the environment to control pests, there is always the possibility that infected animals will be intentionally moved to areas outside the approved release area. In Australia, non-GM myxoma virus was released to control feral nonindigenous rabbits in the early 1950s *(125)*. In 1953, a physician who wanted to control local native European rabbit populations released rabbits infected with the myxoma virus on his estate near Paris. The resulting infections originating from that release devastated wild rabbit populations in Europe, which in turn had an impact on local prey species and caused significant economic damage on rabbit farms in western Europe *(126)*. Although not a GMM pathogen, similar events could lead to the spread of a GMM pathogen beyond its intended control area.

4.2.4. Determining the Impact of Released GMMs on Microbial Ecology

A relevant and complicated component of risk assessment in environmental release of GMMs is profiling the indigenous prerelease microbial populations as a basis to detect changes in biodiversity resulting from the released GMM. Analysis of microbial ecology is a developing science, and the protocols necessary to analyze total microbial populations have yet to be standardized. However, several methods have been established to characterize microbial communities affected by a released GMM *(96)*. Examples include the use of protocols to monitor indigenous enzymatic activity *(127,128)*; and the use of culture techniques *(93,128–131)*, BIOLOG® GN microplates, and other metabolic analysis profiles *(132–135)*, fatty acid profiles *(136–139)* and a wide variety of DNA or ribonucleic acid (RNA) analysis methods *(129,140–153)*. Van Elsas and coworkers *(154)* extensively reviewed these and other methodologies designed to assess the effect of GMMs on diverse microbial ecosystems. Regardless of the method finally selected to analyze the impact of a released GMM on local microbial diversity, it is important to recognize that each protocol has its limitations. Therefore, it currently is not possible to derive a comprehensive picture of all the potential impacts from a released GMM.

Data from release studies have already shown significant differences in microbial communities when comparisons are made between GMM strains and the equivalent non-GMM strains. In one study, *Pseudomonas fluorescens* CHA0 significantly changed soil bacterial populations associated with cucumber roots *(129)*. Likewise, *Pseudomo-*

nas putida GMM has been shown to have an effect on natural fungal flora not targeted specifically by the GMM *(155)*. Impact studies have also been used successfully to analyze the effect of a GM baculovirus on a closed aquatic microbial community *(134)*.

4.2.5. Monitoring GMMs in Environmental Releases

Monitoring GMM cell growth may be accomplished using a variety of methods, including the use of selective and differential media and the use of reporter genes to follow cell growth, immunoblotting, cell profiling, or molecular biology techniques. Many of these methods have been extensively reviewed *(154,156,157)*. Identification of released GMMs using detectable proteins offers both inexpensive and relatively simple protocols to detect microorganisms harboring DNA modifications. β-Galactosidase *(111,115,158–160)*, luciferase *(96,105,161,162)*, *xylE* gene product *(156,163)*, and green fluorescent protein *(164,165)* have all been used effectively for detecting the presence of released GMMs.

Other enzyme markers have been proposed and used as methods to selectively isolate GMMs from the environment. Certainly, the use of antibiotic resistance markers is a feasible method to follow released microorganisms, and such methods have been successful *(41,166)*. However, both US and European agencies governing the use of released GMMs are making substantial attempts to limit, if not eliminate, the use of antibiotic resistance markers in released organisms *(14)*. One of the alternatives under investigation is the use of selective genes such as those that code for metal resistance *(158,167)* and catabolic genes *(168)*.

Immunoblotting techniques have also been developed to monitor released GMMs. One of the advantages of a serological approach is that it permits the detection of both nonviable and viable cells *(169)*. This approach is currently used to detect GM products in foods *(15)*. An alternative method of monitoring that offers similar advantages, although it is generally less sensitive, is lipid profiling of the environmentally released GMM *(159,170)*.

Increasingly, however, molecular DNA or RNA methods are the preferred protocols to monitor released GMMs. In addition to detecting the organisms, nucleic acid-based methods are able to detect the recombinant genes in other hosts, potentially identifying horizontal gene flow. There is a wide selection of methods available, including the use of Southern and Northern hybridization methods, polymerase chain reaction amplification, and gene-chip technologies *(171–174)*. The use of existing sequences, such as ribosomal RNA, highly conserved and unique genes, or specifically engineered unique sequences is considered the best approach for molecular monitoring *(39,156,170)*.

4.2.6. Survival of GMMs and Their DNA in the Environment

A critical consideration in determining safe release of a given GMM is determining whether the GMM is inherently more or less fit to survive in the environment *(46)*. With the advent of recombinant strains, it was generally believed that recombinant organisms would be inherently less fit than indigenous bacterial flora; thus, GMMs would not persist in the environment. Some studies have demonstrated this phenomenon *(175)*. However, there are also data that show GMMs either as stable as non-GM forms *(53,153,176,177)* or with enhanced stability *(160,178)*. Velkov *(47)* reasoned that persistence of GMMs or DNA from GMMs might result from a variety of different cellular processes, including adaptive processes associated with quorum sensing, activation of cellular responses that lead to resistance, and activation of hypermutagenic processes in the cell. Further, these processes might lead to persistence of the DNA itself via plasmid transfer to other species. Evidence has been found that some GMMs can be maintained in the environment for at least 6 years, even in the absence of their symbiotic hosts *(179)*. Studies with recombinant genes in chromosomes or plasmids showed persistence, but generally on the order of a month or less *(40,48,180)*.

Likewise, the genetic stability of the GMM construct is an important consideration when determining risk factors. GM fungi, for example, normally have stable chromosomal constructs *(181)*. However, GM fungal strains may become unstable after selective pressures are removed *(182,183)*. It has been noted that fungi altered as weed control agents may change their host range as a result of a change in the genetic structure of the GM fungi *(184)*. This increased host range may represent a potential hazard in environmental release of GM fungi.

4.3. Horizontal DNA Transfer

One of the major safety concerns surrounding widespread use of GMMs is their ability to exchange DNA with other organisms in an uncontrolled environment *(98)*. In the prerelease evaluation, studies should be undertaken to determine the stability of the construct as defined by its inability to transfer the construct horizontally to other organisms. If any instability is identified in this analysis, release of the organism should be reevaluated. Likewise, during the release, frequent monitoring should be conducted to determine if the construct is stable *(14)*.

It has been well established that bacteria are capable of exchanging DNA between very distant species, even between Gram-positive and Gram-negative organisms *(185–187)*. There is even evidence of horizontal transfer of bacterial genes to eukaryotic organisms *(188)*. Of concern is that, once released, a given GMM may transfer its modified DNA to indigenous species. A short list of bacteria identified as capable of exchanging genetic material under "natural conditions" has been generated *(189)*. These have been broken into four groups: Group I includes members of the genera *Escherichia, Shigella* (excluding *S. dysenteriae*), *Salmonella, Klebsiella, Enterobacter, Citrobacter*, and several *Pseudomonas* species; Group II includes several *Bacillus* species (*Bacillus subtilis, Bacillus licheniformis, Bacillus pumilus, Bacillus globigii, Bacillus niger, Bacillus natto, Bacillus amyloliquefaciens*, and *Bacillus aterrimus*); Group III includes members of the genus *Streptomyces*; and Group IV includes members of the genus *Nocardia*. These studies have shown DNA can be readily exchanged between members of the same group under natural conditions.

Therefore, novel GMMs should be evaluated to determine whether recombinant DNA present in the GMM could be transferred to other species. This analysis needs to be completed prior to an environmental release *(14,190)*. This precaution is necessary to minimize the chance that a transgene may be expressed and produce undesirable consequences from a novel combination of genes. Studies analyzing exchange of GM *Pseudomonas* DNA with other microorganisms in the rhizosphere and spermosphere showed that horizontal transfer could be greatly reduced if the genes are encoded in the chromosome rather than plasmids *(181,191)*. The frequency of transfer increases when the trait in question provides a selective advantage for the host, such as resistance to bacteriophages; acts as a virulence factor; confers additional substrate utilization; or provides a bacterial antibiotic *(14,22)*. Fortunately, the frequency of transmission of DNA, even in "worse case scenarios," is low *(14,180)*. To minimize horizontal DNA transfer further, conjugal plasmids and transposons should be avoided in constructing GMMs used in environmental releases.

4.4. Safety Issues of Foods Derived From GMMs

A quarter of all food products are processed with the aid of microorganisms *(23)*. This includes foods composed of living microorganisms, foods produced through fermentation, and additives that use microbial components. Potentially, many of these foods could benefit from recombinant technology, including improved nutritional value, simplified downstream processing, or increased stability of the food products. However, above all else, foods produced through recombinant technologies must be proven safe.

Several factors need to be taken into account when considering the potential of GMM food safety, including (1) that the GMM is nonpathogenic; (2) whether it will colonize the human gut; (3) the possibility that the GMM will transfer its DNA to indigenous gut flora; (4) that the products produced from the GMM are safe; (5) that the vector components have an approved safe origin; (6) that genetic regulatory elements are safe to use; and (7) that specific foreign genes used in the GMM are safe *(16)*. A few food additives are produced by GMMs, including chymosin, pectinases, and aspartame. However, there are proposals to develop several other GMM foods or GMM-derived food products, notably in the production of cheese and buttermilk *(192)*.

Safety of foods produced via GMMs was studied extensively in the European Union through a joint commission of the FAO and WHO. Collectively, a joint report was generated to develop standards to assess the safety of GMM foods *(24)*. In their deliberations, foods containing viable or nonviable GMMs and those produced by fermentation were considered. This excluded highly purified food additives such as vitamins and GMMs used in the agricultural production of these foods. Additional regulations on novel food products have been published elsewhere *(193)*. As discussed in Section 5.1, GMM food safety in the United States is covered by the FDA, in Australia and New Zealand by the Australia New Zealand Food Authority, and in Canada by Health Canada *(194)*.

To identify potential hazards arising from the use of GMMs, a full accounting needs to be made of the host organism used, DNA donor organism used, specific biotechnology processes used to engineer the GMM, stability of the construct, and the specific genetic modifications used to make the GMM *(24)*. Comments and concerns arising from the initial listing of the components of the GMM should be addressed through examination of the GMM and food product. Once an initial analysis of the GMM is completed, information regarding the impact of the novel GMM and GMM-derived food product on human metabolism, both from direct ingestion and indirect exposure, needs to be reported properly. To make an accurate assessment of the potential risk from the GMM-derived food, a determination must be rendered as to the amount of product to be consumed *(24)*.

Specific attention has been brought to the use of antibiotic resistance genes in developing GMMs for food processing. Because such genes are widely used in processes to develop GMMs, there is a significant concern that the use of such genes could contaminate and transmit antibiotic resistance to normal host flora or even pathogenic organisms. Therefore, it is strongly recommended that the use of antibiotic genes be avoided in developing GMMs for food purposes *(24,59)*. Another concern is that the genetic modifications might activate the production of a toxin not found in the nonmodified strain. It is also conceivable that modifications of the organism can change its nutritional profile, thereby making it a less desirable strain than the nonmodified form. As noted in Section 4.1.3, arguments have also been made that GMMs need to be scrutinized particularly for their ability to cause human allergies, and that an assessment needs to include all populations, including those that individuals who are immune compromised *(24,64)*.

The method widely used to assess GM food safety follows the homologous concepts of substantial equivalence and substantially similar in the European Union and the United States. Through these principles, food safety involving GMOs is determined by directly comparing the GM and non-GM versions of the food product. This methodology has been adopted by the WHO, OECD, FAO, and FDA. The adoption of these concepts is not without its detractors. However, WHO and FAO insist that the substantial equivalence concept is intended to be developed as an initial analysis of the GMO food, not necessarily as a determinative evaluation of its safety *(23)*. This approach has been developed in large part because of the difficulty in applying conventional toxicology to determine the safety of any given GMM food. A typical investigation of a novel GMO food product is likely to include an in vitro analysis of the organism, a detailed analysis of the food product, and an analysis of the consumed product. It has therefore been proposed through FAO and WHO that both the GMO and the resulting food product be appraised separately using the principle of substantial equivalence to evaluate food safety *(24)*.

If living microorganisms themselves are used in the food product, such as is true of many dairy products, there is a reasonable concern regarding the impact of the GMM on the microbial flora of the gut. The organism could potentially transfer its recombinant DNA to indigenous flora of the gut or alternatively may interfere with complex interactions between different microbes. These concerns have been extensively analyzed and studied in lactic acid bacteria *(195)*. It is also possible that recombinant genes may be transferred into pathogenic organisms and convert opportunistic pathogens into pathogenic forms. Opportunistic microorganisms may also become pathogenic if the natural inhibitory effects of the normal microbial flora are altered *(24,26,195)*. Methodologies similar to those previously noted to evaluate complex ecosystems for environmentally released microorganisms may need to be investigated prior to producing viable GMM foods.

For novel GMMs and GMM products that cannot be determined using the substantially similar methodology, it will be necessary to use more extensive toxicological analysis called for through the precautionary principle *(95)*. These assays will normally include general chemical analysis, animal testing, cytotoxic evaluation *(196)*, antinutrition analysis *(26,44)*, and carcinogenic investigation *(14,20)*. For these foods, a case-by-case analysis will be necessary to evaluate human risk from exposure. Foods produced following either philosophy should be further analyzed by monitoring human populations to identify abnormal pathologies in susceptible individuals.

In contrast with the US regulations, since 1977 EU regulations have stipulated that viable GMOs must undergo regulatory approval requiring extensive documentation of the product's safety. Guidelines for safety using GMMs have been established for a variety of different GMMs for food processes; many of these have been published *(16,22,26,30,32,197–199)*. Following the recommendations put forth in the International Life Science Institute Consensus Guidelines, decision trees for assessing the safety of GMMs used in food have been developed. These decision trees are initially used to classify food GMOs into three risk groups termed the Safety Assessment of Food by Equivalence and Similarity Targeting (SAFEST). The different classes of SAFEST GMMs are noted next.

SAFEST Class 1 GMO foods are those in which no foreign DNA has been introduced, and the gene expression pattern is the same as for the unmodified organism. These organisms are considered substantially equivalent to the nonmodified safe microorganism.

SAFEST Class 2 GMO foods are those sufficiently similar to traditional foods. Such products are then assessed on their intended differences, with most of the analysis directed to evaluate the nature and consequences of the genetic differences.

SAFEST Class 3 GMO foods are for novel products for which there are no safe traditional foods to compare the GMM food product. Foods belonging to this category will require the most extensive testing to determine the safety of both the GMM foods and the GMM organism. Presumably, extensive toxicological investigation of the product will need to be undertaken, including the use of animal models. Because these foods will likely be extensively tested using conventional analysis protocols, it is less likely that the doctrine of substantial equivalence would be used in their assessment.

5. Regulations Addressing Safe Uses of GMMs

Listed in this section are general rules and philosophies covered by different international agencies to develop and maintain safe workplaces and safe use of GMMs. It is important that researchers using GMMs adhere closely to safety rules and recommendations when working with any organism that has a perceived risk. Indeed, there is a direct correlation between laboratory personnel who had fewer infections originating from laboratory strains and showed more awareness and concern about infectious agents used in their work, more readily identified hazards in their workplace, and generally maintained an enhanced respect of safety matters than those who generally reported more laboratory accidents *(17,200)*. Although GMM safety regulations vary in different nations, a representative list is noted here. Individuals working with GMMs need to determine the specific agencies and adhere to regulations covering their work.

5.1. United States and GMM Safety Regulations

In the United States, the use of GMMs is controlled by several government agencies, including the EPA, the FDA, the USDA, and the National Institutes of Health (NIH). The EPA generally regulates uses of GMMs that might have a potential impact on the environment; the FDA oversees GMMs used in food and pharmaceutical production;

the USDA regulates GMMs that have an impact on agriculture; and the NIH primarily is responsible for GMMs used in developing or studying issues related to human health. A brief look at each agency's contribution to GMM safety regulation follows.

5.1.1. The Environmental Protection Agency

In the United States, safe use of GMMs in industrial settings such as environmental releases, biofertilizers, and bioremediation is regulated by the EPA *(102)*. In a publication, "Biotechnological Program Under Toxic Substances Control Act," (www.epa.gov/opptintr/biotech/biorule.htm) GMMs such as those used commercially in biotechnology are defined as "intergeneric" and as defining "new" organisms. Before intergeneric organisms are used for commercial means, companies must first submit, 90 days prior to use, a document, "The Microbial Commercial Activity Notice" to the EPA. During this 90-days period, the EPA makes a determination on the document submitted. Likewise, the EPA also evaluates environmental releases of GMMs. At least 60 days prior to field tests, the experimental release application "Biotechnological Program Under Toxic Substances Control Act" (www.epa.gov/biotech_rule/pdf/t8669.pdf) must be properly submitted.

5.1.2. The National Institutes of Health

The NIH likewise has developed strict guidelines for the use of recombinant organisms and enforces its rules under the Office of Recombinant DNA Activities *(201)*. The approach taken through the NIH is to work directly with institutional biosafety committees by developing and institutionalizing standards of containment, both biological and physical. The office has identified four levels of GMM risk groups.

- Risk Group 1: Microbes not associated with disease in healthy adults.
- Risk Group 2: Microbes associated with human diseases that are rarely serious and are generally readily controllable through therapeutic or preventative measures.
- Risk Group 3: Microbes associated with serious human disease that may be controllable through therapeutic or preventive measures.
- Risk Group 4: Microbes associated with serious human diseases that generally lack effective therapeutic or preventative measures.

Using these criteria, the NIH has classified a wide variety of microorganisms into these risk groups. A few examples from each risk group are identified Table 1. The ranking of organisms through these guidelines is designed to maintain the appropriate safe handling of specific GMMs. Several general and specific species of bacteria have been designated as generally exempt from the NIH guidelines for inter- and intraspecies introduction of DNA, provided the appropriate biosafety level for the host is followed (Table 2).

5.1.3. US Department of Agriculture

The USDA's GMM safety regulations are primarily directed at recombinant microorganisms that are plant pathogens. Permits to use such an organism are handled through the Animal and Plant Health Inspection Service (APHIS) of the USDA. APHIS issues two types of permits pertinent to GMMs: those required for field testing of the potential plant pathogen and those required to bring a GMM plant pathogen into the United States or between US states. Permits for environmental testing must document

Table 1
Examples of Pathogenic Microorganisms Classified by Risk Group

Risk Group 1	*Escherichia* K-12, *Bacillus subtilis*, adeno-associated virus types 1 and 4
Risk Group 2	*Bacillus anthracis, Bordetella* spp, *Campylobacter* spp, *Escherichia coli* O157:H7, *Klebsiella* spp, *Listeria, Mycoplasma* spp, *Neisseria gonorrhoeae, Salmonella* spp, *Staphylococcus aureus, Treponema pallidum, Vibrio cholera, Blastomyces dermatitidis*, adenovirus, coronaviruses, hepatitis (A, B, C, D, and E), measles virus, mumps virus, rabies virus, rubella
Risk Group 3	*Brucella, Chlamydia* spp, *Coxiella burnetii, Mycobacterium tuberculosis, Rickettsia* spp, *Yersinia pestis, Coccidiodes immitis, Histoplasma capsulatium,* St. Louis encephalitis, Rift Valley fever virus, yellow fever virus, monkeypox virus, prions, human immunodeficiency virus (HIV), human T-lymphotrophic virus (HTLV), simian immunodeficiency virus (SIV)
Risk Group 4	Lassa virus, Machupo virus, Ebola virus, Marburg virus

Source: From ref. *201.*

Table 2
Examples of Bacteria Exempted From National Institutes of Health Regulatory Guidelines

Escherichia	*Bacillus* spp[a]
Shigella	*Streptomyces* sppa,[b]
Salmonella/Arizona	*Streptococcus* spp[a]
Enterobacter	*Serratia marcescens*
Citrobacter/Levina	*Yersina enterocolitica*
Klebsiella	Erwinia
Pseudomonas spp[a]	

Source: From ref. *201,* Appendix A.
[a]For complete list, see ref. *201.*
[b]Includes members with limited host ranges.

complete information on the organism, including sources and identification of all new genes used, reasons for the study, design of the study, and procedures to prevent dissemination of the organism from the test site. Permits for transportation of GMM plant pathogens require documentation on the organism, sources and identification of all new genes used, and how the organism will be used. APHIS, as part of its safety analysis, prepares an environmental assessment document for field tests. For movement permits, APHIS constructs a preliminary pest risk assessment, contacts the appropriate state department of agriculture, and conducts an on-site inspection of facilities along with state inspectors *(202)*. To simplify the process of future biotechnology regulations, risk assessment and permit issuing will be handled exclusively through the Biotechnology Regulatory Services unit of the APHIS.

5.1.4. US Food and Drug Administration

In 1972, the FDA published policy statements used to regulate foods derived from GMOs. In this document, the FDA determined that GMO foods that are not substantially different from their non-GM counterparts are determined as generally recognized as safe (GRAS). If a substance derived from a GMO is intentionally added to a food and is not determined as GRAS by the FDA, it is considered a food additive. Food additives, unlike GRAS products, are subject to review by FDA prior to use in foods *(59)*. The FDA suggests, but does not require, that comparative structural analysis of the GM protein be compared to known allergens. The FDA recommends that antibiotic marker genes used in the production of foods not contaminate food products. The FDA also urges care that antibiotic resistance enzymes present in the GM foods not reduce the efficacy of oral antibiotics *(59)*. The FDA noted that chymosin, the first GMM food product, was granted approval because the organism and antibiotic resistance gene were destroyed in the manufacturing process, and the products were nontoxic *(203)*.

For drugs produced in GMMs, regulation is not significantly different from those drugs produced in non-GM sources. The source organism and any resistance genes must be noted, and the final product should not show detectable levels of the antibiotic used in the fermentation process. The FDA also regulates GMMs to be used in gene therapy trials *(204)*.

5.2. International Regulations and Safety

Internationally, GMM safety is regulated by a host of national and international agencies. In many circumstances, non-US regulatory agencies have adopted regulations that parallel those in the United States. For example, the Japanese regulations on medically relevant GMMs were developed on principles delineated through the NIH *(201)*, but do differ in specifics of the organisms covered *(205)*. Likewise, the OECD, an international consortium with member states that include 17 European nations, Canada, the United States, Japan, Australia, and New Zealand, develops international safety guidelines for agricultural, industrial, and environmental release of GMMs that parallel regulations in the United States, yet differ in specifics *(206,207)*. Other multinational organizations involved in determining or advising on safety policies of GMMs include the WHO, FAO, and International Food Biotechnology Consortium.

To specifically address EU GMM users' regulations, several directives were issued that pertain to the safe use of GMMs. These directives include the commercialization of GMMs used as plant protection agents (Directive 91/414/CE), the manipulation of GMMs under contained environments (Directives 90/219/CE and amended portions in 94/51/CE and 98/81/CE), and the deliberate release of GMMs into the environment (90/220/CE) *(14)*.

6. General GMM Safety Considerations

Specific protocols and equipment are necessary to use GMMs safely in research and production facilities. Although reviewed in detail elsewhere *(17)*, a general outline is provided here. The emphasis here is on (1) developing safe practices in the GMM facility; (2) developing control structures to prevent aerosols; (3) methods to contain GMMs; and (4) methods to protect personnel.

6.1. Containment Equipment

Fermenters, centrifuges, and centrifuge bottles are the primary containment barriers used to prevent the dispersal of microorganisms *(208)*. Biological safety cabinets of the various classes are used to provide varying degrees of protection. Class I cabinets are designed with open hoods with inward air flow, Class II cabinets are laminar flow hoods designed with inward flow with the supplied HEPA (high-efficiency particulate air) filter, and Class III cabinets or Glove boxes are designed to provide entirely enclosed systems *(209)*. Selection of equipment should be appropriate for the level of work done in the laboratory or facility.

6.2. Containment Facilities

Containment facilities provide the physical workplace for personnel using GMMs. The facilities should be designed to provide protection of those workers and prevent dissemination of the GMM beyond the immediate facilities into the environment. Included in this area are the physical barriers in the facility controlling air movement, differential air pressures used to contain GMMs, and equipment to treat GMM-contaminated wastes. To contain GMMs that represent different hazard levels, four classes of containment facilities have been developed. The following classes represent a combination of NIH and Japanese containment structures *(89,205)*:

- Biosafety Level 1 (BL-1); P1 (Japan): These containment systems are designed for use of GMM organisms that do not cause human disease and that work with organisms identified as NIH Risk Group 1. Biological safety cabinets are not required; work may be conducted in open laboratories using nonporous bench tops. Decontamination of work surfaces takes place daily and after spills. All contaminated liquid or solid materials are decontaminated before reuse or disposal. General laboratory practices include the use of mechanical pipetting devices and the wearing of protective coats, which should only be used in the laboratory. Eating, drinking, and smoking are prohibited, as is the storage of food in the laboratory.
- Biosafety Level 2 (BL-2); P2 (Japan): These containment systems are designed for GMMs or their DNA derived from NIH Risk Group 2 organisms. These are GMMs that pose some level of identifiable risk. Generally, work must be carried out in biological cabinets or chemical fume hoods, but work that does not generate aerosols may still be conducted on the open bench. Other regulations are similar to those of BL-1. Laboratory should be posted as BL-2.
- Biosafety Level 3 (BL-3); P3 (Japan): These are containment systems designed for GMMs and their DNA included in NIH Risk Group 3. Organisms at BL-3 are associated with significant risk to personnel. The facilities contain physical barriers, including sealed walls, floors, and ceilings. A biosafety laminator flow hood or glove box is used when manipulating viable cultures. There should be limited access to the facilities. Airflow is regulated to produce "negative pressure" within the facility and is appropriately discharged outside the facility. Lab coats or gowns should be autoclaved before laundering. A hazard sign needs to be posted identifying the class of organisms used. It is suggested that baseline serum samples be stored for persons at risk; periodic sampling may be collected to determine exposure. Although this principle was noted in Japanese protocols and not in NIH documents, it seems a logical precaution and so is noted here.

- Biosafety Level 4 (BL-4); P4 (Japan): These facilities represent the highest level of containment and are designed to protect personnel from extremely hazardous NIH Risk Group 4 organisms. These facilities are designed to be isolated completely from other parts of the facility, including physical barriers, ventilation, and waste treatment. Personnel change clothes and shower as they enter and exit the facility *(89)*.

Other general practices should be followed when using GMMs. Microbial cultures, of course, should be maintained and manipulated using aseptic technique to minimize the possibility of contamination and therefore cross contamination of the GMM DNA *(210)*. For maintenance, cultures should be properly stored frozen, cryogenically in or over liquid nitrogen or in a conventional freezer. Specimens can also be safely stored in a lyophilized form, but care must be taken to ensure the lyophilizer itself does not become contaminated with the GMM *(89)*.

Management of waste streams is of particular concern in the biotechnology industry. GMMs can be effectively controlled through conventional methods to eliminate microorganisms, such as proper autoclaving procedures. However, these techniques are typically ineffectual at breaking down DNA to the monomer level. Most DNA, notably chromosomal DNA, will typically be degraded rapidly by DNases in the environment *(21,211,212)*. However, some DNA is stable in fragmented form or as supercoiled forms, such as plasmids, for extended periods of time. The method suggested to minimize hazards from escaped GMM DNA is to use DNA exclusively from GMMs classified as nonhazardous and that do not contain mobile DNA. Finally, a full accounting of product production, recovery and processing, waste management, and accident reports should be completely documented. Methodologies for decontamination materials and accidents have been reviewed *(21,89)*.

6.3. Large-Scale Fermentation of GMMs

The OECD Council has laid out several principles to minimize general risk from GMMs used for large-scale industrial purposes. These principles, based on those developed for use of organisms in small-scale production, represent sound management of GMMs to minimize potential risks involved. Further, to be considered safe by the OECD, the microorganisms must have several traits that are deemed essential. The GMM (1) must be nonpathogenic; (2) must not harbor known viruses or co-contaminating bacteria; (3) must have been extensively used safely in a non-GM form for industrial purposes; and (4) must be unable to grow outside its industrial setting. The foreign DNA used in the host should be of limited size to minimize the inclusion of nonessential DNA; it should not provide additional stability to the construct unless such stability is essential for the construct's function; it should not permit increased mobility of the construct; and it should not confer resistance to other organisms that do not already possess the resistance. Finally, the GMMs themselves should not contain any deleterious properties.

To implement industrial GMMs safely, a series of principles termed the good industrial large-scale practice principles has been developed:

1. Exposure of GMMs and GMM products should be kept at levels appropriate for the organism used, the process developed, and the product produced.
2. Dissemination of the organism must be maintained through appropriate preventive engineering protocols and equipment. Use personal protective devices and clothing as needed.

3. Keep control equipment properly maintained through appropriate testing. Evaluate control protocols frequently to match the intrinsic nature of the GMM, the product produced, and the process used.
4. Monitor for the presence of viable organisms and their molecules outside its controlled environment.
5. Personnel involved in production and handling of GMMs need to receive proper training and have sufficient experience. This training and experience needs to be documented appropriately to ensure the safety of the production of the GMM.
6. A biological safety committee that consults with external regulatory committees should be established. This safety committee distributes its findings to worker representatives.
7. To ensure a philosophy of safety at the facility, a code of safe practice should be developed and implemented.

6.4. Small-Scale Field Release

The OECD has also developed safety practices, termed good developmental principles, for basic and applied research involving development of GMMs and for small-scale research field studies *(207)*. The good developmental principles are similar to the good industrial large-scale practice principles and are similar to previous practices described elsewhere *(207)*. The following is a summary and interpretation of the OECD principles:

1. Minimize levels of GMMs and GMM products at levels appropriate for specific field experiments.
2. Prevent the dissemination of the organism beyond the test area through appropriate protocols and equipment.
3. GMMs need to be monitored within the research site both during and after the experiment. Safe protocols should be developed to control the GMM at any step in the process to prevent harmful environmental effects.
4. Monitor for the presence of GMMs and their molecules outside the test area.
5. If GMMs are detected outside the test area, control methods must be implemented to prevent further contamination.
6. Appropriate protocols must be developed to terminate the experiment and properly dispose of waste generated in the experiment.
7. Personnel involved in production and handling of GMMs need to receive proper training and have sufficient experience to handle the GMMs safely.
8. Appropriate documentation needs to collected and maintained for all experimental trials.

6.5. Development of Suicide GMMs

One of the concerns regarding GMM releases is the possibility that they will linger in the environment long after their desired activity is completed. A further concern is that these organisms will continue to multiply and leave the release site. One method to prevent this outcome is to design a bactericidal mechanism into the GMM. However, the bactericidal trait introduced into the GMM needs to be sufficiently stable to minimize the possibility of revertants *(213)*.

One promising direction in preventing the controlled growth of released GMMs is the adaptation of the TOL benzoate mineralization pathway for suicide activation. By utilizing the *gef* family of genes, it has been proven possible to develop GMMs that will undergo controlled cell death *(213–217)*. Three members of this gene family, including the *E. coli hok, relF*, and *gef* genes, effectively cause cell death by activating a cascade in the cell that leads to disruption in the cell's membrane potential, causing an influx of periplasmic RNase into the cytoplasm. The constitutively expressed *hok* (host killing) is normally blocked by the antisense RNA gene *sok* (suppression of killing). To use this system effectively in a GMM, *sok* is deleted or mutated to prevent its expression, and an inducible promoter replaces the *hok* constitutive promoter. Alternative suicide systems have been developed based on the *E. coli relF* gene *(218)* and streptavidin-based system *(219)*. Both systems were developed for use in *P. putida*, an organism more relevant than *E. coli* in environmental release. The *E. coli relF* system has been tested in both seawater and soil models *(218)*.

Additional suicide systems have been developed for this purpose *(220)*. However, all such induced-suicide systems remain highly ineffective. Even under laboratory conditions, there is still a significant survival rate (10^4) for the *hok* system. The *relF* and streptavidin-based suicide systems have reported cell resistance in the 10^6 to 10^8 range *(218)*. Although the *relF* and streptavidin-based systems have proven significantly more effective than the *hok* systems in suicide activation, their optimum efficiencies were obtained using isopropyl-β-D-thiogalactopyranoside induction, which would be impractical in an environmental release. Other induction methods may be developed to activate these killer genes in environmental systems. For example, linking these suicide systems to stress-induced control systems *(157)* or activation using the depletion of a substrate, such as a pollutant, *(221)* would be reasonable alternatives.

Alternative killing methods have also been developed. For example, the use of bacteriophages specific for a given GMM is potentially an effective way to control released organisms. Studies using the bacteriophage PhiR2f against GM *P. fluorescens* showed a 1000-fold reduction of cells in simulated soil environments *(39)*. In another approach, phage-resistant and rapid-ripening lactic cocci used in the production of cheese have been engineered with lysin, which autolyses the culture after the stationary phase *(29)*. This technology might be adapted in other organisms for environmental release. Finally, recombinant technologies have been used to engineer suicide fungal pathogens such as *Fusarium*, which is used as for biocontrol of parasitic broomrape weeds. Specifically, the fungus is engineered to be asporogenic *(222)*. These fungal GMMs are constructed through deletions in sporulation genes and are engineered as such to prevent the organism from spreading beyond the release site.

7. Ethical Issues and GMMs

7.1. Introduction to Ethical Issues

Ethical issues involving GMMs are both complex and contentious, involving parties with different attitudes and understanding of the issues *(223,224)*. Reiss *(225)* argued that not all ethical arguments are equal, and that ethical conclusions need to be based on reason, established ethical principles, and general consensus. The use of GMMs focuses in general terms on issues that look at the potential consequences of using GMMs (for example, damage to an ecosystem, introduction of antibiotic resistance genes) and that are intrinsically wrong ("polluting" the world with GMMs or their DNA). Researchers and end users of GMMs generally develop safety protocols to address the potential consequences of the genetic modifications present in GMMs and usually do not identify GMMs themselves as intrinsically wrong. This in turn drives much of the conflict generated between the different sides in the GMM debate.

Some ethical issues primarily have an impact on personnel and institutions using the technology. For example, are GMMs truly patentable? At what point does the control of recombinant technologies through patents and litigation begin to impact seriously the very science from which they were designed? When GMMs are constructed in universities or other research institutions receiving public funds, who ultimately controls the profits and intellectual property derived from the GMMs? And, does a corporation have the right to withhold technologies based on intellectual property that could legitimately help developing nations?

Alternatively, there are ethical issues that are of more concern to the general population. Should people be eating GMM-derived foods? Just how "safe" are GMM foods? Should the environment be polluted with GMMs and GMM DNA? Could an environmental release be catastrophic? Should a person be compelled to accept GMM products even if they have strong personal convictions to the contrary? What will be the consequence of those who use non-GMMs and must now document that their products are GMM free? Who will have control and who will regulate safety issues? On balance, it also needs to be asked, if GMMs have this incredible potential and have demonstrated such little risk, is society overly cautious and missing out on the potential benefits afforded by GMMs? An accounting of many of these issues has been outlined elsewhere *(226)*. Because it is not possible to address all of these ethical issues, a subset of them is discussed.

How do ethical issues involving GMMs develop? The impact that GMMs have on the environment is an excellent example. The introduction of GMMs or their DNA into the environment will likely have some impact on the local microbial flora and, if not properly controlled or monitored, potentially may cause harm outside the control area. Such an event might introduce an undesirable gene, such as an antibiotic resistance gene, into the environment, which in turn might have an impact on human health through increased antibiotic resistance. It is then conceivable that the gene could be picked up by a pathogen, ultimately resulting in harm or death to individuals. It is easy to understand why an environmental release leading to increased antibiotic resistance of pathogens and human death is deemed immoral. How these concerns are addressed by GMM developers and regulators will ultimately determine how widely accepted GMMs will become *(223)*.

Other examples of GMM use less clearly demonstrate "harm." Which ethical issues result from actions that have no direct bearing on humans? For example, releasing a large number of GMMs might locally disrupt a natural ecosystem by interrupting a specific soil predator–prey relationship. If the action changes the soil, somehow making it less fertile, it can be said to damage the soil's "instrumental value." This phrase refers to the value of the nonhuman world in terms of its usefulness to humans *(12)*. Even if the disruption has no direct bearing on human health, human agriculture, or other human activity, it still can be acknowledged as causing a change in the intrinsic value of the site. In other words, inherent value can be derived from the nonhuman world itself *(12)*. Concerns for the intrinsic value of the earth in combination with the instrumental value drive much of the ethical debate regarding environmental issues.

In approaching issues of bioethics and GMOs, it is important to realize that this field of knowledge is unlikely to produce an ultimate resolution that will satisfy all parties. Reiss *(225)* established a framework that delineates a general standard for "ethical conclusions" by suggesting that they should be based on reason, use well-established ethical principles, and be derived from a general consensus. Although these may be difficult to formulate, a consensus on how to define these three principles should serve as a useful guideline to evaluate GMMs.

7.2. Public Concerns and Governmental Philosophies of GMMs

Pharmaceutical products made from GMMs are generally well received and do not receive much condemnation *(14)*. This is not generally true for GM foods *(224)*. Surveys conducted in 1999 to 2000 in Switzerland, the United Kingdom, and the United States showed a marked difference in US and European responses to GM food products. Whereas in the United States only 2% of the public felt that GM foods were a potential risk, 59% of Europeans shared this view *(227)*. To a great extent, in Europe food is not simply the fuel that drives bodies; it has a cultural value as well *(33)*. It has also been speculated that the difference in European and American views is based on wider support in the United States of agencies that regulate GMMs, such as the FDA. By contrast, the European experience includes a series of unrelated serious biotechnology incidents (noted in this section) that have led to deep public mistrust of EU regulatory agencies. In both the United States and the European Union, approaches to convince reluctant populations as to the true merit and safety of GMMs and GMM products need to be conducted via a well-meaning and thorough dialog with the public rather than the paternal approach frequently used in this debate *(33)*.

Another major problem currently facing the US vs European approaches to GM food safety issues are the conflicting philosophies of how safety should be determined. In the United States, GM foods are generally considered safe using the principle of substantial equivalence in that the GM food is essentially the same as the unmodified form. The FDA is far more supportive of GM foods than EU or Japanese granting agencies. This is indicative of the relative high percentage of GRAS status granted to the US GM food producers. The European model currently follows the precautionary principle, suggesting that a GM food or other GM food product must first be determined safe before it is released into the market *(33,34)*. Not surprisingly, in the United States, significantly fewer regulations detailing the specific use of GMMs have been generated than in Europe *(59)*. In 1996, the FDA produced a document, *Safety Assurance of Foods Derived by Modern Biotechnology in the United States*, that outlines this philosophy:

> "Based on our present knowledge of developments in agricultural research, we believe that most of the substances that are being introduced into food by genetic modification have been safely consumed as food or are substantially similar to such substances. Therefore, we do not anticipate that most newly added substances in bioengineered foods will require premarket approvals *(59,* p.3*)*."

This statement is not without substantial supportive data. The FDA rules are the result of intensive analysis of GM foods over the last 25 years. To date, despite extensive analysis, no GM food product brought to the market has shown adverse health effects. It is also clearly in the best interests of biotechnology companies to maintain the safety of their products as documented that "no other foods in history have been tested and observed as diligently as the foods developed from modern biotechnology" *(52,* p.225*)*. Although the current FDA regulations do not require extensive testing, the FDA does specifically address concerns about limiting the use of antibiotic resistance genes in food products. The FDA noted that fermentation products, such as chymosin, which are made with GMMs containing antibiotic resistance genes, must demonstrate that they are free of "transforming DNA" *(59)*.

The cautious EU/Japanese approach arguably stifles research and production of GMM products that may be beneficial. The US model leaves itself open to criticism of a laissez-faire policy regarding GMM regulation, leaving too much control of safety in the hands of the industries that develop GMMs. However, an alternative safety model, similar to those developed for evaluating pharmaceuticals, lies between these two extremes *(33)*. In this approach, initial safety analysis requires physicochemical, biological, pharmacological, toxicological, and clinical testing of the product; this analysis is conducted and financed by the developing industry. This process is followed up through government epidemiological analysis, which acts as a population monitor to detect undesirable side effects. Finally, national agencies that specifically monitor the GMMs evaluate the product over the long term *(33)*.

In a similar approach, the OECD has developed the concept of *familiarity* in utilizing GMMs. Familiarity is based on overall knowledge acquired from (1) the host organism itself; (2) the environment in which the GMM is to be used; (3) the life cycle of the organism; and (4) criteria used in its construction *(15,42)*. Through familiarity, researchers and regulatory agencies can develop risk assessment based on previous knowledge and experience *(30)*. Using this approach, industrial GMMs can adopt the concept of familiarity as "an extended history of safe industrial use" *(206)* and agricultural GMMs can adopt the phrase of "an extended history of safe agricultural use" to develop safe GMMs *(42,* p.16*)*.

Other concerns and issues of bioethics also spring from recombinant technologies. In spite of almost 30 years of developing GMMs, there is still a significant level of distrust in the general population of this technology *(52)*. Although generally accepted in the production of medicines, such as recombinant vaccines and other therapeutic proteins, there is widespread fear of GMMs used as foods or in the production of food additives *(228)*. In part, this is a reflection of serious mistakes made in nonrecombinant biotechnology. Events such as mad cow disease and hoof and mouth disease outbreaks centered primarily in the United Kingdom, dioxin contamination of meats and poultry in Belgium, and the Starlink corn gene contamination in the United States and Mexico have convinced a large part of the population that biotechnology and, by association, recombinant technologies are inherently unsafe *(13,14,223)*. This connection of unrelated events may seem counterintuitive to those familiar with their actual causation. Nonetheless, these examples are frequently used in arguments directed against GM products.

Likewise, GM agricultural biotechnology has become engrained with the politics of globalization and the power of multinational agriculture industries *(229)*. Although a significantly larger issue in Europe, there are vocal advocates in North America and Latin America who likewise fear the development of GM food products *(229)*. If GMMs are to be developed to meet their full potential, there must be adherence to safe protocols, appropriate analysis of potential risk factors must be identified, and concerns regarding their use must be transparently addressed. It is this final point that may be the hardest to fulfill; paternal assurances as to the safety of these technologies will not bring wider acceptance of their introduction *(34,230)*.

Despite the difficulties of gaining general acceptance by the public, examples exist of the general acceptance of GMM products *(223)*. These include the production of recombinant medicines through fermentation technologies and the use of GM viruses in gene therapy trials. The use of GMMs to produce pharmaceutical products has greatly reduced the cost of these drugs, making them more generally available. Further, because of the purification protocols used in producing drugs, GMM pharmaceuticals are inherently going to have the same risks associated with them as the non-GMM versions.

Novel GMM therapeutics that do not have complementary non-GMM versions, such as gene therapy, are another issue. The single human death and two documented leukemia cases resulting from gene therapy trials demonstrate the inherent risk of live GM viral vaccines. The FDA and European regulation agencies, as of September 2002, put a halt to the X-linked form of severe combined immune deficiency disease trials *(87)*. Obviously, clinical trials are inherently never risk free. Therefore, medical ethics committees at both the institutional and the national levels as well as the gene therapy patients themselves (or their guardians) must have sufficient understanding of the inherent risks of the therapy before such treatment is ever initiated *(86)*.

Another example of the general acceptance of GMMs is in the production of "vegetarian cheese." In the traditional process to make cheese, rennin containing the enzyme chymosin is extracted from the stomachs of calves. To make the enzyme more readily available, the chymosin gene was cloned into an *Aspergillus* expression vector, resulting in the production of recombinant chymosin *(192,231)*. Chymosin, which is a product of recombinant technology, is consumed. It is interesting to speculate why this product is so widely accepted and yet so many other GMM derived foods are rejected. Is it because it seems so much less humane to isolate the enzyme from a calf than from a fungus? Was it the marketing of the enzyme as vegetarian that brought it some level of consumer acceptance? Or, is it simply that consumers do not recognize it as the product of a GMM?

Another issue that relates patterns of concern about GMMs by the general public is the poor communication and misunderstanding that exists among the biotechnology industries, the public, and the media *(229,232)*. Whether a process or product is deemed safe is hardly important if the public does not have confidence that the scientific process used to evaluate the product is independent of industrial interests *(95)*. If the public perception is that multinational corporations, motivated by profit, deduce and report findings on the safety of their GMM product in a biased manner, it hardly seems likely that the process itself will provide assurances about the product's safety to a skeptical public. The public needs to be informed about the real risks, if any, of a given GMM product. An informed public should at least understand that the GMM product has been extensively analyzed, certainly for human harm, well before it is ever brought into the market, and that the process has been carried through without bias *(95,233)*. Communicating to the public true risk is perhaps best expressed briefly through the phrase "to produce the appropriate level of concern and action" *(234)*.

In identifying and communicating the risks of recombinant technologies, we are confronted with a truism: The scientific community evaluates safety and risk of these processes through interpretation of data, assigning probabilities and consequence based on data itself. Often, the scientific community evaluates public concern as not based on scientific data and therefore irrelevant to these issues. It is important to understand that, although public fears may have been devised using different methodologies, different standards of evidence, and different values, their concerns are every bit as rational and deserving of consideration as a scientific approach *(232,235)*. It is not possible or even necessary to try to convince everyone that a given procedure is entirely safe. Rather, it is critical that, through open dialog, it is possible to communicate that the assessment process, determined through a rational scientific approach, has been given a transparent account by all concerned, and the final evaluative outcome represents an honest conclusion of that process *(233)*.

7.3. Labeling of Foods and Food Products Derived From GMMs

In the United States, labeling is not required for any GM foods or GM food products. US citizens have indicated a preference for labeling that identifies products as GM foods *(228)*. If so, why does the FDA so earnestly resist the labeling process? FDA documents point out that labeling a product as being with or without GM DNA would become prohibitively complicated *(59)*. For example, to determine that a product is "GM free," documentation would need to be compiled and followed regarding the strains used, the food processing facilities used, and all transportation equipment used to ensure the noncontamination of the product *(59)*. It is imaginable that a process such as the production of acetic acid by GMM and non-GMM forms would require separate fermenters, separate transportation vehicles, and separate process equipment to be able to identify one product as a GMM product and the other as GM free *(52,59)*. Such a complicated and cumbersome process would undoubtedly increase the cost of the product. Finally, it is also believed that including labels on products warning the public about the presence of GMMs when there are no known hazards associated with the product may only serve to unnecessarily frighten consumers *(52)*.

The alternative, not labeling products as containing recombinant DNA, is unacceptable to many individuals, including those who profess a wish to maintain a GM-free diet. Many people practice forms of environmentalism that are arguably religious in nature *(236,237)*. Therefore, it might be further argued that these environmentalists' convictions against eating GMM foods are equally compelling as religious' prohibitions. This is true whether the GMM product is determined to be completely safe for both human consumption and environmental use *(232)*. However, the argument leads to the question, should individual consumers have the right to know and control their dietary intake even if a GMM product is deemed completely safe? This concern has led to mobilization of a large number of interest groups in the United States and Europe determined to require labeling on GM foods *(43,238,239)*. Already, governments in Europe and Asia have either recommended or enacted labeling requirements *(43)*. Recent government actions, including a resolution passed in the US Congress in defense of US GM products, *(240)* will likely only intensify what is already a divisive issue between the United States and the European Union.

A compelling reason to label GM foods is to help identify potential food allergens in rare susceptible individuals *(183)*. It seems unlikely that a GM food product capable of producing severe allergies would become established in a food product. This is largely because of the current level of allergen analysis conducted on GM foods. However, as pointed out by several watch groups concerned about GM issues, without proper labeling of products, if a problem were to exist with allergens in a small percentage of the population, no database will be available to make an epidemiological analysis of the food product *(43)*. In a reply to concerns raised by the Union of Concerned Scientists, the FDA indicated that the agency was reviewing this issue *(241)*.

The labeling issues are complex and divisive. Although dialog and consensus between concerned parties is the best approach to resolve some of these issues, it is most likely that resolution will be accomplished either through legislation or through national and international courts. Although not specifically involving GMM-derived foods, several lawsuits involving GM foods have been litigated against US regulatory agencies and biotechnology companies. Current class action suits are spearheaded by over 600 plaintiffs, including environmental groups such as the Union of Concerned Scientists, Greenpeace, and the Sierra Club *(242)*.

GM food labeling also has an impact on international trade. In a compromise designed to head off an international trade embargo between nations over GM labeling, the Cartagena Protocol on Biosafety Treaty was developed *(28)*. Signed in Montreal by more than 130 countries, this treaty bridges the extremes between not labeling GM products and the trade barriers likely to affect world GM producers. Specifically, the United States agreed to label foods derived from GMOs as "maybe" containing recombinant DNA products. Future requirements may be added to strengthen this reporting mechanism. In return, the protocol allows wider use of GM products in other nations without unilateral embargos threatened against exporters.

7.4. Ethical Issues of GMMs and Environmental Release

Two separate incidents are useful case studies regarding the ethics of released GMMs. In 1987, plant pathologist Gary Strobel of Montana State University conducted an unauthorized release of a mutated strain of *Pseudomonas syringe*, a GMM engineered to control the pathogenic fungi *Ophiostoma ulmi*, the causative agent of Dutch elm disease. He specifically violated a 1986 EPA rule delineating a 90-day notice prior to initiating field releases of GMMs *(243)*. In statements released by him to the committee, Dr. Strobel noted the *Pseudomonas* strain had not been genetically engineered in that it was a nonpathogenic transposon-marked organism. It was his contention that, by definition, the NIH guidelines regarding recombinant microorganisms did not apply *(244)*. In a separate incident, an accidental release of a GM virus containing genes from hepatitis C viruses and Dengue fever occurred at London's Imperial College. The accident was a result of inadequate containment *(245)*. Although no perceived harm occurred to human health or the environment from either of these incidents, public perception of these events helped galvanize concerns of legitimate and properly regulated releases of GMMs.

Potential uses of GMMs include examples by which the organisms will need to be released into an uncontrolled environment to be effective. The use of GMMs in pest control, bioremediation, wildlife and livestock vaccination, metal extraction, and crop yield improvement all suggest that the use of released GMMs will represent a growing part of the biotechnology industry *(58)*. However, will such environmental GMM releases alter, in some fundamental way, the ecology of the systems into which they are exposed?

This question is certainly one of the most contested of all GMM issues, in large part because of the huge stakes involved. On one side of the issue are the biotechnology and agricultural interests, which see a potential panacea of benefits from controlled releases of GMMs *(179)*. On the other side of the issue are concerned environmental groups, which fear GMMs are inherently dangerous because of their very nature as organisms. The fear is that, once released, a GMM cannot be controlled if an undesirable event takes place *(50)*.

Both sides have compelling arguments for their views. Parties interested in release of GMMs can note an array of safety mechanisms developed that maintain the process as safe. Further, the benefits of GMM releases potentially help society in numerous ways. Parties opposed to such releases point out that the history of human colonization of the planet is a lesson in what can go wrong when nonindigenous species are placed in novel environments *(246)*.

It is easy to understand the source of these fears in evaluating the damage caused through the intentional or accidental releases of numerous nonindigenous organisms into the environment. For example, the establishment of human-released organisms such as kudzu plants and European starlings in the Eastern United States; mosquitoes, mongooses, and pigs in Hawaii; European rabbits, Cane toads, and domestic mice in Australia; and the brown snake on the island of Guam and the damage these organisms have inflicted in their respective environments underscore that concern. It is hard to imagine completely what the consequences might be of an uncontrolled GMM. Speculation of potential harm caused by a GMM release includes impacts on human health, crop damage, or damage to indigenous microbial communities *(10,18,89,190)*.

Regardless of the perspective on the issue, there must be one starting point in agreement: The ecology of vast areas of the world has been irreversibly changed through human activities. Ever-increasing demands on cropland and "pristine" environments to maintain substance for a growing world population and the needs to remediate lands already polluted beyond the capacities of natural systems require novel answers. Many of those potential answers will undoubtedly be devised from the ability to manipulate genes in microorganisms that are then released into the environment *(58)*.

It was generally believed, and some data support the fact, that most GMMs are effectively less fit to survive in complex ecosystems *(46)*. Even if true, like all organisms, GMMs do evolve, and some have been shown to persist for extended periods of time in harsh environments despite the best attempts to engineer biological and physical methods to constrain them. Further, because of the complex nature of these environments, it is virtually impossible to design experiments that will accurately assess all the parameters of whether a released GMM or its DNA will persist in the environment. Even if it were possible to design complex release studies for each GMM, the cost associated with such studies would be prohibitive.

Lenski argued *(46)*, however, that there is a reasonable approach to this conundrum. Lenski suggested that GMM applications appreciably similar to previous releases of GMMs or non-GMMs be used as a baseline to assess their risk. This approach would permit more intensive risk analysis of those applications evaluated as of greater risk.

Using this approach, a detailed assessment would be required for GMM applications that are novel and do not have comparative non-GMM releases.

Releasing GMMs into the environment also has definite social implications that invoke other ethical considerations *(95,224)*. Whereas research in a laboratory takes place in a confined, and therefore private, domain, releasing GMMs into the environment enters the realm of public involvement. It is inescapable that people do fear the technology. What responsibility do the GMM user and developer have to address those fears when analytical responses are ineffective? This is a difficult and important issue that is likely to elude simple answers.

7.5. GMMs and Patent Issues

Is it possible actually to patent a living GMM? That was the question put before the US Supreme Court in June 1980. Ananda Chakrabarty applied for a patent for a recombinant *Pseudomonas* that had been engineered to disperse oil slicks more effectively than the non-GM form. Several parties resisted the granting of this patent, including the US Patent Office, the USDA, and several individuals opposed to the process. By a 5-to-4 judgment, the US Supreme Court granted the patent to Chakrabarty on the grounds that the GMM was in fact a human-made invention and not a product of nature *(247,248)*. Despite this ruling, there are still moral considerations that need to be addressed in patent law, both in the United States and internationally. What moral code is used to address the granting of a given patent? What cultural norm should a court uphold to make such a determination? Court challenges based on the interpretation of the morality of a given patent are increasingly used as a method of choice by those opposed to the technology to block the granting of the patent *(226)*.

There are sound reasons for developing strong patent laws in nations. Patents provide protection of development costs and, certainly for companies, a way to receive a reasonable return on their investment *(248)*. Basically, all a patent provides is a way through the courts to find remedy of unauthorized use of the product of the patent *(235)*. Through the creation of a monopoly under the protection of the law, patents provide value for discoveries and thus incentives for private company investment.

A major concern that has developed from the patent process is the effect that patents have on science *(235)*. Notably, in the United States and in the United Kingdom, the undeniable monetary returns offered by the granting of patents have had an impact on the science conducted *(230)*. The process does provide funds for active research projects and is often used in the United States for general institutional needs. However, the patent process often imposes secrecy on the scientific process, including that of public institutions *(249)*.

Increasingly, the goal of research in many laboratories has become the creation of intellectual property. Should it not be the philosophical goal of publicly funded research that science is an enterprise that engages in free exchange of reason and thought? Further, issues of secrecy permeate the patenting process. In the United States, a patent must be filed within 1 year of the publication of the product; in the United Kingdom, the patent must be filed by the time the researcher publishes. Either way, the process stifles open scientific dialog of the findings, generally for fear of losing control of the patent. These and additional concerns stemming from patents were addressed recently by the Royal Society in the United Kingdom *(249)*.

Finally, once granted, a patent acts as a monopoly, entitling the inventor to exclusive rights in charging fees or exclusive use of that process or product *(230)*. That prohibition, through legal or economic pressures, may interfere with promising lines of research. Such arguments come full circle when it is realized that public funding through grants is often the catalyst that makes the research possible. However, because the patents are only granted as rights to prevent the unauthorized use of the product, it can be argued that the act of receiving a patent is not in itself immoral *(226)*.

7.6. Biological Warfare and GMMs

How does one conceive that biological weapons are more or less moral than conventional weapons? Since the September 11 attacks in New York City and Washington, DC, and the subsequent mailings of anthrax that resulted in several US deaths, this question has been part of active debate in the United States *(250)*. In fact, despite the US anthrax deaths, it is still believed by many that the effective use of bioengineered pathogens as weapons is likely quite low *(251)*. However, biological weapons pose many obvious advantages to their users, including their relative ease of production, relatively low cost, ease of concealment, and civilian vulnerability to bioweapons *(252,253)*. Although the development of GMM as bioweapons is more difficult to accomplish than development of non-GMM weapons, the fact that genomic databases provide details on the specific pathogenicity of genes may render methods to develop novel pathogens more efficient *(254)*. Many of the general issues regarding biological weapons have been reviewed elsewhere *(253,254)*.

The BWC treaty *(255)* banning the development and use of microorganisms as weaponized agents should have been the cornerstone in preventing the continued development of GMM-derived bioweapons. The threat of GMM-derived bioweapons was first addressed to the United Nations by Joshua Lederberg, discoverer of bacterial conjugation *(256)*. Unfortunately, it has become clear that microorganisms have been intentionally modified for weapons use, apparently first in the former Soviet Union *(254)* and later in other nations.

It has been suggested that enhancements provided by recombination technology could be used to deliver toxin genes (i.e., anthrax toxin, myelin toxin), antibiotic resistance genes (i.e., penicillin or tetracycline resistance), genes to change the mode of infection of a pathogen (such as addition of a respiratory mode), or genes to make microbes resistant to vaccination *(91)*. A short list of microbes implicated in GMM-derived bioweapons includes *Bacillus anthrax, Yersinia pestis, Francisella tularensis*, and smallpox *(91)*. Clearly, it is possible to ponder other destructive GMMs directed at humans as well, as demonstrated in the inadvertent creation of lethal GM mouse pox *(81,257)*.

Other forms of biological warfare are conceivable. Plant pathogen genes could be cloned into hosts normally associated with crops, or any number of pathogens could be engineered against livestock *(258–260)*. There are even suggestions that recombinant technologies currently used in biodegradation be adapted to consume metals, lubricants, or plastics that might be directed at opposing force weapons *(261)*. If true, such weapons would arguably not only be in violation of the BWC treaty by their very nature, but also a microbe developed and used in such a capacity could be isolated and turned against its developer, although presumably suicide genes would be engineered into such strains. Regardless, such weapons directed at the fabric of industrialization could have far-reaching consequences in a modern world.

This misguided predilection to engineer microorganisms as weapons requires reasonable countermeasures. For example, it will likely be necessary in the foreseeable future to develop novel vaccines as a defense against GM viruses or GM bacteria. Although the extent of countermeasure development is unclear, in the Reagan administration active research into this question was conducted *(256)*. The United States has refused to support a stronger BWC, partly because of concerns about confidential business information and partly because of national security concerns *(245)*. It is reasonable to speculate that these national security concerns involve countermeasures developed against known bioweapons using much the same technologies that it took to develop them *(255)*.

7.7. The "Do Nothing" Principle

Many would argue, notably Jeremy Rifkin *(36)*, that the potential hazards of recombinant technologies are so vast that they should not be attempted, and the technology should not be approved. This argument, to ban a technology, has its consequences as well. There are serious problems, notably in pollution and food production or improvement, for which GMMs can provide obvious solutions.

For example, soils contaminated with xenobiotic compounds can be extremely recalcitrant. To remediate sites that are extensively contaminated, soils may have to be removed physically, treated with solvents or detergents first to remove the compound, and then be destroyed by physical methods, such as incineration or chemical neutralization. Such protocols potentially expose workers or local communities to aerosols and dusts containing the compound. A GMM that can effectively mineralize a toxic compound *in situ* would almost certainly be less expensive than conventional cleanup methods and expose cleanup crews to less of the pollutant.

The same principle is true in developing GMMs to make soil more fertile. It has been suggested that, over the next 40 years, food production will have to more than double to meet projected world populations *(52,230,262)*. Not a complete answer, but GMMs will likely compose part of the solution.

The concept of not using GMMs has other consequences as well. Species such as mice and rabbit populations periodically increase to produce significant plagues in Australia, resulting in major damage to regional farms *(124,263)*. The use of conventional methods to control these pests is simply inadequate to meet the staggering need both in realistically controlling their numbers and in limiting the harm conventional control techniques (poisoning, physical traps) might have on indigenous species *(263)*. The use of GM viruses has been proposed as a method to target and control individual pest species. However, as noted in Section 4.1.4, this approach comes with its own set of risks *(124)*. The ethical dilemma is which approach to take: allow destructive plagues to go unchecked, continue to use ineffective and environmentally damaging pest control methods, or look at targeted GMM control methods. Each comes with its own ethical consequences. If substantial data are gathered that the GMM control method is specific to its target organism and is effective in safely controlling pests, then in certain circumstances, GMMs may well become the preferred method to control pests.

7.8. Other GMM Ethical Issues

Finally, 3 billion years of selection and evolution have resulted in a biosphere full of microorganisms of incredible complexity and diversity *(12)*. Should the natural evolutionary processes that developed living organisms and therefore the integrity of individual species be respected? In other words, simply because there is the intellectual capacity to do so, is it appropriate to manipulate any organism of choice? These perhaps are ridiculous questions to ponder because humankind has been manipulating the genes of organisms for millennia using conventional genetic methods. Further, because there can be no return to a time when GMMs were not part of life on this planet, it is likely these questions represent moot concerns. However, not to appreciate and understand the relevance of this capability to manipulate life at the molecular level and the consequences of this ability are to deny what may be humanity's most profound impact on life on earth.

References

1. Biotechnology Industry Organization. (2003) Guide to Biotechnology: Ethics. Available online at: http://www.bio.org/er/ethics.asp. Accessed June 2003.
2. Hartl, D. L. (1985) Engineered organisms in the environment: inferences from population genetics. In *Engineered Organisms in the Environment: Scientific Issues* (Halvorson, H. O., Pramer, D., and Rogul, M., eds.). ASM, Washington, DC, pp. 83–88.
3. Barinaga, M. (2000) Asilomar revisited: lessons for today? *Science* **287,** 1584–1585.
4. Pramer, D. (1985) Orientation. In *Engineered Organisms in the Environment: Scientific Issues* (Halvorson, H. O., Pramer, D., and Rogul, M., eds.). ASM, Washington, DC, p. 1.
5. Lindow, S.E. (1985) Ecology of *Pseudomonas syringae* relevant to the field use of ice$^-$ deletion mutants construction in vitro for plant frost control. In *Engineered Organisms in the Environment: Scientific Issues* (Halvorson, H. O., Pramer, D., and Rogul, M., eds.). ASM, Washington, DC, pp. 23–35.
6. Moss, B. and Buller, M. L. (1985) Vaccinia virus vectors: potential use as live recombinant virus vaccines. In *Engineered Organisms in the Environment: Scientific Issues* (Halvorson, H. O., Pramer, D., and Rogul, M., eds.). ASM, Washington, DC, pp. 36–39.
7. Watrud, L. S., Perlak, F. J., Tran, M.-H., et al. (1985) Cloning of the *Bacillus thuringiensis* subsp *kurstaki* delta into *Pseudomonas fluorescens*: molecular biology and ecology of an engineered microbial pesticide. In *Engineered Organisms in the Environment: Scientific Issues* (Halvorson, H. O., Pramer, D., and Rogul, M., eds.). ASM, Washington, DC, pp. 40–46.
8. Slater, J.H. (1985) Gene transfer in microbial communities. In *Engineered Organisms in the Environment: Scientific Issues* (Halvorson, H. O., Pramer, D., and Rogul, M., eds.). ASM, Washington, DC, pp. 89–98.
9. Suter, G.W., II. (1985) Application of environmental risk analysis to engineered organisms. In *Engineered Organisms in the Environment: Scientific Issues* (Halvorson, H. O., Pramer, D., and Rogul, M., eds.). ASM, Washington, DC, pp. 211–219.
10. Whipps, J. M., De Leij, F. A. A. M., Lynch, J. M., and Bailey, M. J. (1998) Risk assessment with the release of genetically modified *Pseudomonas fluorescens* in the field. Biological control of fungal and bacterial plant pathogens. *IOBC Bull.* **21,** 199–204.
11. Steinhauser, K.G. (2001) Environmental risks of chemicals and genetically modified organisms: a comparison. Part I: Classification and characterization of risks posed by chemicals and GMOs. *Environ. Sci. Pollut. Res. Int.* **8,** 120–126.
12. Southgate, C. (2002) Introduction to environmental ethics. In *Bioethics for Scientists* (Bryant, J., La Velle, L. B., and Searle, J., eds.). John Wiley and Sons, New York, pp. 39–55.

13. Ratcliff, J. (2001) Genetically modified organisms in animal feed—a European perspective. In *Concepts in Pig Science 2001. The Third Annual Turtle Lake Pig Science Conference* (Lyons, T. P. and Cole, D. J. A., eds.). Nottingham University Press, Nottingham, UK, pp. 39–45.
14. Migheli, Q. (2001) Genetically modified biocontrol agents: environmental impact and risk analysis. *J. Plant Pathol.* **83,** 47–56.
15. Migheli, Q. (2001) The deliberate release of genetically modified biocontrol agents. III: Towards an alternative risk assessment paradigm. *Agro-Food-Industry Hi-Technol.* **12,** 23–25.
16. Lingren, S. (1999) Biosafety aspects of genetically modified lactic acid bacteria in EU legislation. *Int. Dairy J.* **9,** 37–41.
17. Liberman, D. F., Fink, R., and Schaefer, F. (1999) Biosafety and biotechnology. In *Manual of Industrial Microbiology and Biotechnology, 2nd ed.* (Demain, A. L. and Davis, J. E., eds.). ASM, Washington, DC, pp. 300–309.
18. Landis, W. G., Lenart, L. A., and Spromberg, J. A. (2000) Dynamics of horizontal gene transfer and the ecological risk assessment of genetically engineered organisms. *Hum. Ecol. Risk Assess.* **6,** 875–899.
19. Doblhoff-Dier, O., Bachmayer, H., Bennett, A., et al. (1999) Safe biotechnology 9: values in risk assessment for the environmental application of microorganisms. *Trends Biotechnol.* **17,** 307–311.
20. Cockburn, A. (2001) Assuring the safety of genetically modified (GM) foods: the importance of an holistic, integrative approach. *J. Biotechnol.* **98,** 79–106.
21. Doblhoff-Dier, O., Bachmayer, H., Bennett, A., et al. (2000) Safe biotechnology 10: DNA content of biotechnological process waste. *Trends Biotechnol.* **18,** 141–146.
22. Food and Agriculture Organization. (1996) *Biotechnology and Food Safety. Report of a Joint FAO/WHO Consultation.* UN Food and Agriculture Organization, Rome.
23. Food and Agriculture Organization/World Health Organization. (2000) *Safety Aspects of Genetically Modified Food of Plant Origin.* World Health Organization, Geneva, Switzerland.
24. Food and Agriculture Organization/World Health Organization. (2001) Safety assessment of foods derived from genetically modified microorganisms. Report of a Joint FAO/WHO Expert Consultation on Foods Derived from Biotechnology. *Microb. Ecol. Health Dis.* Sept., 197–211.
25. Frommer, W., Archer, L., Boon, B., et al. (1992) Safe biotechnology 4: recommendations for safety levels for biotechnological operations with microorganisms that cause diseases in plants. *Appl. Microbiol. Biotechnol.* **38**, 139–140.
26. International Life Science Institute. (1999) Europe Novel Foods Task Force. *Microb. Ecol. Health Dis.* **11**, 198–207.
27. Cartagena Protocol on Biosafety Treaty. (2000) Available on-line at: http://www.biodiversity.org/ratification.asp. Accessed June 2003.
28. Lelieveld, H. L. M., Bachmayer, H., Boon, B., et al. (1997) Safe biotechnology 8: transport of infectious and biological materials. *Appl. Microbiol. Biotechnol.* **48**, 135–140.
29. Millis, N. (1992) Second International Symposium on the Biosafety Results of Genetically Modified Plants and Microorganisms. *Australas Biotechnol.* **2**, 237–239.
30. Organization for Economic Cooperation and Development. (1993) *Safety Evaluation of Foods Produced by Modern Technology—Concepts and Principles.* Organization for Economic Cooperation and Development, Paris.
31. Lammerts van Bueren, E. (Ed.). (1997) *Proceedings of the Ifgene Conference, The Future of DNA.* Kluwer, Dordrecht, The Netherlands.
32. World Health Organization. (1991) *Strategies for Assessing the Safety of Food Produced by Biotechnology. Report of Joint FAO/WHO Consultation.* World Health Organization, Geneva, Switzerland.

33. Meningaud, J.-P., Moutel, G., and Hervé, C. (2001) Ethical acceptability, health policy and foods biotechnology based foods: is there a third way between the precaution principle and an overly enthusiastic dissemination of GMO? *Med. Law* **20**, 133–141.
34. Giampietro, M. (2002) The precautionary principle and ecological hazards of genetically modified organisms. *R. Swed. Acad. Sci.* **31**, 466–470.
35. Milewski, E. and Talbot, B. (1983) Proposals involving field testing of recombinant DNA containing organisms. *Recomb. DNA Tech. Bull.* **6**, 141–145.
36. Palfreman, J. (2001) Interview with Jeremy Rifkin. In *Harvest of Fear*. Palfreman Film Group. Available on-line at: www.pbs.org/wgbh/harvest/interviews/Rifkin.html. Accessed June 2003.
37. Maykuth, A. (1986) Genetic wonders to come: some see boon, others calamity. *Philadelphia Inquirer*. Available on-line at: http://maykuth.com/Archives/gene86.htm. Accessed June 2003.
38. Tappeser, B., Jäger, M., and Eckelkamp, C. (2002) Survival, persistence, transfer: the fate of genetically modified microorganisms and recombinant DNA in different environments. In *Genetically Engineered Organisms, Assessing Environmental and Human Health Effects* (Letourneau, D. and Burrows, B. E., eds.). CRC, Boca Raton, FL, pp. 223–250.
39. Smit, E., Wolters, A. C., Lee, H., Trevors, J. T., and van Elsas, J. D. (1996) Interactions between a genetically marked *Pseudomonas fluorescens* strain and bacteriophage PhiR2f in soil: effects of nutrients, alginate encapsulation, and the wheat rhizosphere. *Microb. Ecol.* **31**, 125–140.
40. Henschke, R. B., Henschke, E. J., and Schmidt, F. R. (1991) Monitoring survival and gene transfer in soil microcosms of recombinant *Escherichia coli* designed to represent an industrial production strain. *Appl. Microbiol. Biotechnol.* **35**, 247–252.
41. Herron, P. R., Toth, I. K., Heilig, G. H. I., Akkermans, A. D. L., Karagouni, A., and Wellington, E. M. H. (1998) Selective effect of antibiotics on survival and gene transfer of *Streptomyces* in soil. *Soil Biol. Biochem.* **30**, 673–677.
42. Zadoks, J. C. (1998) Risk analysis of beneficial microorganisms—wild type and genetically modified. In *Proceedings on Microbial Plant Protection Products—Workshop on the Scientific Basis for Risk Assessment*. Swedish National Chemicals Inspectorate, Stockholm, Sweden, pp. 9–38.
43. Haslberger, A. G. (2000) Policy forum: genetic technologies. Monitoring and labeling for genetically modified products. *Science* **287**, 431–432.
44. Schmidt, C. W. (1999) Caught in the middle: should the World Trade Organization settle environmental disputes? *Environ. Health Perspect.* **107**, A562–A564.
45. Brooke-Taylor, S. (2001) Practical approaches to risk assessment. *Biomed. Environ. Sci.* **14**, 14–20.
46. Lenski, R. E. (1993) Evaluating the fate of genetically modified microorganisms in the environment: are they inherently less fit? *Experientia* **49**, 201–209.
47. Velkov, V. V. (2001) Stress-induced evolution and the biosafety of genetically modified microorganisms released into the environment. *J. Biosci.* **26**, 667–683.
48. Lee, G.-H. and Stotzky, G. (1999) Transformation and survival of donor, recipient, and transformants of *Bacillus subtilis* in vitro and in soil. *Soil Biol. Biotechnol.* **31**, 1499–1508.
49. Dwyer, D. F., Hooper, S. W., Rojo, F., and Timmis, K. N. (1988) Fate of genetically-engineered bacteria in activated sludge microcosms. *Schriftenr Ver Wasser Boden Lufthyg* **78**, 267–276.
50. Greenpeace. (2001) Man Made Bacteria on the Loose. Available on-line at: http://www.greenpeace.org/~geneng/reports/bio/rhizsummary.pdf. Accessed June 2003.
51. Union of Concerned Scientists. (2003) Web site. Available on-line at: http://www.ucsusa.org/food_and_environment/biotechnology/index.cfm?pageID=9. Accessed June 2003.

52. Grabowski, G. (2001) Food industry perspective on safety and labeling of biotechnology. In *Genetically Modified Organisms in Agriculture* (Nelson, G., ed.). Academic, New York, pp. 225–231.
53. Bott, T. L. and Kaplan, L. A. (2002) Autecological properties of 3-chlorobenzoate-degrading bacteria and their population dynamics when introduced into sediments. *Microb. Ecol.* **43**, 199–216.
54. Hefle, S. L., Nordlee, J. A., and Taylor, S. L. (1996) Allergenic foods. *Crit. Rev. Food Sci. Nutr.* **36**, S69–S89.
55. Renault, P. (2002) Genetically modified lactic acid bacteria: applications to food or health and risk assessment. *Biochimie* **84**, 1073–1087.
56. Bratoeva, M. and Raitchev, G. (1991) A case of shigellosis caused by *Shigella dysenteriae* type 1 *Shigella flexneri* type 5B. *J. Hyg. Epidemiol. Microbiol. Immunol.* **35**, 35–39.
57. Yamamoto, T., Honda, T., Miwatani, T., and Yakota, T. (1984) A virulence plasmid in *Escherichia coli* enterotoxigenic for humans: intergenetic transfer and expression. *J. Infect. Dis.* **150**, 688–698.
58. Henderson, L. and Gatewood, D. M. (2000) Release of genetically engineered microorganisms in the environment in the United States. In *Microbial Interactions in Agriculture and Forestry* (Rao, N. S. and Dommergues, Y. R., eds.). Science Publishers, Enfield, NH, pp. 83–109.
59. Food and Drug Administration. (1996) *Safety Assurance of Foods Derived by Modern Biotechnology in the United States.* Center for Food Safety and Applied Nutrition. Available online at http://www.cfscan.fda.gov/~/rd/biojap96.html. Accessed June 2003.
60. Kimber, I., Kerkvliet, N. I., Taylor, S. L., Astwood, J. D., Sarlo, K., and Dearman, R. J. (1999) Toxicology of proteins allergenicity: prediction and characterization. *Toxicol. Sci.* **48**, 157–162.
61. Sampson, H. A. (1999) Food allergy. Part I: immunopathogenesis and clinical disorders. *J. Allergy Clin. Immunol.* **103**, 717–728.
62. Young, E., Stoneham, M. D., Petruckevitch, A., Barton, J., and Rona, R. (1994) A population study of food intolerance. *Lancet* **343**, 1127–1130.
63. Hourihane, J. O. (1998) Prevalence and severity of food allergy-need for control. *Allergy* **53**, 84–88.
64. Lack, G., Chapman, M., Kalsheker, N., King, V., Robinson, C., and Venables, K. (2002) Report on the potential allergenicity of genetically modified organisms and their products. *Clin. Exp. Allergy* **32**, 1131–1143.
65. Nordlee, J. A., Taylor, S. L., Townsend, J. A., Thomas, L. A., and Bush, R. K. (1996) Identification of a Brazil nut allergen in transgenic soybeans. *N. Engl. J. Med.* **334**, 688–692.
66. Gendel, S. M. (1998) Sequence database for assessing the potential allergenicity of proteins used in transgenic foods. *Adv. Food Nutr. Res.* **42**, 63–92.
67. Astwood, J. D., Leach, J. N., and Fuchs, R. L. (1996) Stability of food allergens to digestion in vitro. *Nat. Biotechnol.* **14**, 1269–1273.
68. Miller, K., Meredith, C., Selo, I., and Wal, J. M. (1999) Allergy to bovine β-lactoglobulin: specificity of immunoglobulin E generated in the Brown Norway rat to tryptic and synthetic peptides. *Clin. Exp. Allergy* **29**, 1696–1704.
69. Hilton, J., Dearman, R. J., Satter, N., Basketter, D. A., and Kimber, I. (1997) Characteristics of antibody responses induced in mice by protein allergens. *Food Chem. Toxicol.* **35**, 1209–1218.
70. Metcalfe, D. D., Astwood, J. D., Townsend, R., Sampson, H. A., Taylor, S. L., and Fuchs, R. L. (1996) Assessment of the allergenic potential of foods derived from genetically engineered crop plants. *Crit. Rev. Food Sci. Nutr.* **36**, S165–S186.
71. Sangodkar, U., Chapman, P., and Chakrabarty, A. M. (1988) Cloning, physical mapping and expression of chromosomal genes specifying degradation of the herbicide 2,4,5-T by *Pseudomonas cepacia* AC1100. *Gene* **71**, 267–277.

72. Homma, Y., Sato, Z., Hirayama, F., Kanno, K., Sirahama, H., and Suzui, T. (1989) Production of antibodies by *Pseudomonas cepacia* as an agent for biological control of soilborne pathogens. *Soil Biol. Biochem.* **21**, 723–728.
73. McLoughlin, T. J., Quinn, J. P., Bettermann, A., and Bookland, R. (1992) *Pseudomonas cepacia* suppression of sunflower wilt fungus and role of antifungal compounds in controlling the disease. *Appl. Environ. Microbiol.* **58**, 1760–1763.
74. Holmes, A., Govan, J., and Goldstein, R. (1998) Agricultural use of *Burkholderia* (*Pseudomonas*) *cepacia*: a threat to human health? *Emerg. Infect. Dis.* **4**, 221–227.
75. Goldmann, D. and Klinger, J. (1986) *Pseudomonas cepacia*: biology, mechanisms of virulence, epidemiology. *J. Pediatr.* **108**, 806–812.
76. Thomassen, M. J., Demko, C. A., Klinger, J. D., and Stern, R. C. (1985) *Pseudomonas cepacia* colonization among patients with cystic fibrosis. A new opportunist. *Am. Rev. Respir. Dis.* **131**, 791–796.
77. Sajjan, U. S., Sun, L., Goldstein, R., Forstner, J. F. (1995) Cable (cbl) type II pili of cystic fibrosis-associated *Burkholderia* (*Pseudomonas*) *cepacia*: nucleotide sequence of the *cblA* major subunit pilin gene and novel morphology of the assembled appendage fibers. *J. Bacteriol.* **177**, 1030–1038.
78. Govan, J. R. and Deretic, V. (1996) Microbial pathogenesis in cystic fibrosis: mucoid *Pseudomonas aeruginosa* and *Burkholderia cepacia*. *Microbiol. Rev.* **60**, 539–574.
79. Reddy, M. (1997) *Status on Commercial Development of* Burkholderia cepacia *for Biological Control of Fungal Pathogens and Growth Enhancement of Conifer Seedlings for a Global Market.* US Forest Service General Technical Report PNW 389, Washington, DC, pp. 235–244.
80. Av-Gay, Y. (1999) Uncontrolled release of harmful microorganisms. *Science* **284**, 1621.
81. Jackson, R. J., Ramsay, A. J., Christensen, C. D., Beaton, S., Hall, D. F., and Ramshaw, I. A. (2001) Expression of mouse interleukin-4 by a recombinant ectromelia virus suppresses cytolytic lymphocyte responses and overcomes genetic resistance to mousepox. *J. Virol.* **75**, 1205–1210.
82. Hentges, D. J., Petschow, B. W., Dougherty, S. H., and Marsh, W. W. (1995) Animal models to assess the pathogenicity of genetically modified microorganisms for humans. *Microb. Ecol. Health Dis.* **8**, S23–S26.
83. Giannoukakis, N. and Trucco, M. (2003) Gene therapy technology applied to disorders of glucose metabolism: promise, achievements, and prospects. *Biotechniques* **35**, 122–145.
84. Floeth, F. W., Langen, K. J., Reifenberger, G., and Weber, F. (2003) Tumor-free survival of 7 years after gene therapy for recurrent glioblastoma. *Neurology* **61**, 270–271.
85. Hege, K. M. and Carbone, D. P. (2003) Lung cancer vaccines and gene therapy. *Lung Cancer* **41**, S103–S113.
86. Dettweiler, U. and Simon, P. (2001) Points to consider for ethics committees in human gene therapy trials. *Bioethics* **15**, 491–500.
87. American Society of Gene Therapy Report. (2003) American Society of Gene Therapy Responds to a Second Case of Leukemia Seen in a Clinical Trial of Gene Therapy for Immune Deficiency. Available on-line at: http//www.asgt.org/press_releases/01132003.html. Accessed June 2003.
88. Pike, R. M. (1979) Laboratory-associated infections: incidence, fatalities, causes and prevention. *Annu. Rev. Microbiol.* **33**, 41–66.
89. Liberman, D. F., Israeli, E., and Fink, R. (1991) Risk assessment of biological hazards in the biotechnology industry. *Occup. Med. State of the Art Rev.* **6**, 285–299.
90. Wedum, A. G., Barkley, W. E., and Hellman, A. (1972) Handling of infectious agents. *J. Am. Vet. Med. Assoc.* **161**, 1557–1567.
91. Alibek, K. (1999) *Biohazard.* Random House, New York, pp. 40–42, 155–157, 160–161, 166–167, 259–260, 336.

92. Migheli, Q. (2001) The deliberate release of genetically modified biocontrol agents. I: Is it possible to harmonize safety, emotions and rationality? *Agro-Food-Industry Hi-Tech.* **12,** 41–42.
93. De Leij, F. A. A. M., Sutton, E. J., Whipps, J. M., and Lynch, J. M. (1994) Effect of a genetically modified *Pseudomonas aureofaciens* on indigenous microbial populations of wheat. *FEMS Microbiol. Ecol.* **13**, 249–257.
94. Office of Technology. (1981) *Impacts of Applied Genetics: Microorganisms, Plants, and Animals.* US Government Printing Office, Washington, DC.
95. Steinhauser, K. G. (2001) Environmental risks of chemicals and genetically modified organisms: a comparison. Part II: Sustainability and precaution in risk assessment and risk management. *Environ. Sci. Pollut. Res. Int.* **8**, 222–226.
96. Popova, L. Y., Pechurkin, N. S., Maksimova, E. E., et al. (1999) Experimental microcosms as models of natural ecosystems for monitoring survival of genetically modified microorganisms. *Life Support Biosph. Sci.* **6**, 193–197.
97. Molin, S. (1992) Designing microbes for release into the environment. *Sci. Prog.* **76**, 139–148.
98. Sharples, F. E. (1983) Spread of organisms with novel genotypes: thoughts from an ecological perspective. *Recomb. DNA Tech. Bull.* **6**, 43–56.
99. Cairns, J. and Orvos, D. R. (1992) Establishing environment hazards of genetically engineered microorganisms. *Rev. Environ. Contam. Toxicol.* **124,** 19–39.
100. Kozdrój, J. (1999) Impact of introduced *Pseudomonas fluorescens* mutants on indigenous rhizosphere microflora of bean. *J. Sci. Health* **34,** 435–459.
101. Käppeli, O. and Auberson, L. (1998) Planned releases of genetically modified organisms into the environment: the evolution of safety considerations. *Chimia* **52,** 137–142.
102. EPA Fact Sheet. (2003) Microbial Products of Biotechnology: Final Regulations Under the Toxic Substances Control Act Summary. Available on-line at: http://www.epa.gov/opptintr/biotech/fs-001.htm. Accessed June 2003.
103. Vázquez, M. M., Barea, J. M., and Azcón, R. (2002) Influence of arbuscular mycorrhizae and a genetically modified strain of *Sinorhizobium* on growth, nitrate reductase activity and protein content in shoots and roots of *Medicago sativa* as affected by nitrogen concentrations. *Soil Biol. Biotechnol.* **34,** 899–905.
104. Organization for Economic Cooperation and Development. (1997) *Consensus Document on Information Used in the Assessment of Environmental Applications Involving* Pseudomonas. OECD Series on the Harmonization of Regulatory Oversight in Biotechnology. Organization for Economic Cooperation and Development, Paris.
105. Ramos, C., Molina, L., Mølbak, L., Ramos, J. L., and Molin, S. (2000) A bioluminescent derivative of *Pseudomonas putida* KT2440 for deliberate release into the environment. *FEMS Microbiol. Ecol.* **34,** 91–102.
106. Nuti, M. P., Squartini, A., Giacomini, A., Casella, S., and Corich, V. (1994) The use of genetically modified organisms (GMOs) in the environment: biosafety results of field tests and key scientific issues. In *Proceedings of the Sixth European Congress on Biotechnology*, Vol. 2 (Alberghina, L., Frontali, L., and Sensi, P., eds.). Elsevier Science, Amsterdam, The Netherlands, pp. 1291–1296.
107. Turner, J. T., Lampel, J. S., Stearman, R. S., Sundin, G. W., Gunyuzlu, P., and Anderson, J. J. (1991) Stability of the δ-endotoxin gene from *Bacillus thuringiensis subsp kurstaki* in a recombinant strain of *Clavibacter xyli* subsp *cynodontis*. *Appl. Environ. Microbiol.* **57,** 3522–3528.
108. Drouault, S., Juste, C., Marteau, P., Renault, P., and Corthier, G. (2002) Oral treatment with Lactococcus lactis expressing *Staphylococcus hyicus* lipase enhances lipid digestion in pigs with induced pancreatic insufficiency. *Appl. Environ. Microbiol.* **68,** 3166–3168.
109. Bonning, B. C. and Hammock, B. D. (1996) Development of recombinant baculovirus for insect control. *Annu. Rev. Entomol.* **41,** 191–210.

110. Daniell, T. J., Davy, M. L., and Smith, R. J. (2000) Development of a genetically modified bacteriophage for use in tracing sources of pollution. *J. Appl. Microbiol.* **88,** 860–869.
111. Gillespie, K. M., Angle, J. S., and Hill, R. L. (1995) Runoff losses of Pseudomonas aureofaciens (lacZY) from soil. *FEMS Microbiol. Ecol.* **17,** 239–246.
112. Hekman, W. E., Heijnen, C. E., Burgers, S. L. G. E., van Veen, J. A, and van Elsas, J. D. (1995) Transport of bacterial inoculants through intact cores of two different soils as affected by water percolation and presence of wheat plants. *Microbiol. Ecol.* **16,** 143–157.
113. Snyder, W. E., Tonkyn, D. W., and Kluepfel, D. A. (1999) Transmission of a genetically engineered rhizobacterium by grasshoppers in the laboratory and in the field. Ecol. Appl. 9, 245–253.
114. Lilley, A. K., Hails, R. S., Cory, J. S., and Bailey, M. J. (1997) The dispersal and establishment of pseudomonad population in the phyllosphere of sugar beet by phytophagous caterpillars. *FEMS Microbiol. Ecol.* **24,** 151.
115. Kluepfel, D. A., Lamb, T. G., Synder, B., and Tonkyn, D. W. (1994) Six years of field testing a *lacZY* modified fluorescent pseudomonad. In *Proceedings of the Third International Symposium on the Biosafety Results of Field Tests of Genetically Modified Plants and Microorganisms* (Jones, D. D., ed.). University of California Press, Oakland, CA, pp. 169–176.
116. Clegg, C. D., Anderson, J. M., Lappinscott, H. M., van Elsas, J. D., and Jolly, J. M. (1995) Interactions of a genetically modified *Pseudomonas fluorescens* with the soil-feeding earthworm *Octolasion cyaneum (Lumbricidae). Soil Biol. Biochem.* **27,** 1423–1429.
117. Kolter, R., Siegle, D. A., and Tormo, A. (1993) The stationary phase of the bacterial life cycle. *Annu. Rev. Microbiol.* **47,** 855–874.
118. Terré, J., Chappuis, G., Lombard, M., and Desmettre, P. (1996) Eradication of rabies, using a rec-DNA vaccine. *J. Biotechnol.* **46,** 155–157.
119. Rupprecht, C.E. (2003) *Rabies Vaccines: Past, Present and Future.* Sixth Annual Conference on Vaccine Research, Arlington, VA, May 5–7. 2003.
120. van Loon, L. C. (2000) Helping plants to defend themselves: biocontrol by disease-suppressing Rhizobacteria. In *Phytosfere'99—Highlights in European Plant Biotechnology Research and Technology Transfer* (de Vries, G. E. and Metzlaff, K., eds.). Elsevier Science, Amsterdam, The Netherlands, pp. 203–212.
121. Tyndale-Biscoe, C. H. (1994). Virus-vectored immunocontraception of feral mammals. *Reprod. Fertil. Dev.* **6**, 281–287.
122. Chambers, L. K., Lawson, M. A., and Hinds, L. A. (1999) Biological control of rodents—the case for fertility control using immunocontraception. In *Ecologically-based Management of Rodent Pests* (Singleton, G. R., Hinds, L. A., Leirs, H., and Zhang, Z., eds.). Australian Centre for International Agricultural Research, Canberra, pp. 215–242.
123. Singleton, G. R., Hinds, L. A., Lawson, M. A., and Pech, R. P. (2001) Strategies for management of rodents: prospects for fertility control using immunocontraceptive vaccines. In *Advances in Vertebrate Pest Management II* (Pelz, H. J., Cowan, D. P., and Feare, C. J., eds.). Filander-Verlag, Fürth, Germany, pp. 301–318.
124. Williams, C. K. (2002) Risk assessment for release of genetically modified organisms: a virus to reduce the fertility of introduced wild mice, *Mus domesticus. Reproduction,* **Suppl. 60**, 81–88.
125. Hayes, R. A. and Richardson, B. J. (2001) Biological control of the rabbit in Australia: lessons not learned? *Trends Microbiol.* **9**, 459–460.
126. Flowerdew, J. R., Trout, R. C., and Ross J. (1992) Myxomatosis: population dynamics of rabbits (*Oryctolagus cuniculus* Linnaeus, 1758) and ecological effects in the United Kingdom. *Rev. Sci. Tech.* **11**, 1109–1113.
127. Jones, R. A., Broder, M. W., and Stotzky, G. (1991) Effects of genetically engineered microorganisms on nitrogen transformations and nitrogen-transforming microbial populations in soil. *Appl. Environ. Microbiol.* **57**, 3212–3219.
128. Doyle, J. D., Short, K. A., Stotzky, G., King, R. J., Seidler, R. J., and Olsen, R. H. (1991) Ecologically significant effects of *Pseudomonas putida* PPO301(pRO103), genetically engi-

neered to degrade 2,4-dichlorophenoxyacetate, on microbial populations and processes in soil. *Can. J. Microbiol.* **37**, 682–691.
129. Natsch, A., Keel, C., Hebecker, N., Laasik, E., and Défago, G. (1998) Impact of *Pseudomonas fluorescens* strain CHA0 and a derivative with improved biocontrol activity on the culturable resident bacterial community on cucumber roots. *FEMS Microbiol. Ecol.* **27**, 365–380.
130. Orvos, D. R., Lacy, G. H., and Cairns, J., Jr. (1990) Genetically-engineered *Erwinia carotovora*: survival, intraspecific competition, and effects upon selected bacterial genera. *Appl. Environ. Microbiol.* **56**, 1689–1694.
131. White, D., Crosbie, J. D., Atkinson, D., and Killham, K. (1994) Effect of an introduced inoculum on soil microbial diversity. *FEMS Microbiol. Ecol.* **14**, 169–178.
132. Legard, D. E., McQuilken, M. P., Whipps, J. M., et al. (1994) Studies of seasonal changes in the microbial populations on the phyllosphere of spring wheat as a prelude to the release of a genetically modified microorganism. *Agric. Ecosys. Environ.* **50**, 87–110.
133. Gagliardi, J. V., Buyer, J. S., Angle, J. S., and Russek-Cohen, E. (2001) Structural and functional analysis of whole-soil microbial communities for risk and efficacy testing following microbial inoculation of wheat roots in diverse soils. *Soil Biol. Biotechnol.* **33**, 25–40.
134. Kreutzweiser, D., England, L., Shepherd, J., Conklin, J., and Holmes, S. (2001) Comparative effects of a genetically engineered insect virus and a growth-regulating insecticide on microbial communities in aquatic microcosms. *Ecotoxicol. Environ. Saf.* **48**, 85–98.
135. Richardson, R. E., James, C. A., Bhupathiraju, V. K., and Alvarez-Cohen, L. (2002) Microbial activity in soils following steam treatment. *Biodegradation* **13**, 285–295.
136. Buyer, J. S. and Drinkwater, L. E. (1997) Comparison of substrate utilization assay and fatty acid analysis of soil microbial communities. *J. Microbiol. Methods* **30**, 3–11.
137. Haack, S. K., Garchow, H., Odelson, D. A., Forney, L. J., and Klug, M. J. (1994) Accuracy, reproducibility, and interpretation of fatty acid methyl ester profiles of model bacterial communities. *Appl. Environ. Microbiol.* **60**, 2483–2493.
138. Frostegard, A., Baath, E., and Tunlid, A. (1993) Shifts in the structure of soil microbial communities in limed forests as revealed by phospholipid fatty acid analysis. *Soil Biol. Biochem.* **25**, 723–730.
139. Zelles, L., Bai, Q. Y., Beck, T., and Beese, F. (1992) Signature fatty acids in phospholipids as lipooliogosaccharides as indicators of microbial biomass and community structure in agricultural soils. *Soil Biol. Biochem.* **24**, 317–323.
140. Mills, D. K., Fitzgerald, K., Litchfield, C. D., and Gillevet, P. M. (2003) A comparison of DNA profiling techniques for monitoring nutrient impact on microbial community composition during bioremediation of petroleum-contaminated soils. *J. Microbiol. Methods* **54**, 57–74.
141. Polz, M. F., Bertilsson, S., Acinas, S. G., and Hunt, D. (2003) A(r)Ray of hope in analysis of the function and diversity of microbial communities. *Biol. Bull.* **204**, 196–199.
142. Greene, E. A. and Voordouw, G. (2003) Analysis of environmental microbial communities by reverse sample genome probing. *J. Microbiol. Methods* **53**, 211–219.
143. Stin, O. C, Carnahan, A., Singh, R., et al. (2003) Characterization of microbial communities from coastal waters using microarrays. *Environ. Monit. Assess.* **81**, 327–336.
144. Ramirez-Moreno, S., Martinez-Alonso, M. R., Mendez-Alvarez, S., Esteve, I., and Gaju, N. (2003) Seasonal population changes in the restriction fragment length polymorphism (RFLP) patterns from PCR-amplified 16S rRNA genes of predominant ribotypes in microbial mat samples from the Ebro Delta (Spain). *Curr. Microbiol.* **46**, 190–198.
145. Yi, S., Tay, J. H., Maszenan, A. M., and Tay, S. T. (2003) A culture-independent approach for studying microbial diversity in aerobic granules. *Water Sci. Technol.* **47**, 283–290.
146. Casamayor, E. O., Massana, R., Benlloch, S., et al. (2002) Changes in archaeal, bacterial and eukaryal assemblages along a salinity gradient by comparison of genetic fingerprinting methods in a multipond solar saltern. *Environ. Microbiol.* **4**, 338–438.

147. Abed, R. M., Safi, N. M., Koster, J., et al. (2002) Microbial diversity of a heavily polluted microbial mat and its community changes following degradation of petroleum compounds. *Appl. Environ. Microbiol.* **68**, 1674–1683.
148. Ibekwe, A. M., Papiernik, S. K., Gan, J., Yates, S. R., Yang, C. H, and Crowley, D. E. (2001) Impact of fumigants on soil microbial communities. *Appl. Environ. Microbiol.* **67**, 3245–3257.
149. Engelen, B., Meinken, K., von Wintzingerode, F., Heuer, H., Malkomes, H. P., and Backhaus, H. (1998) Monitoring impact of a pesticide treatment on bacterial soil communities by metabolic and genetic fingerprinting in addition to conventional testing procedures. *Appl. Environ. Microbiol.* **64**, 2814–2821.
150. Stotzky, G. (1997) DNA in the environment: ecological, and therefore societal, implications. In *Proceedings of the Ifgene Conference, The Future of DNA* (Lammerts van Bueren, E., ed.). Kluwer, Dordrecht, The Netherlands, pp. 55–77.
151. Felske, A., Engelen, B., Nubel, U., and Backhaus, H. (1996) Direct ribosome isolation from soil to extract bacterial rRNA for community analysis. *Appl. Environ. Microbiol.* **62**, 4162–4167.
152. Lee, D. H., Zo, Y. G., and Kim, S. J. (1996) Non radioactive method to study genetic profiles of natural bacterial communities by PCR-single-strand-conformation polymorphism. *Appl. Environ. Microbiol.* **62**, 3112–3120.
153. Liu, W. T., Marsh, T. L., Cheng, H., and Forney, L. J. (1997) Characterization of microbial diversity by determining terminal restriction fragment length polymorphisms of genes encoding 16S rRNA. *Appl. Environ. Microbiol.* **63**, 4516–4522.
154. van Elsas, J. D., Duarte, G. F., Rosado, A. S., and Smalla, K. (1998) Microbiological and molecular biological methods for monitoring microbial inoculants and their effects in the soil environment. *J. Microbiol. Methods* **32**, 133–154.
155. Glandorf, D. C. M., Verheggen, P., Jansen, T., et al. (2001) Effect of genetically modified *Pseudomonas putida* WCS358r on the fungal rhizosphere microflora of field-grown wheat. *Appl. Environ. Microbiol.* **67**, 3371–3378.
156. Prosser, J. I. (1994) Molecular marker systems for detection of genetically engineered microorganisms in the environment. *Microbiology* **140**, 5–17.
157. Lee, H. (2003) A Hypothetical Scenario in Environmental Biodegradation/Biotransformation. Available on-line at: http://www.uoguelph.ca/~hlee/418chap4.htm. Accessed June 2003.
158. Corich, V., Giacomini, A., Vian, P., et al. (2001) Aspects of marker/reporter stability and selectively in soil microbiology. *Microb. Ecol.* **41**, 333–340.
159. Germida, J. J., Siciliano, S. D., and Seib, A. M. (1998) Phenotypic plasticity of *Pseudomonas aureofaciens* (*lacZY*) introduced into and recovered from field and laboratory microcosm soils. *FEMS Microbiol. Ecol.* **27**, 133–139.
160. Naseby, D. C. and Lynch, J. M. (1998) Impact of wild-type and genetically modified *Pseudomonas fluorescens* on soil enzyme activities and microbial population structure in the rhizosphere of pea. *Mol. Ecol.* **7**, 617–625.
161. Drahos, D. J., Hemming, B. C., and McPherson, S. (1986) Tracking recombinant organisms in the environment: β-galactosidase as a selectable non-antibiotic marker for fluorescent pseudomonads. *Biotechnology* (NY) **4**, 439–443.
162. Shaw, J. J., Dane, F., Geiger, D., and Kloepper, J. W. (1992) Use of bioluminescence for detection of genetically engineered microorganisms released into the environment. *Appl. Environ. Microbiol.* **58**, 267–273.
163. Jansson, J. K. (1995) Tracking genetically modified microorganisms in nature. *Curr. Opin. Biotechnol.* **6**, 275–283.
164. Leung, K. T., So, J. S., Kostrzynska, M., Lee, H., and Trevors, J. T. (2000) Using a green fluorescent protein gene-labeled p-nitrophenol-degrading *Moraxella* strain to examine the protective effect of alginate encapsulation against protozoan grazing. *J. Microbiol. Methods* **39**, 205–211.

165. Kostrzynska, M., Leung, K. T., Lee, H., and Trevors, J. T. (2002) Green fluorescent protein based biosensor for detecting SOS-inducing activity of genotoxic compounds. *J. Microbiol. Methods* **48**, 43–51.
166. Colwell, R. R., Somerville, C., Knight, I., and Straube, W. (1988) Detection and monitoring of genetically-engineered microorganisms. In *The Release of Genetically Engineered Microorganisms* (Sussmann, M., Collins, G. H., Skinner, F. A., and Stewart-Tall, D. E., eds.). Academic Press, London, pp. 47–60.
167. Iwasaki, K., Uchiyama, H., and Yagi, O. (1993) Survival and impact of genetically engineered *Pseudomonas putida* harboring mercury resistance gene in aquatic microcosms. *Biosci. Biotechnol. Biochem.* **57**, 1264–1269.
168. Hwang, I. and Farrand, S. K. (1994) A novel gene tag for identifying microorganisms released into the environment. *Appl. Environ. Microbiol.* **60**, 913–920.
169. Olsen, P. E. and Rice, W. A. (1989) *Rhizobium* strain identification and quantification in commercial inoculants by immunoblot analysis. *Appl. Environ. Microbiol.* **55**, 520–522.
170. Böttger, E. C. (1996) Approaches for identification of microorganisms. *ASM News* **62**, 247–250.
171. Chaudhry, G. R., Toranzos, G. A., and Bhatti, A. R. (1989) Novel method for monitoring genetically engineered microorganisms in the environment. *Appl. Environ. Microbiol.* **55**, 1301–1304.
172. Steffan, R. J. and Atlas, R. M. (1988) DNA amplification to enhance detection of genetically engineered bacteria in environmental samples. *Appl. Environ. Microbiol.* **54**, 2185–2191.
173. Steffan, R. J. and Atlas, R. M. (1991) Polymerase chain reaction: applications in environmental microbiology. *Annu. Rev. Microbiol.* **45**, 137–161.
174. van Overbeek, L. S., van Veen, J. A., and van Elsas, J. D. (1997) Induced reporter gene activity, enhanced stress resistance, and competitive ability of a genetically modified *Pseudomonas fluorescens* strain released into a field plot planted with wheat. *Appl. Environ. Microbiol.* **63**, 1965–1973.
175. McClure, N. C., Frey, J. C., and Weightman, A. J. (1991) Survival and catabolic activity of natural and genetically engineered bacteria in a laboratory-scale activated-sludge unit. *Appl. Environ. Microbiol.* **57**, 366–373.
176. Glandorf, D. C. M., Verheggen, P., Jansen, T., et al. (1998) Field release of genetically-modified *Pseudomonas putida* WCS358r to study effects on the indigenous soil microflora. In *Past, Present and Future Considerations in Risk Assessment When Using GMOs* (de Vries, G. E., ed.), CCRO Workshop Proceedings, Noordwijkerhout, The Netherlands, pp. 41–46.
177. Short, K. A., Doyle, J. D., King, R. J., Seidler, R. J., Stotzky, G., and Olsen, R. H. (1991) Effects of 2,4-dichlorophenol, a metabolite of a genetically engineered bacterium, and 2,4-dichlorophenoxyacetate on some microorganism-mediated ecological processes in soil. *Appl. Environ. Microbiol.* **57**, 412–418.
178. Velicer, G. J. (1999) Pleiotrophic effects of adaptation to a single carbon source for growth on alternative substrates. *Appl. Environ. Microbiol.* **65**, 264–269.
179. Morrissey, J. P., Walsh, U. F., O'Donnell, A., Moenne-Loccoz, Y., and O'Gara, F. (2002) Exploitation of genetically modified inoculants for industrial ecology applications. *Antonie Van Leeuwenhoek* **81**, 599–606.
180. Genthner, F. J., Campbell, R. P., and Pritchard, P. H. (1992) Use of a novel plasmid to monitor the fate of a genetically engineered *Pseudomonas putida* strain. *Mol. Ecol.* **1**, 137–143.
181. van Elsas, J. D. and Migheli, Q. (1999) Evaluation of risk related to the release of biocontrol agents active against plant pathogen. In *Integrated Pest and Disease Management in Greenhouse Crops* (Albajes, R., Gullino, M. L., van Lenteren, J. C., and Elad, Y., eds.). Kluwer, Dordrecht, The Netherlands, pp. 337–393.

182. Leslie, J. F. and Dickman, M. B. (1991) Fate of DNA encoding hygromycin resistance after meiosis in transformed strains of *Gibberella fujikuroi* (*Fusarium monilforme*). *Appl. Environ. Microbiol.* **57**, 1423–1429.
183. Keller, N. P., Bergstrom, G. C., and Yoder, O. C. (1991) Mitotic stability of transforming DNA is determined by its chromosomal configuration in the fungus *Cochliobolus heterostrophus. Curr. Genet.* **19**, 227–233.
184. Kistler, H. C. (1991) Genetic manipulation of plant pathogenic fungi. In *Microbial Control of Weeds* (TeBeest, D. O., ed.), Chapman and Hall, London, pp. 152–170.
185. Mazodier, P. and Davies, J. (1991) Gene transfer between distantly related bacteria. *Annu. Rev. Genet.* **25**, 147–171.
186. Gormley, E. P. and Davies, J. (1991) Transfer of plasmid RSF1010 by conjugation from *Escherichia coli* to *Streptomyces lividans* and *Mycobacterium smegmatis. J. Bacteriol.* **173**, 6705–6708.
187. Amabile-Cuevas, C. F. and Chicurel, M. E. (1993) Horizontal gene transfer. *Am. Sci.* **81**, 332–341.
188. Heinemann, J. A. and Sprague, G. F., Jr. (1989) Bacterial conjugative plasmids mobilize DNA transfer between bacteria and yeast. *Nature* **340**, 205–209.
189. Ministry of Education, Science, and Culture (1991) *Guidelines for Recombinant DNA. Experiments.* Tokyo, Japan.
190. Pretty, J. (2001) The rapid emergence of genetic modification in world agriculture: contested risk and benefits. *Environ. Conserv.* **28**, 248–262.
191. Sengeløv, G., Kristensen, K. J., Sørensen, A. H., Kroer, N., and Sørensen, S. J. (2001) Effect of genomic location on horizontal transfer of a recombinant gene cassette between *Pseudomonas* strains in the rhizosphere and spermosphere of barley seedlings. *Curr. Microbiol.* **42**, 160–167.
192. Henriksen, C. M., Nilsson, D., Hansen, D., Hansen, S., and Johansen, E. (1999) Industrial applications of genetically modified microorganisms; gene technology at Chr. Hansen A/S. *Int. Dairy J.* **9**, 17–23.
193. European Commission. (1997) Regulation (EC) No. 258/97 concerning novel foods and novel food ingredients. Official J. Eur. Commun. ***L43**, 1–7.*
194. Brooke-Taylor, S. (2001) Practical approaches to risk assessment. *Biomed. Environ. Sci.* **14,** 14–20.
195. Mattila-Sandholm, T., Mättö, J., and Saarela, M. (1999) Lactic acid bacteria with health claims—interactions and interference with gastrointestinal flora. *Int. Dairy J.* **9,** 25–35.
196. Yang, C.-Y., Pang, J.-C., Kao, S.-S., and Tsen, H.-Y. (2003) Enterotoxigenicity and cytotoxicity of Bacillus thuringiensis strains and development of a process for Cry1Ac production. J. *Agric. Food Chem.* **51,** 100–105.
197. International Food Biotechnology Council. (1990) Biotechnologies and food: assuring the safety of foods produced by genetic modification. *Reg. Tox. Pharm.* **12,** 1–193.
198. Hertel, C., Probst, A. J., Cavadini, C., Meding, E., and Hammes, W. P. (1995) Safety assessment of genetically modified microorganisms applied in meat fermentations. *Syst. Appl. Microbiol.* **18,** 469–476.
199. Jonas, D., Antignac, E., Antoine, J., et al. (1996) The safety assessment of novel foods. *Food Chem. Toxicol.* **34,** 931–940.
200. Martin, J. C. (1980). Behavior factors in laboratory safety: personnel characteristics and modification of unsafe acts. In *Laboratory Safety: Theory and Practice* (Fuscaldo, A. A., Erlick, B. J., and Hindman, B., eds.). Academic Press, New York, pp. 321–342.
201. National Institutes of Health. (1998). Guidelines for Research Involving Recombinant Molecules. Available on-line at: http://www.niehs.nih.gov/odhsb/biosafe/nih/rdna-apr98.pdf. www.aphis.usda.gov/brs/index.html. Accessed June 2003.

202. APHIS document. (2003) Agricultural Biotechnology. Permitting Notification and Regulation. Available on-line at: http://www.aphis.usda.gov/brs/index.html. Accessed June 2003.
203. CFSAN handout. (1995) FDA's Policy for Foods Derived by Biotechnology. US Food and Drug Administration. Available on-line at: http//www.cfsan.fda.gov/~1rd/biopoly.html.
204. Food and Drug Administration. (1998) Guidance for Human Somatic Cell Therapy and Gene Therapy. Available on-line at: http//www.fda.gov/cber/gdlns/somgene.htm. Accessed June 2003.
205. Tomita, F. (1994) Regulation and safety of recombinant microorganisms. *Bioprocess Technol.* **19**, 15–26.
206. Organization for Economic Cooperation and Development. (1986) *Recombinant DNA Safety Considerations for Industrial Agricultural and Environmental Appliactions of Organisms Derived by Recombinant DNA Technique*. Bulletin. Organization for Economic Cooperation and Development, Paris.
207. Organization for Economic Cooperation and Development. (1992) *Safety Considerations for Biotechnology*. Organization for Economic Cooperation and Development, Paris.
208. Center for Disease Control. (1974) *Classification of Etiological Agents on the Basis of Hazard*. 4th ed. US Department of Health, Education and Welfare, Public Health Service, Atlanta, GA.
209. National Institutes of Health. (1979) *Laboratory Safety Monograph*. National Institutes of Health, Bethesda, MD.
210. Cooney, C.L. (1983) Bioreactors: design and operation. *Science* **219**, 728–733.
211. Lorenz, M. G., Gerjets, D., and Wackernagel, W. (1991) Release of transforming plasmid and chromosomal DNA from two cultured soil bacteria. *Arch. Microbiol.* **156**, 319–326.
212. Romanowski, G., Lorenz, M. G., and Wackernagel, W. (1993) Plasmid DNA in a groundwater aquifer microcosm-adsorption, DNase resistance and natural genetic transformation of *Bacillus subtilis*. *Mol. Ecol.* **2**, 171–181.
213. Ramos, J. L., Andersson, P., Jensen, L. B., et al. (1995) Suicide microbes on the loose. Biological containment could decrease the uncertainties surrounding the deliberate release of recombinant microorganisms. *Biotechnology* **13**, 35–37.
214. Contreras, A., Molin, S., and Ramos, J. L. (1991) Conditional-suicide containment system for bacteria which mineralize aromatics. *Appl. Environ. Microbiol.* **57**, 1504–1508.
215. Molin, S., Klemm, P., Poulsen, L. K., Biehl, H., Gerdes, K., and Andersson, P. (1987) Conditional suicide system for containment of bacteria and plasmids. *Biotechnology* **5**, 1315–1318.
216. Jensen, L. B., Ramos, J. L., Kaneva, Z., and Molin, S. (1993) A substrate-dependent biological containment system for *Pseudomonas putida* based on the *Escherichia coli gef* gene. *Appl. Environ. Microbiol.* **59**, 3713–3717.
217. Bej, A. K., Perlin, M. H., and Atlas, R. M. (1988) Model suicide vector for containment of genetically engineered microorganisms. *Appl. Environ. Microbiol.* **54**, 2472–2477.
218. Knudsen, S., Saadbye, P., Hansen, L. H., et al. (1995) Development and testing of improved suicide functions for biological containment of bacteria. *Appl. Environ. Microbiol.* **61**, 985–991.
219. Kaplan, D. L., Mello, C., Sano, T., Cantor, C., and Smith, C. (1999) Streptavidin-based containment systems for genetically engineered microorganisms. *Biomol. Eng.* **16**, 135–140.
220. Molin, S., Boe, L., Jensen, L. B., Kristensen, C. S., Givskov, M., Ramos, J. L., and Bej, A. K. (1993) Suicidal genetic elements and their use in biological containment of bacteria. *Annu. Rev. Microbiol.* **47**, 139–166.
221. Ronchel, M. C., Ramos, C., Jensen, L. B., Molin, S., and Ramos, J. L. (1995) Construction and behavior of biologically contained bacteria for environmental applications in bioremediation. *Appl. Environ. Microbiol.* **61**, 2990–2994.
222. Amsellem, Z., Barghouthi, S., Cohen, B., et al. (2001) Recent advances in the biocontrol of *Orobanche* (broomrape) species. *BioControl* **46**, 211–228.

223. Braum, R. (1994) People's concerns about biotechnology: some problems and some solutions. *J. Biotechnol.* **98**, 3–8.
224. Rehmann-Sutter, C. (1993) Nature in the laboratory-nature as a laboratory. Considerations about the ethics of release experiments. *Experientia* **49**, 190–200.
225. Reiss, M. J. (2002) Introduction to ethics and bioethics. In *Bioethics for Scientists* (Bryant, J, La Velle, L. B, and Searle, J., eds.). John Wiley and Sons, New York, pp. 3–17.
226. Crespi, R. S. (2000) An analysis of moral issues affecting patenting inventions in the life sciences: a European perspective. *Sci. Eng. Ethics* **6**, 157–180.
227. Bodenmüller, K. (2001) Health-Relevant and Environmental Aspects of Different Farming Systems: Organic, Conventional and Genetic Engineering. InterNutrition - Swiss Association for Research and Nutrition, Switzerland. Available on-line at: http://www.internutrition.ch/in-news/mediainfo/index_f.html. Accessed June 2003.
228. Trautman, T. D. (2000) Risk communication—the perceptions and realities. *Food Addit. Contam.* **18**, 1130–1134.
229. Palfreman, J. (2001) Frontline and Nova (co-production) *Harvest of Fear*. Palfreman Film Group. WGBH, Boston, MA.
230. Pellizzoni, L. (2001) Democracy and the governance of uncertainty—the case of agricultural gene technologies. *J. Hazard Mater.* **86**, 205–222.
231. Hughes, S. (2002) The patenting of genes for agricultural biotechnology. In *Bioethics for Scientists* (Bryant, J., La Velle, L. B., and Searle, J., eds.). John Wiley and Sons, New York, pp. 153–170.
232. Anderson, W. A. (2000) The future relationship between the media, the food industry and the consumer. *Br. Med. Bull.* **56**, 254–268.
233. Beringer, J. (1999) Keeping watch over genetically modified crops and foods. *Lancet* **353**, 605–606.
234. Minnesota Extension Service. (1990) *Food, Agriculture, and Nutrition Forum IV, Food Safety: Developing Communication Strategies*. University of Minnesota, Minneapolis. Minnesota Extension Service, Minneapolis.
235. Sandman, P. M. (1989) Hazard vs outrage in the public perception of risk. In *Effective Risk Communication: The Role and Responsibility of Government and Non Government Organizations* (Covello, V. T., McCallum, D. B., and Pavlova, M. T., eds.). Plenum, New York, pp. 45–49.
236. Bharathan, G., Chandrashekaran, S., May, T., and Bryant, J. (2002) Crop biotechnology in developing countries. In *Bioethics for Scientists* (Bryant, J., La Velle, L. B, and Searle, J., eds.). John Wiley and Sons, New York, pp. 171–196.
237. Hayden, T. (ed.). (1996) *The Lost Gospel of Earth: A Call for Renewing Nature, Spirit, and Politics*. Sierra Club Books, San Francisco, CA.
238. Beardsley, T. (2000) Rules of the game. *Sci. Am.* **282**, 42–43.
239. Mitchell, P. (2003) Europe angers United States with strict GM labeling. *Nat. Biotechnol.* **21**, 6.
240. US House of Representatives. (2003) 108th Congress, 1st Session, Roll Call 256, *Congr. Rec.* H. Res. 252.
241. Union of Concerned Scientists. (2002). Food and Environment Comments on genetically modified food. Available on-line at: http://www.ucsusa.org/food_and_environment/biotechnology_archive/page.cfm?pageID=381. Accessed June 2003.
242. Knight, D. (1999) Agriculture: Costs and Benefits of Biotechnology. InterPress Services. Available on-line at: http://www.ifpri.org/media/innews/052099.htm. June 10, 2003.
243. Roberts, L. (1987) MSU faults Strobel for Dutch elm test. *Science* **237**, 1286.
244. Strobel, G. (1987) I have acted in good faith. *Scientist* **1**, 11.
245. Dorey, E. (2001) US rejects stronger bioweapons treaty. *Nat. Biotechnol.* **19**, 793.

246. Simberloff, D. (1985) Predicting ecological effects of novel entities: evidence from higher organisms. In *Engineered Organisms in the Environment: Scientific Issues* (Halvorson, H. O., Pramer, D., and Rogul, M., eds.). ASM, Washington, DC, pp. 152–161.
247. *Diamond vs Chakrabarty*. (1980) 206 US Patent Q. 193 US patent 4,259,444.
248. Jenner, M. W. (2002) Biotechnology crops—a producer's perspective. In *Genetically Modified Organisms in Agriculture* (Nelson, G., ed.). Academic Press, New York, pp. 151–156.
249. Couzin, J. (2003) Report deplores growth in academic patenting. *Science* **300**, 406.
250. US Senate. (2003) *Biological, Chemical, Radiological Weapons Countermeasure Research Act*. US Senate, 108th Congress Bill S.666.IS.
251. Rose, G. (1999) It could happen here: facing the new terrorism. *For. Affairs* **78**, 131–137.
252. Coughlin, T. R. (2003) Biological Weapons; Malignant Biology. Available on-line at: http://www.centrexcorporation.com/Downloads/BIOLOGICAL%20WEAPONS%20TRC.pdf. Accessed June 2003.
253. Fidler, D. P. (1999) Facing the global challenges posed by biological weapons. *Microbes Infect.* **1**, 1059–1066.
254. Fraser, C. M. and Dando, M. R. (2001) Genomics and future biological weapons: the need for preventive action by the biomedical community. *Nat. Genet.* **29**, 253–256.
255. Biological Weapons Convention. (1972) *Int. Legal Materials* **11**, 309–315.
256. Miller, J., Engelberg, S., and Broad, W. (2001) *Germs—Biological Weapons and America's Secret War*. Simon and Schuster, New York, pp. 80–84.
257. Nowak, R. (2001) Disaster in the making: an engineered mouse virus leaves us one step away from the ultimate bioweapon. *New Sci.* **169**, 4–5.
258. Ban, J. (2000) Agricultural biological warfare: an overview. *Arena* **9**, 1–8.
259. Pearson, J. E. (2000) *Biological Agents as Potential Weapons Against Animals. Biological Warfare Technical Brief.* Office International des Epizooties, Paris, p. 3.
260. Dudley, J. P. and Woodford, M. H. (2002) Bioweapons, bioterrorism, and biodiversity: potential impacts of biological weapons attacks on agricultural and biological diversity. *Rev. Sci. Off. Int. Epiz.* **21**, 125–137.
261. Mackenzie, D. (2002) US non-lethal weapon report suppressed. *New Sci.*. Available on-line at: http://www.newscientist.com/news/news.jsp?id=ns99992254. Accessed June 2003.
262. Gasser, C. S. and Fraley, R. (1992) Transgenic Crops. *Sci. Am.* **266**, 34–39.
264. Pech, R. P., Hood, G. M., Singleton, G. R., Salmon, E., Forrester, R. I., and Brown, P. R. (1999) Models for predicting plagues of house mice (*Mus domesticus*) in Australia. In *Ecologically-Based Management of Rodent Pests* (Singleton, G. R., Hinds, L. A., Leirs, H., and Zhang, Z., eds.). Australian Centre for International Agricultural Research, Canberra, pp. 81–112.

III

MAMMALIAN GMOs

5

Large-Scale Exogenous Protein Production in Higher Animal Cells

William Whitford

1. Introduction

Recombinant proteins and polypeptides are now produced in a variety of ways, from direct chemical synthesis, to expression of cloned genes in in vitro systems, to exogenous gene expression in a wide variety of in vivo systems. In vivo expression systems include those based in fungal, bacterial, yeast, plant, insect, and mammalian cells, as well as those based in transgenic plants and animals.

With the choice of so many systems and with so many factors involved in the selection of any particular one, system selection has become a study in itself. Broadly, factors to consider include the (1) size and structure of the gene to be expressed; (2) required posttranscription and translation events; (3) size and structure of the final gene product; (4) required product fidelity; (5) required amount of product and term of production; (6) regulatory concerns; (7) required final product purity (including purification systems available); (8) required product molecular homogeneity; (9) ultimate product use; (10) production team expertise; (11) acceptable lead times and start-up costs; (12) acceptable production costs, and (13) schedule and financial tolerance for failure.

A simple approach to getting started in cell-based expression has been described *(1)*. It is suggested that products can be assigned to groups such as (1) small peptides (<80 amino acids), which can be most efficiently expressed as fusion proteins in prokaryotes; (2) polypeptides, which are normally secreted proteins (80–500 amino acids), can be expressed in almost any system, but preferably as a secreted product in yeast or higher animal cells; (3) large, secreted, and cell surface proteins (>500 amino acids), which are most efficiently produced in mammalian systems; and (4) large, but nonsecreted proteins, for which the choice of system depends mainly on their particular characteristics. A good review of the posttranslational functions provided by the basic systems was presented by Grey and Subramanian *(2)*.

From: *The GMO Handbook: Genetically Modified Animals, Microbes, and Plants in Biotechnology*
Edited by: S. R. Parekh © Humana Press Inc., Totowa, NJ

However, as more expression systems in a number of diverse cell types and species have become practical, the choice of system for those who desire to consider all of the options has become daunting. Beyond the number of new choices available, each existing system continually evolves and incorporates newly discovered improvements. For example, recombinant insect cell production hosts are now providing additional, or previously absent, posttranslational power to that system *(3)*.

It has been slightly over 20 years since techniques allowing the production of protein in a cloned gene in an animal cell emerged. At that time, it was a new and exotic technology practiced in research laboratories. Since then, a number of systems have become standard practice. In fact, embarking on an animal cell expression project has evolved from a sophisticated and intimidating prospect, to a common and routine tool, and now back to an intimidating prospect. Why is it intimidating again? It is not because a researcher wonders if the gene can be successfully expressed, but because one wonders about which one of the dozens of available systems will give the best results. It is now possible to get lost in the vast number of disparate systems readily, even commercially, available. Presented here is an introduction to production in those higher animal cell systems that have become the most popular in larger-scale applications.

In addressing large-scale production, bench-top expression systems used primarily in verification of cloning product and in research applications are excluded, such as the analysis of (1) expression on cell activities; (2) copy deoxyribonucleic acid (cDNA) library gene products; (3) glycoprotein structure and carbohydrate moieties; (4) protein-folding structure and processing, as well as most transient expression systems in general. Systems that have become widely applied in larger-scale production of biomedically active proteins, antibodies, and antibody fragments are highlighted.

By limiting the topic to higher animal cells; fungal, bacteria, yeast, and plant technologies are excluded, and a group (insect and mammals) with similar biochemical, culture, and production characteristics is distinguished. These eukaryotic cells have advanced and related systems of protein expression and processing. That the cells are in culture excludes the field of transgenic animal production.

Looking at exogenous proteins as the product of interest excludes other products from large-scale culture, such as viral vectors or the cells themselves. Furthermore, activities prior and subsequent to production are omitted. These activities include (1) expression vector and promoter selection; (2) gene identification, isolation, manipulation, and cloning; (3) cell transfection, clone selection, and expansion; and (4) raw product handling, analysis, and purification.

1.1. History of the Field

Looking at the current state of any technology, its principal systems can seem quite obvious. This can be true of the current capabilities to use animal cells to produce exogenous products. However, the history of cellular-based protein production is a story of step-by-step accomplishments. When it is considered that humans have been maintaining living animal cells outside the body since the end of the 19th century, the reality of the gradual evolution of the science of the cellular-based production of selected and modified recombinant proteins is better appreciated.

Milestones in this history include the first observation of cells dividing in vitro in 1907 *(4)*. In 1928, the first system for the propagation of virus in cultured tissue was reported *(5)*. During those times, high concentrations of animal fluids were the main source of nutrition, but in 1955, a relatively nutritionally complete synthetic growth medium was presented *(6)*. It was about that time that the first pharmaceutically relevant product, the Salk poliovirus vaccine, was produced in primary monkey kidney cells *(7)*.

Even the Nobel Prize-worthy invention in 1975 of the production of monoclonal antibodies (MAb) *(8)*, as significant as it was at the time, was really just one step toward the current state of the art. Hybridoma technology is essentially a way of immortalizing an antibody-producing mouse splenocyte to allow its maintenance in culture and production of the exact antibody it had been making in the mouse. Before that accomplishment, systems to allow for the continuous culture of animal cells had to be developed, and biochemical means of cloning and selection had to be understood. Moving from there to the current capability of pharmaceutical-scale production of human or humanized antibody required 30 years of yet more incremental discoveries. The realization of cellular-based, large-scale recombinant protein production required numerous independent technologies. From molecular biology to bioreactor technology, many distinct disciplines contributed to this history of gradual development.

1.2. Products From Cell-Based Systems

A number of products from cultured living organisms are now used in such areas as industrial research, and medical diagnosis and therapy. These products range from simple molecules to the most complex, including multimeric proteins, lipids, collagen, polyglycans, viruses, and cells. The term *recombinant protein biologicals* can be used to specify high molecular weight polypeptides with biochemical activity. Production in animal cells is a major system in use to produce many protein biologicals for therapeutic, diagnostic, and manufacturing purposes. The types of products now in use include such proteins as MAb and fragments, human growth factors, interleukins, interferons, erythropoietin, plasminogen activators, enzymes and blood clotting factors (Table 1).

1.3. Advantages of Higher Animal Cells

Each system of production has been pursued because of its unique set of features. Factors in the choice of a host for production include system efficiency and scalability, clonal stability, economic issues, and characteristics of the product desired. These characteristics include required posttranslational modifications such as propeptide proteolytic processing, glycosylation and carbohydrate trimming, γ-carboxylation, hydroxylation, sulfation, phosphorylation, fatty acid acylation, and subunit assembly. The time required to establish higher animal systems and the cost to operate them are greater than in many other systems. Therefore, their main advantage is their more authentic protein processing of, for example, secreted glycoproteins of mammalian or human origin.

1.4. Production Scale

The concepts of "large" or "small" scale can be relative depending on the context, purpose, and system of expression. In discussing eukaryotic expression systems, it is practical to refer to work in 3- to 300-mL lots as small scale. This is because the common formats for such scale are T- and shake flasks. Large-scale expression, or production, can refer to expression in 0.3- to 20-L lots. This is the range supported by the

Table 1
Recent Pharmaceutical Products Manufactured Using Transgenic Higher Animal Cells

Product	Manufacturer	Launch	Product	Disease	Host Cell
Remicade™	Centocor	1998	MAb	Rheumatoid arthritis	Sp2/0
Reopro™	Centocor	1998	MAb	Myocardial Infarction	Sp2/0
Rituxan™	IDEC	1998	MAb	Non-Hodgkin's lymphoma	CHO
Synagis™	MedImmune	1998	MAb	Respiratory syncytial virus	NS0
Simulect™	Novartis	1998	MAb	Organ rejection	Sp2/0
Zenapax™	Roche	1998	MAb	Organ rejection	NS0
Enbrel™	Amgen	1998	MAb fusion	Rheumatoid arthritis	CHO
Herceptin™	Genentech	1998	MAb	Breast cancer	CHO
NovoSeven™	Novo	1999	Factor VIIa	Hemophilia	BHK
Campath™	Berlex	2001	MAb	Cancer	CHO
Amevive™	Biogen	2003	Fc fusion	Psoriasis	CHO
Fabrazyme™	Genzyme	2003	Enzyme	Fabry's disease	CHO
Aldurazyme™	Biomarin	2003	Enzyme	Mucopolysaccharidosis	CHO
Xolair™	Genentech	2003	MAb	Asthma	CHO
Humira™	Abbott	2003	MAb	Rheumatoid Arthritis	CHO
Arastin™	Genentech	2003	MAb	Cancer	CHO

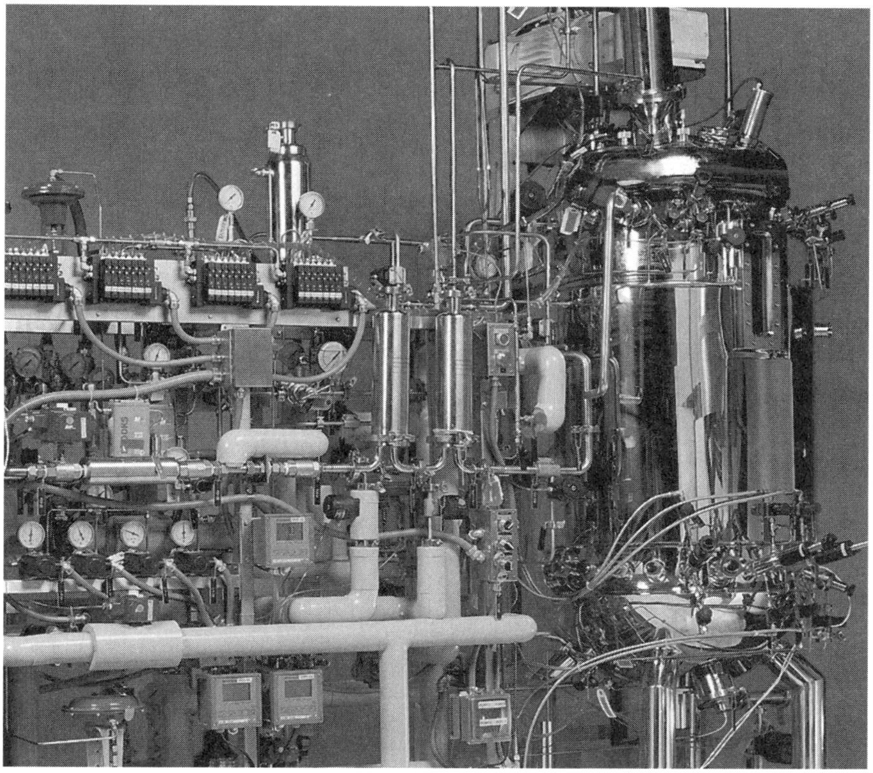

Fig. 1. This sterilizable-in-place and clean-in-place 500-L industrial bioreactor features retractable probe housings and redundant probes and filters. (Courtesy of New Brunswick Scientific, Edison, NJ.)

newer high surface area, multiple stacked-plate monolayer systems as well as roller bottle assemblies, shake and Fernbach flasks, and bench-top or lab-scale bioreactors. Very large-scale production of 20 to 15,000 L is accomplished in stand-alone bioreactors such as those employed by manufacturers of protein biologicals for commercial products (Fig. 1).

Because this chapter addresses the large-scale production of recombinant proteins, it introduces systems that became popular in large- to very large-scale, although the techniques presented address primarily only large-scale implementations. Escalating from large- to very large-scale requires its own set of engineering disciplines and is largely outside the scope of this work.

2. Mammalian Systems

2.1. Overview

There have been many developments in cell-based expression technology since the early days of HAT medium-selected hybridomas producing mouse antibody in ascites culture. Today, researchers can construct transfectoma by direct cloning methods that promote high and stable expression of a multimeric recombinant protein under the control of multiple or inducible promoters, and they can accomplish production in a 15,000-L bioreactor using an animal component-free medium.

Tens of mammalian cell types are used in the production of recombinant product, including such well-known lines as BHK-21 and SP2/0 *(9)*. However, stable expression systems from Chinese hamster ovary (CHO) and NS0, more for empirical than theoretical reasons, have become the most popular choices for very large-scale production (Table 1). Robust growth in suspension culture, fidelity in product processing, a sound regulatory pedigree, and success in a variety of fusion, transfection, selection, and production approaches are properties that make them practical candidates for construction of transgenic derivatives and expression of product. Their propensity to produce stable clones of high expression efficiency means that, once established, a clone can be expected to provide a consistent product over many passages.

Cell fusion efficiency was one of the early properties selected in deriving myeloma lines. Today, their transfection efficiency in processing exogenous DNA is key. As mammalian cells, they have the systems for efficient and relatively authentic posttranslational processing of medically important products. The fact that both can present a nonadherent phenotype in various media and environments allows efficient large-scale suspension culture.

Levels of expression in individual constructs vary greatly both between and among each line, and have been increasing steadily. Specific productivity now considered successful is on the order of 10–400 mg of secreted protein per 10^9 cells per day. Volumetric levels range from 2 to 200 mg/L for recombinant proteins and from 200 to 2000 mg/L for recombinant antibodies; remarkably, levels of up to 6 g/L have been reported.

2.2. Vectors, Promoters, and Selection Systems

It is beyond the scope of this chapter to review the disciplines of vector and promoter selection, gene cloning and transfection, and producer clone selection. However, the results of these systems employed to generate and maintain the recombinant line can have an impact on the handling of the cultures and methods of protein production. Therefore, some elements of transgenic line production and popular selection marker systems are mentioned.

The functions of selectable marker systems in the generation of transgenic cells are to select successfully transformed cells and often to provide selective pressure to maintain or amplify the inserted gene *(10)*. Having an essential gene with a product that is required for cell proliferation linked to the gene of interest in the expression vector determines that only cells undergoing successful transformation events will dominate the culture. In many cases, the selectable marker is also a way to increase the copy number of that gene in the host. It may involve an exogenous enzyme activity, such as in markers expressing product that inactivates a toxin or blocks a toxic event. Some of these dominant selectable markers are very familiar, such as those used in antibiotic selection systems common in prokaryotic culture.

There are two kinds of systems that employ endogenous markers: (1) positive selection, which uses salvage pathway mutant cells with drugs that block normal biosynthetic pathways; and (2) negative selection, which uses toxic analogs of biomolecule intermediates (e.g., nucleotides) that are incorporated into the final product only if the salvage pathway is used. Some selection systems or vectors require an appropriate culture medium or cellular host possessing a corollary phenotype. For example, common applications of Tk vectors require a Tk^- host cell to provide for this positive selection. Others, such as neo, depend on toxic additives (e.g., geneticin) to a normal culture, with the exogenous gene product (e.g., aminoglycoside phosphotransferase II) providing the rescue by degrading the toxin. This system may therefore be used in a wide range of hosts and media.

Any number of selection systems may be discovered in the literature. Some of the more popular include those presented in Table 2. There are good reasons for considering most of the various systems available; however, those desiring to begin in large-scale production can usually obtain acceptable results with commercially available vectors that support a few popular systems. GS (glutamine synthetase), DHFR (dihydrofolate reductase), HPT (hygromycin phosphotransferase), Hgprt (hypoxanthine–guanine phosphoribosyltransferase), and G418 (aminoglycoside-3'-phospho-transferase) have been employed widely and work well in the majority of cases. In stable expression systems, one or more copies of the structural region of the gene of interest, along with relevant DNA-based signals for transcription by RNA polymerase II, are integrated into the chromosomal DNA of the host cell. Vectors to accomplish this contain minimally, a promoter and various promoter-associated elements, promoter distal and proximal cloning sites, an origin of replication for *Escherichia coli*, a polyadenylation sequence, and possibly a eukaryotic origin of replication. Commonly employed constitutive promoters include the SV40 early, adenovirus late, and the cytomegalovirus (CMV) early.

2.2.1. Available Systems

Plasmid vectors of various origins, selectable markers, enhancers, and cloning features with expression driven by a number of promoters have been designed and described in the literature. Many choices are commercially available from a number of vendors. For example, Promega (Madison, WI) offers vectors for both transient and stable expression of recombinant proteins in mammalian cells. The pSI and pCI vectors can be used for high levels of transient expression because each of these vectors contains a strong viral promoter–enhancer, an optimized chimeric intron, and the SV40 late polyadenylation signal. The pSI vector contains the SV40 enhancer–promoter region, a viral promoter that is known to express strongly in many different cell lines. The enhancer–promoter region also allows episomal replication in cell lines that express the SV40 large T antigen (such as COS cells), which can lead to even higher levels of expression. The pCI vector contains the CMV immediate early promoter and enhancer regions, which can promote strong constitutive expression in many different cell types *(11)*. The chimeric intron contains β-globin and immunoglobulin G (IgG) sequences, which have been optimized to match consensus donor and acceptor sites *(12)*. The presence of an intron flanking the protein-coding region can help increase levels of gene expression *(13)*. The SV40 late polyadenylation signal is known to enhance stability of messenger RNA (mRNA) transcripts *(14)*. In addition, both the pCI and pSI vectors have a T7 RNA polymerase promoter upstream of the multiple cloning region, which allows generation of RNA transcripts in vitro.

For the engineering of stably transformed lines, Promega offers two vectors containing the neomycin phosphotransferase gene, the pCI–neo and pTargeT™ vectors. Here,

Table 2
Common Markers Used in Animal Cell Systems for Selection of Genetic Transformation and Amplification of Gene Copy Number and Product Expression

Marker	Activity	Mechanism
	Endogenous marker	
Ada	Adenosine deaminase; makes inosine from adenosine	Xyl-A (9-β-D-xylofuranosyl adenosine) is toxic adenosine analog; ADA detoxifies added Xyl-A
Aprt	Adenine phosphoriboyl-transferase; makes adenosine 5'-monophosphate from adenine	Positive selection: Adenine and azaserine block dATP synthesis; only cells using salvage pathway survive
Cad	Carbamyl phosphate synthase; aspartate transcarbamylase; dihydro-oroatase; provides initial steps in uridine biosysthesis	Positive selection: PALA (*N*-phosphonacetyl-L-aspartate) blocks the aspartate transcarbamylase activity of CAD
Dhfr	Dihydrofolate reductase; converts folate to dihydrofolate, then to tetrahydrofolate	Positive selection: DHFR is required for nucleotide or amino acid biosynthesis, so selection is in nucleotide-free medium
Hgprt	Hypoxanthine–guanine phosphoribosyltransferase; converts hypoxanthine to IMP and guanine to GMP	Positive selection: Hypoxanthine and aminopterin block IMP synthesis; only cells using salvage pathway survive.
Tk	Thymidine kinase; makes dTMP from thymidine	Positive selection: Thymidine and aminopterin to block dTTP synthesis; only cells using salvage pathway survive.
	Dominant selectable markers	
AS	Asparagine synthase	Only bacterial enzyme uses ammonia as amide donor; cells transformed with AS survive asparagine-free medium containing toxic glutamine analog albizziin
ble	Glycopeptide-binding protein	Provides resistance to antibiotics bleomycin and pheomycin
gpt	Guanine-xanthine	Similar to mammalian *hgprt*, but has xanthine phosphoribosyl-transferase activity; allows aminopterin/mycophenolic acid
hisD	Histidinol dehydrogenase	Provides resistance to histindinol
hpt	Hygromycin phosphotransferase	Provides resistance to hygromycin-B
neo	Neomycin phosphotransferase	Provides resistance to aminoglycoside antibiotics
pac	Puromycin *N*-acetyltransferase	Provides resistance to puromycin
trpB	Tryptophan synthesis	Provides resistance to indole
	Markers for copy number amplification	
Ada	Adenosine deaminase	Deoxycoformycin
AS	Asparagine synthase	β-Aspartylhydroxamate
Cad	Aspartate transcarbamylase	*N*-Phosphonacetyl-L-aspartate
Dhfr	Dihydrofolate reductase	Methotrexate
gpt	Xanthine–guanine phosphoribosyltransferase	Mycophenolic acid
GS	Glutamine synthetase	Methionine sulfoximine
Hgprt	Hypoxanthine–guanine phosphoriboosyltransferase	Aminopterin
Impdh	Inosine monophosphate dehydrogenase	Mycophenolic acid
Mt-1	Metallothionein 1	Cd^{2+}
Mres	Multidrug resistance: P-glycoprotein 170 gene	Adriamycin, colchicine, others
Odc	Ornithine decarboxylase	Difluoromethylornithine
Rnr	Ribonucleotide reductase	Hydroxyurea
tk⁻	Thymidine kinase	Aminopterin
Umps	Uridine monophosphate synthase	Pyrazofurin

the presence of the neo gene allows selection of transfected plasmids using the antibiotic G418. The pCI–neo vector has the same features as the pCI vector, a CMV immediate early enhancer–promoter for high levels of protein expression, a chimeric intron, and the SV40 late polyadenylation signal. Expression of the neomycin resistance gene is driven by the SV40 promoter–enhancer, which also allows the plasmid to be replicated episomally in cell lines containing T antigen.

The pTargeT Mammalian Expression Vector System is a convenient system for cloning polymerase chain reaction (PCR) products and for expression of cloned PCR products in mammalian cells. Nearly identical to the pCI–neo vector, the pTargeT vector contains a multiple cloning region modified to allow T/A cloning of PCR-generated inserts. To prepare the vector, pTargeT is digested to create a blunt end, and thymidine is added to the 3' ends of the vector to create a T overhang. The T overhang allows easy cloning of PCR fragments containing a 3'-A overhang into the pTargeT vector.

Many other vectors employing a number of diverse features have been developed, including many with unique advantages. For example, in most systems, the successful introduction of the DNA into a cell, as well as the site and number of plasmids incorporated, is a rather random event. However, methods that enable site-specific or targeted integration of the genes of interest into more transcriptionally active loci (hot spots) have been developed *(15,16)*. Regardless of the system chosen, most require selection of cells that have gone through a productively successful genetic transformation. Even exceptions to this, such as recombinase-mediated cassette exchange, require the use of a marker at some stage of recombination *(17)*. Two of the more popular systems, the Dhfr and GS, are introduced here.

2.2.2. GS Selection System

Glutamine synthetase (GS) produces glutamine from ammonia and glutamic acid. Normal cells will obtain required glutamine either through the action of GS or by absorption from the ambient growth medium. GS^- cells cultured in media deficient in glutamine will starve for it unless an exogenous source of GS is present. Methionine sulfoximine (MSX), an inhibitor of GS, is used to select for cells in the culture that have increased copies or transcriptional levels of GS. Through the stepwise increase of MSX concentration in the medium, cells harboring incremental copies of the GS gene will be selected. If the GS gene is linked to the gene of interest for production, concomitant increases in that gene will be achieved. A number of variations of this basic system are popular in selecting for cells that have undergone a successful transfection event and increasing the copy number of exogenous genes in the resulting recombinant lines.

2.2.3. Dhfr Selection System

Dihydrofolate reductase (DHFR) is required for several reactions in *de novo* and salvage nucleotide–DNA biosynthesis. It is one of many enzymes in this pathway used as a marker. As one of the oldest and most used systems, a number of variations in the employment of this enzyme have been developed. Normal cells will obtain required nucleotide precursors either through the action of DHFR or by absorption of them from the ambient growth medium. $Dhfr^-$ cells cultured in media deficient in nucleotides will starve for them unless an exogenous source of Dhfr is present. Methotrexate, an inhibitor of DHFR, is used to select for cells in the culture that have increased copies or transcriptional levels of Dhfr. By gradually increasing the concentration of methotrexate in the medium, cells harboring greater numbers of the Dhfr and desired product genes will be selected.

2.3. Cell Systems

Suitable cell lines for stable production of foreign proteins display a number of common characteristics, including (1) a propensity to readily grow, unclumped, in suspension cultures; (2) high rates of productive and stable integration of heterologous DNA; (3) a capacity for higher posttranslational modifications; (4) robust growth and high levels of production in a variety of media; and (5) adaptability to a variety of derivatization, selection, and production environments *(18)*.

Mammalian cells share many metabolic processes and display many similar characteristics in protein expression. Glycoforms of products from both CHO and NS0 display many characteristics in common with human counterparts, including O- and N-linked carbohydrates with bi-, tri-, and tetraantennary sialiated branches. However, there are species and cell-line-determined differences that can have a significant impact on their performance in production systems. For example, differences in the glycosylation of a given protein as expressed in various mammalian systems have been reported *(19)*. One specific difference noted is that human glycoprotein processing employs both α2,3- and α2,6-sialyltransferase; both CHO and BHK lack the α2,6 enzyme.

Technology exists to improve host posttranslational function through genetic transformation providing genes with a function either to augment endogenous pathways or to knock out genes, forcing increased exogenous gene expression or altered processing. Examples include the introduction of such protein-folding active agents as PDI, BiP, and GRP78 *(20)* and the examination of various glycosyltransferases *(21)*. In other work, a CHO line was engineered to express α2,6-sialyltransferase *(22)*, but although the exogenous erythropoietin expressed in it did contain α2,6-linked sialic acids, it showed lower bioactivity. Although much promise exists, these newer transgenic improvements have yet to gain general acceptance.

Many characteristics of each of the more commonly used cell lines, and even of some of their specialized derivatives, have been observed and reported over the years. However, the fact that each transgenic line constructed for protein production has been clonally derived means that it has been selected for some particular characteristics, such as high levels of production. In this process other characteristics can be unintentionally co-selected, which determines nutritional and performance features in the cloned line quite divergent from those of the parent line. It is common to observe two lines derived from the same parent line, genetically transformed with the same vector and harboring the gene for a similar protein, display significant differences in metabolic demand and other performance once they are assessed in large-scale culture.

2.3.1. NS0

Early murine myeloma lines were derived over 30 years ago from induced mouse plasmacytomas and were of particular interest at the time as stable immortalized lines producing antibody. As the goals for hybridoma applications advanced, requirements for the myeloma lines increased. Robust cell lines were developed that were inhibited in their capacity to produce antibody and that provided an acceptable frequency of high-yield monoclonal producers on fusion with B-lymphocytes. NS0 is now one of many mouse myeloma cell lines available for the construction of hybridoma and transfectoma.

NS0 originated as an IgG secretor, yet its current lack of endogenous antibody or fragment production makes it particularly efficient in the processing and secretion of exogenous antibody. An especially strong incentive for employing an NS0 line is the development of an NS0-based GS selection system. From the original MOPC-21 line, 289/16 (NS1/1), and then P3-NS1/1-1Ag4.1 (NS1) were isolated. NS0, isolated from NS1 over 20 yr ago, has been subcloned and otherwise modified in its protracted history at many locations. Here, the ECACC no. 85110503 characteristics are referred to as nominal, referring to some of the many variant lines as appropriate.

2.3.2. Chinese Hamster Ovary

Chinese hamster ovary (CHO) cells were established from the biopsy of an ovary of a Chinese hamster over 30 years ago *(23)*, and they are the origin of each of the many derivative lines developed since that time. The popular CHO-KI (ATCC CCL-61) and CHO-S were established in 1968 and 1989, respectively, directly from the original CHO line. DX B11 (DUKX B11) was derived from the K1 line in 1978 and has been the basis for a number of popular lines for expression, including Dhfr and adenosine deaminase-selectable strains *(24)*. The currently popular DG44 was derived from the CHO pro3$^-$ (CHO Toronto). Gibco CHO-S was established from the LANL line, which is a direct descendant of the original line.

2.4. Culture–Production Systems

2.4.1. NS0

Whether used in the construction of hybridomas or transfectomas or in transient expression systems, NS0 has become a popular line for the production of MAb and recombinant proteins *(25)*. The selection–amplification systems reported successful with this line include GS (MSX), Dhfr (MTX), APH (G418), HPT (hygromycin), Iapt (histidinol), Hgprt (MPA), and 3-ketosteroid reductase (cholesterol). The GS system *(26)* is currently the most popular with large-scale producers. Promoters/expression vectors used in producing stable and transient expression abound, with the strong promoters from CMV, respiratory syncytial virus, and SV40 the most common. Such implementations have been used as producers of MAb, from murine to fully human, as well as nonantibody recombinant proteins *(27)*. As mammalian cells, mouse myeloma cells provide all of the generic posttranslational processing steps required for authentic human product processing. However, the effects of derivative and especially murine-specific glycan motifs, cell fusion, recombination, overexpression level, as well as the particular nutritional and culture environment do affect the way and extent to which any specific process is applied to the product.

The origin of NS0 dictates that it has the capacity for efficient production and secretion of antibody and polymeric proteins. The absence of Ig chain or fragment expression in this parent cell line is a feature contributing to its popularity *(28)*. Protease activity in medium following cell culture is of particular importance when employing serum-free medium (SFM). Both absent and significant protease activity from NS0 derivative culture have been reported, but the majority describe no significant product degradation in application.

Cholesterol is required by all animal cells for a number of functions, including the maintenance of membrane fluidity. Most cultured animal cells can produce what they require from the most elementary nutrient precursors. Nevertheless, it is true that some cultured cells fully capable of endogenous cholesterol generation grow more efficiently with cholesterol supplementation. As the need for expression in serum-free environments developed, it was discovered that NS1 and derivatives (including NS0) had become auxotrophs for cholesterol *(29,30)*. The biochemical step of deficiency has been identified as the demethylation of lanosterol to C-29 sterols. This means that any precursor to cholesterol above lanosterol will be ineffective as a metabolic component. Therefore, a functional supplement to the culture medium must be in the area of lathosterol to cholesterol.

Apoptosis is known to be an issue in the culture of murine myeloma and derivatives, especially under conditions of environmental or nutritional stress. NS0 demonstrate a particular susceptibility in this regard *(31)*. A number of factors could contribute to this including its lack of Hsp70 expression potential and variability in membrane cholesterol concentration *(32)*. Reported inducers of the response include (1) glucose, glutamine, phosphate, oxygen, and essential amino acid starvation; (2) temperature, pH, and osmolality changes; and (3) shear stress and metabolic by-product buildup.

On the other hand, it has also been reported that the specific molecular onset of apoptosis in hybridoma is actually associated with cell proliferation and full metabolic activity rather than with the decline of cell viability *(33,34)*. GADD153 expression has been observed as either a trigger, or at least an indicator, of NS0 apoptosis in response to environmental stress *(35,36)*. Transfection-based expression of the mitochondrial cytochrome-*c* active Bcl-2 has been repeatedly shown to reduce stress-induced apoptosis in this line greatly and increase overall MAb production *(37,38)*. Inducible expression of the p21 (CIPI) cyclin-dependent kinase inhibitor also appears to do the same *(39)*. Some protein hydrolysates reportedly provide an antiapoptotic effect in the SFM culture of hybridoma, whereas serum and any associated growth factors do not *(40)*. Finally, it has been observed that maintenance of ambient cholesterol levels by sequential supplementation may reduce apoptosis in suspension culture *(personal observation)*.

Shear force sensitivity is of concern in the suspension culture of NS0. Although each cell line exhibits its own particular assortment of characteristics in response to hydrodynamic stress, it is often convenient to generalize by referring to shear-sensitive or shear-tolerant lines *(41)*. This appears not to be the case with NS0. Personal observations, and those of premier commercial producers, indicate a particular issue with NS0 in this respect. NS0, in media supplemented with pluronic acid and even low concentrations of serum or protein, yields somewhat shear-tolerant cultures. However, when moved to bioreactor scale, specific, apparently shear force-induced culture impairment can occur, especially as sparge or perfusion (or other cell separation) systems are implemented *(42,43)*. Serum- and protein-free NS0 cultures are apparently particularly sensitive to some cell separation systems used in perfusion bioreactor culture.

2.4.1.1. Clonal Derivatives

As with most cell lines that develop wide popularity, and especially over a long period of time, NS0 exists as a number of distinct subclone lines. These lines have been generated by a number of means to support particular functions, growth media, culture conditions, and selective agents. In addition, each step of derivatization will further alter properties of the line. Of course, the generation of hybridoma, in introducing entire lymphoid chromosomes, can introduce significant changes of phenotype. But, even a transiently transfected line may exhibit distinct properties based on the metabolic demands of vector expression, any required selective pressure, or properties of the new gene products. In any event, significant and divergent properties have been reported, particularly with individual MOPC-21 derivatives *(30,44,45)*. Transgenic producers derived from the newer GS NS0 lines from Lonza® are an example. Not only must the culture environment be modified to accommodate the GS selection system specifically, but also these lines display significant metabolic characteristics requiring specific attention for performance optimization. A number of individual NS0 transformants have been extensively characterized in high-efficiency production. Although the producers of biomedical products often maintain this information as intellectual property, published reports do exist that provide insight into individual clonal metabolic patterns to be anticipated. Maintenance of production through tens of passages, even in the absence of selection pressure, has been observed, and the stability of derivatives through cryogenic preservation has been documented.

2.4.1.2. NS0-Specific Demands

Subclones and derivatives of NS0, because of the diversity of generation and maintenance, complicate the identification of NS0 media requirements. One culturist's transfectoma from a GS selection line will likely have quite divergent media requirements from another's ECACC-derived line *(46)*. This should be kept in mind when applying prescribed culture media formulations and protocols to any particular line and is especially important when applying available MAb production media, supplements, and methods, as these are often designed and tested in non-NS0 hybridoma cultures.

Autocrine and cytokine responses by myeloma and hybridoma have been explored for decades, and the various cocktails recommended are numerous and diverse. NS1 and derivatives have been shown by growth medium fractionation and add-back experimentation to be totally unresponsive to, or non-secretors of, endogenous cytokines. Similar experiments have implied their existence and function, particularly in hybridoma lines. Interleukin 6 has been shown to be an effector in promoting certain strains derived from distant precursors of NS0 as well as hybridoma from yet other precursors.

NS1 and NS0 cultures have been variously reported as both unresponsive to and obligate for insulin or insulin-like growth factor 1 (IGF-1). Successful (albeit not optimized) performance has been observed in the culture of NS0 in quite elementary chemically defined and protein- or peptide-free media, including in the absence of selenium, transferrin, or insulin or IGF-1 *(47)*. Studies demonstrated a distinct insulin requirement for GS NS0, which may be supplanted by zinc chloride ($ZnCl_2$). Furthermore, many chemically defined and protein-free growth media have been very successfully applied to NS0s and derivatives, achieving doubling times of 22 hours or better, cell densities above 5×10^6/mL in batch culture, and production levels in multiples of serum-containing controls *(48,49)*.

Shear protectants are required for all suspension culture applications. Serum-supplemented media and serum-free formulations high in protein can provide sufficient protection for most applications. Common nonprotein protectants, such as pluronic acid, provide some protection, but some NS0 derivatives remain shear sensitive to aggressive sparging and cell separation techniques in bioreactor applications even at nominal levels of supplementation *(personal observation)*. Increasing pluronic acid to 0.2% w/v seems to provide some additional protection, but shear stress remains an issue in serum- and, especially, protein-free applications.

Amino acid, vitamin, trace element, and other ion concentration or ratio optima for various myeloma and derivatives, including NS0, have been reported for decades. General basal conditions as well as high-density feeding and perfusion approaches for any particular application can be established from them. However, review of these formulations, as well as direct experimentation, reveals significant subclone, derivative, basal medium formulation, and culture mode-induced variation in these requirements. For example, it is not uncommon to see an NS0 reported to require additional supplementation of a particular amino acid in an application and to discover that your culture either is not utilizing it or actually is producing it *(48)*.

Apoptosis is of concern, especially in high-density or production-scale culture modes. Approaches to reducing the problem include avoiding particular nutritional and environmental stresses. It appears that this issue also is greatly affected by the diversity of lines and applications employed. Factors in media formulation or supplementation that can be addressed include design to avoid glucose, glutamine, and essential amino acid starvation; pH and osmolality extremes; and shear stress and by-product generation *(41,50,51)*. Chemical additives that have been shown to reduce or inhibit apoptosis may be a consideration here as well. These include bongkrekic acid, cyclosporin A, pyrrolidine dithiocarbamate, *N*-acetylcysteine, as well as the caspase inhibitors Z-VAD-FMK and Ac-DEVD-CHO.

Iron transport is always an issue in the development of any SFM. Added transferrin will normally replace the transport potential provided by serum. In protein-free formulations, a variety of chelators and added iron have been shown to support most cultured cells, although many have a preference for particular complexes. Optimal NS0 performance can be obtained in this way, although it appears from some studies that NS0 has a distinct phenotype in this respect compared to other myelomas. A media composition-dependent relationship has been observed such that a transferrin replacement chemistry that works well in one formulation with a particular clone may not in other basal formulations or clones.

Glutamine supplementation is an absolute requirement for cells that cannot produce it on their own, but are obligate for an exogenous source. However, the GS genes allowing glutamine production from glutamate and asparagine are available to NS0 from transfection or fusion with other cells. Some NS0-derived hybridomas have been observed to possess a full glutamine synthesis function. It is of note that some SFMs identified as "without glutamine" in fact contain trace levels from other ingredients (e.g., hydrolysates) that, although not sufficient for full nutrient support, may be significant when employing the GS selection system.

Lipids, because of their very limited solubility in aqueous media, are a special issue in culture media supplementation. This is usually solved by dispersing the lipids in a microfilterable suspension using a number of approaches. The fact that NS0 requires such a high concentration of a sterol lipid (2 to 6 mg/L of cholesterol) is a particular challenge to supplementation. Lipid supplementation issues include solubility, filterability, dispersion, physical stability, cell delivery kinetics, and container vessel adhesion. Many technologies are available for this supplementation, including the use of vesicles, emulsions, microemulsions, carrier proteins, and carrier polymers such as cyclodextrin.

Vesicles of phospholipids in the lamellar phase can be constructed in the appropriate size, and will carry significant levels of cholesterol intercalated between their acyl chains. However, issues remain in both the control of the cell delivery kinetics and in providing sufficient physical stability to the preparations to allow for a practical shelf life *(52)*. The same issue limits the use of emulsions, which are prone to such destabilization as coalescence. Carrier proteins such as bovine serum albumin (BSA) work well, but are both prohibited in protein-free formulations and require high BSA concentrations at the cholesterol levels required by NS0. Carrier polymers such as cyclodextrin provide a qualified solution *(53)*. Although cholesterol loads efficiently into certain cyclodextrins and the resultant complex is filterable, stable, and innocuous, the nature of the association leads to instability in high dilution, determining a requirement for many supplements to be added throughout culture *(personal observation)*. A commercially available cyclodextrin-based supplement reported to be stable in high dilution is proscribed from supplementation in fed-batch protocols by the manufacturer. A medium composition-dependent relationship exists with the means of cholesterol supplementation. Notably, well-established high-cholesterol serum extracts have been shown in multiple laboratories to work well with some SFM, but not at all with others *(personal observation)*.

N-Glycan motifs applied to proteins expressed in NS0 that are murine specific (e.g., Galα1, 3Galβ1, 4GlcNAc) are of concern to some. Attempts to modify *N*-glycan processing by manipulation of intracellular nucleotide sugar content by adding precursors to the growth medium yielded mixed results, affecting some residue ratios or sequences and not others *(54)*.

2.4.2. Chinese Hamster Ovary

Recombinant protein expression in CHO cells began about 1975 because they were readily available, well characterized, easily transfectable and subcloned, grew and produced well in low serum concentrations, and displayed higher and robust posttranslational processes. Since that time, the number of both CHO producer subclones and transgenic approaches has increased remarkably. Specific information regarding the biochemical and genetic differences between common CHO lines, surprisingly, is difficult to obtain. Those desiring to contribute to a casual repository of such information may email trish_benton@pacbell.net.

A number of CHO derivatives have been developed that provide specific features useful to producers of recombinant proteins. For example, DG44 are dhfr$^-$, CHO N3 are AS$^-$, and CHO C55.7 are ODC$^-$. They are therefore valuable for use in the Dhfr, asparagine synthetase, and ornithine decarboxylase systems, respectively. Some lines provide more than one selection factor. For example, DX B11 is both dhfr$^-$ and Ada$^-$ and therefore can be used in either system.

One of the complaints about the CHO system is that it was difficult to obtain the levels of expression expected in other systems. Not too many years ago, levels of 2–20 mg/L of product were considered acceptable. Popular promoters such as hCMV are not very efficient in CHO, and several rounds of gene copy number amplification are usually required to obtain currently acceptable levels (0.1–1.0 g/L) of expression. The Dhfr system is one way to drive the copy number to 50 or more, and quite acceptable levels of production may be obtained this way. However, such levels of copy number are difficult to maintain without continued use of methotrexate, which can be an issue in large-scale production.

CHO cells possess all of the basic systems of higher posttranslational modification, for instance, proteolytic propeptide processing, *N*- and *O*-linked glycosylation, carbohydrate trimming, γ-carboxylation of glutamic acid residues, hydroxylation of aspartic acid and asparagine residues, sulfation of tyrosine residues, phosphorylation, fatty acid acylation, and multimeric protein assembly. However, when addressing the production of some (e.g., human) glycoproteins, they (1) are inefficient at some specific substeps; (2) lack the capacity for some steps entirely; or (3) provide some additional activities, any of which can result in nonauthentic product. Although human (and other) cells contain both α2-6- and α2-3-sialyltransferase, CHO contains only the α2-6 *(55)*. Normal human processing results in the addition of *N*-acetylneuraminic acid, whereas CHO has been reported also to add, to varying degrees, the human oncofetal *N*-glycolylneuraminic acid *(56)*. Efforts to correct these deficiencies have had mixed results. For example, a line engineered to express α2-3-sialyltransferase *(22)* and one to express additional β1,4-galactosyltransferase and α2-3-sialyltransferase *(57)*, although biochemically successful, yielded products of dubious improvement functionally. However, this work is clearly the direction of the future, and current efforts to improve the functionality of product through cells engineered to provide new, or augmented, processing steps will likely result in lines providing more authentic proteins.

CHO cells are known to have a very unstable karyotype because of chromosomal rearrangements arising from translocations and homologous recombination, especially in response to amplification procedures *(58)*. Numerous instances of instability of production, especially in the absence of selective pressure, have been reported. Loss of productivity in the extended culture of dhfr$^-$ CHO has been determined to be because of a loss of transgene copy number *(59)*. The region of incorporation of incremental gene copies is suspected to be a factor in the degree of transgene stability.

In Dhfr applications, gradual increases in MTX have been reported to effect incorporation into telomeric regions, and this is suspected to result in more stable lines *(60)*. Increase in the concentration of MTX from 0 to 20 μ*M*, and reaching 500 μ*M* over a period of many weeks, has been suggested to yield the most productive derivatives. Any basal level of Dhfr activity in the parent line also affects the efficacy of MTX amplification. MTX gradient experimentation of the CHO-K1 cell line (RCB 0285) transfected with pSV2-GS/hGM-CSF, constructed from pSV2-dhfr and cultivated in medium supplemented with 10% FBS (fetal bovine serum) with added glutamate *(61)*, demonstrated optimal results beginning with no MTX. However, the means of producing high levels of expression from CHO are still developing *(62)*. A new vector utilizing flanking regions of the Chinese hamster EF-1α gene has been reported that provides for high levels of expression of MAb without the requirement of methotrexate-driven copy number amplification *(63)*.

2.4.3. Episomally Based Systems

From the early work with hybridoma to the current genetically transformed lines, large-scale production is based on genes integrated in the host chromosome constitutively producing the product of interest or, in the case of newer inducible promoters, at least continuously maintaining the genetic capability of production. However, there are two systems available that, although used for years in small scale, because of recent and significant improvements are just now becoming interesting in large-scale applications *(10)*.

2.4.3.1. Transient Expression Systems

Long-established systems of small-scale transient expression have traditionally been the system of choice for the production of total amounts of less than 1 mg of protein. The number of systems available for transient expression has been expanding beyond the popular COS cell and vaccinia virus. Some can be supplied in large-scale production and therefore are mentioned here. Advantages of transient systems include (1) simplicity of expression vector construction; (2) short, high levels of production; (3) genetic consistency and stability; (4) flexibility of host cell lines; (5) flexibility in type of product produced; and (6) flexibility of combinations of coexpressed product or subunits.

Transient expression systems rely on the high-efficiency introduction of a suitable vector into the cells comprising a culture, followed by the immediate expression of product from that vector. Therefore, the introduced DNA does not have to integrate into the host chromosome or be otherwise maintained in the cell. The system is temporary because when a culture has produced product and dies, the process must begin again.

There are two distinct methods of transient expression; one employs plasmid vectors, and the other is viral, with the plasmid vector systems relying on the more scalable means of transfection or electroporation. Both electroporation and transfection have been successfully employed to the 10-L scale. Those transfection methods reported successful include calcium phosphate precipitation and polyethyleneimine. Although most cell lines would be theoretical candidates for this approach, good results have been reported in HEK293, COS, and BHK, with the most promise generally held for HEK293. Transfection efficiencies exceed 70%, and expressed product levels for nonantibody products generally range from 1 to 10 mg/L.

Relative expression levels from transient systems are regarded to be higher than stably transformed lines; however, absolute levels for a particular construct in a particular line are hard to predict. Improvements in systems themselves and in large-scale applications are reported. New promoters, recombinant hosts expressing helper functions, and means to amplify the copy number and increase persistence of the plasmids in the host hold the promise of significant improvement in product yield. Viral vector systems have been developed in Semliki Forest virus, adenovirus, and vaccinia virus. These seem to hold the most promise for regular large-scale production as each demonstrates one or more of the following advantages: (1) cell-based production of the vector; (2) high levels of expression; (3) scalability to 1000 L, and (4) wide host range *(64–66)*.

2.4.3.2. Episomally Maintained Vectors

Self-replicating extrachromosomal expression vectors stably maintained in a culture of eukaryotic cells possess, at least theoretically, a number of advantages over the currently popular integrated vectors *(10)*. Here, although the plasmids are not integrated, they are stably maintained in the cell indefinitely, similar to bacterial systems. In various implementations, maintenance of a transformed line for over a year and plasmid copy numbers of over 9000 have been reported. New strategies in controlling the runaway replication of SV40-derived vectors are allowing the development of one approach. Three other viruses with an origin of replication and other features that hold promise are bovine papillomavirus, BK virus, and Epstein–Barr virus (EBV).

Advantages for these systems lie in their high expression levels, expression controllability, and versatility of host. The fact that the expression of product is independent of positional effects and copy number variances means that a constant and moderate level of expression is maintained throughout passage and cloning. The fact that the genes of interest can be maintained at high and controllable copy numbers means that a commensurate level of expression can be achieved. An additional advantage for some applications (e.g., expression-based cloning) is that, as the exogenous genes are maintained episomally, they can easily be separated from the chromosomal DNA.

2.5. Cell Culture Techniques

The techniques of culturing the vast number of mammalian cell types in the many media and formats available today have been introduced elsewhere *(67)*. An introduction to the handling of suspension cultures is presented here for those with a background in the basics of cell culture.

2.5.1. Culture Basics

2.5.1.1. Shake Culture

The most popular format for the maintenance and small-scale study of cultures of both CHO and NS0 is shake, or shaker, culture. This system provides a convenient means of suspension culture that produces as close to optimal conditions for growth and production as can be easily obtained in the laboratory. One popular implementation is the use of 500-mL nominal capacity disposable Erlenmeyer cell culture flasks with a platform orbital shaker installed in 37°C jacketed incubators. Suspension-adapted cultures may be seeded in 100–200 mL of medium at concentrations from 2 to 6×10^5 cells/mL. Cultures are incubated at speeds between 80 and 125 rpm.

This system may be used with any medium as long as SFM and protein-free media are supplemented with a shear protectant. Flask sizes are available to support culture volumes from 15 to1500 mL. The largest flasks are listed in vendor catalogs as Fernbach flasks. Such cultures support a wide variety of applications, including the single-culture seeding of bioreactors up to 20 L.

2.5.1.2. Culture Passage

Seeding density, doubling time, and peak culture density vary greatly even between individual CHO and NS0 clones. However, even if little is known about the behavior of the clone, its basic properties in the medium and culture format chosen may be established in 1 to 2 weeks. Seeding may be attempted at a few densities between 1×10^5

and 1×10^6 cells/mL, with the expectation of a doubling time from 22 to 42 hours, peak viabilities by trypan blue of 80 to 99%, and peak counts reaching 1×10^6 to 1×10^7 cells/mL. Cultures should be monitored daily and split while the growth rate and viability are still high.

The optimal density for culture passage can be simply determined by allowing a culture under nominal conditions to progress to greatly reduced viability and plotting the peak cell density. Then, the standard culture passage point will be established at between 1 day prior to, or at, that peak density. Some cells, in some formats and media, are sensitive to culturing to peak density, and care must be taken to avoid it. Others are more robust and can withstand even extended maintenance at peak densities. The optimal density at which to seed cultures can be determined by setting up parallel cultures at incrementally diminished densities and observing the minimal density at which the culture survives or does not lag excessively.

2.5.2. Culture Media

For many cells, the most robust, forgiving, and easiest medium to employ is a classical basal medium supplemented with a high-quality FBS. In addition, most clones are established and come to the production facility in such a medium. The recommended media for general maintenance of CHO cells is DMEM-F12 supplemented with 10% FBS. NS0 maintains well in RPMI 1640 supplemented with 10% FBS. Serum-supplemented media of many varieties have been shown to support good product secretion. However, the issues of increased performance; serum availability, cost and lot consistency; and regulatory compliance have directed many culturists to SFM formulations (Table 3). NS0 and CHO can be adapted to a number of commercially available cell culture media. Both lines have been reported to perform well in a variety of custom-formulated and commercially available SFM as long as some cell line- and clone-determined restrictions and media supplementation are observed.

Downstream issues for media selection have actually been reduced of late, especially for SFM culture. Newer adsorption and affinity chemistries and resins and simplified purification schemes with increased specificity provide robust purification methods with fewer contraindicated raw material components. Most commercially available SFM formulations are easily handled by existing purification schemes.

2.5.2.1. Media Platforms

Serum-dependent media were the first synthetic media developed, beginning 50 years ago. These media provide all of the glucose (and other energy sources), amino acids, major inorganic salts, vitamins, and buffering required by the cells. The serum provides sterols, fatty acids, lipid vehicles, growth factors, protein-based shear force protection, and additional trace elements and vitamins not provided by the media. There are differences in the number of sera available. Sera generated from various species and at various stages of animal development determine the relative concentration of ingredients active in cell growth promotion and in product processing. The skill and care employed in the collection and handling affects the overall quality in many ways, chiefly in avoiding contamination from other elements of the animal and in reducing hemolysis of red blood cells during processing. Also, there are a number of fortified and processed sera available that may provide additional benefits to the culture at a reduced cost.

Table 3
Well-Established Biological Protein Products Manufactured Using Transgenic Higher Animal Cells

BLA Name	Manufacturer	Molecule	Cell Line	Method	Medium
Therapeutics					
Antihemophilic factor/ReFacto™; control of hemorrhagic episodes	Genetics Institute	Recombinant	CHO	Continuous perfusion bioreactor	Serum free (human albumin and recombinant insulin)
Coagulation factor VIIa/NovoSeven™; control of Bleeding: type A/B	Nova Nordisk A/S	Recombinant glycoprotein	BHK	Not revealed	Serum containing
Infliximab/Remicade™; treatment of Crohn's disease	Centocor	MAb, mouse/ human chimeric $IgG_{1\text{-}k}$	NSO	Stirred-tank bioreactor, suspension, perfusion culture (spin filter)	Serum free (hydrolysates, bovine transferrin/albumin, insulin, and Excyte™)
Daclizumab/Zenapax™; prophylaxis of acute organ rejection	Hoffman LaRoche	MAb, mouse/ human chimeric IgG_1	SP2/0	Not revealed	Not revealed
Diagnostics					
Capromab pendetide/ProstaScint™; Cancer imaging agent	Cytogen	MAb, murine $IgG_{1\text{-}k}$	Murine hybridoma	Hollow-fiber bioreactor	Serum free (BSA albumin and and transferrin)
Imciromab pentetate/Myoscint™; Cardiac imaging agent	Centocor	MAb, murine $IgG_{2\text{-}k}$ FAb fragment	Murine hybridoma	Not revealed	Serum containing (gentamicin)

SFM have been engineered to provide the activities of serum without the need for supplementation. This is accomplished by adding a number of components that provide those serum functions to a basal media formulation. These media can be categorized in many ways, including by the original basal formula from which they were built, the cell lines that they support, or the nature of the additional ingredients.

Protein-free media contain nonproteinaceous ingredients that provide the specific functions of, and replace, proteins. These heterogeneous ingredients range from hydrolysates of animal tissue to recombinant peptides, synthetic molecules, and organic salts. The definition of a protein is often stretched here to include quite large polypeptides. Demand for an absolutely protein-free medium has been reduced in deference to the (1) more robust and specific purification techniques available; (2) availability of non-animal-derived proteins; and (3) the fact that, regardless of the initial formulation, harvest media from cell culture contains some cellular protein anyway. Nevertheless, there is a shift from the use of protein because of the availability of protein-free formulations and cost considerations.

Chemically defined media contain only ingredients that are either molecularly homogeneous or mixtures with constituents that can be identified and their concentrations fully characterized. Obviously, some definition regarding chemical classifications and concentration limits must be accepted in this designation. The impetus for this category is mainly reproducibility in performance and purification and transportability of formulation.

Animal component-free media can contain anything that does not come directly from an animal. Precise definition of this category is ongoing and is driven by the intention of the category to avoid the possibility of contamination by infectious agents. Issues remaining include (1) the definition of an animal (e.g., yeast); (2) the proximate origin of the material (e.g., vs a derivitization of an ultimately animal-derived material); (3) raw material traceability (e.g., how far back to trace and what is acceptable certification) of the "nonanimal" component; and (4) acceptable limits of the material's exposure to, and contamination by, materials containing animal product (e.g., human processors and manufacturing or packaging materials).

Liquid and powder formats are available for many media. Powder provides a number of advantages, especially for large-volume consumers. Maximum lot sizes for a liquid product typically reach 10,000 L. Powder lots, depending on the specific weight of the formulation, reach over 600,000 L. The shelf life of liquid product ranges from 6 months to 2 years, although most powders are validated to 3 years or more. The price of liquid media ranges from \$10 to \$45 per liter; the same formulas in powder range from \$2 to \$10 per liter. Another potential feature of powder is the opportunity for slight customization at the time of hydration.

The main disadvantages of powder include the user's requirements for (1) availability of large quantities of high-quality water; (2) expertise, facilities, and procedures for hydration, filtration, and quality control testing; and (3) materials and facilities for large-volume liquid storage. The main advantage of liquid is convenience, and for large-scale operators, this is especially true as better and larger package sizes have become available. The largest liquid packaging available for liquid media (900 L) is from HyClone® (Logan UT). Another advantage of liquid is the ability to rely on the expertise and certifications of the vendor facilities for processing and quality control.

Supplementation of a standard medium prior to use is often required to provide components that are (1) too labile to be included in the original formulation (e.g., glutamine); (2) optional for standard use (e.g., antibiotics); (3) special application requirements (e.g., antifoams); (4) vector-determined selection agents (e.g., hygromycin); (5) cell line-specific requirements (e.g., cholesterol); or (6) extensions of standard nutrient complements discovered to improve culture performance. The last category introduces the supplementation approaches defined in fed-batch culture (*see* Section 2.6.3).

Most producers are free to choose any type of medium and supplementation based on cost, production efficiency, and other proximate inherent performance features. The current trend for pharmaceutical manufacturers in the United States in selecting media for new endeavors is, if feasible, to stipulate a serum-free, animal component-free (or at least reduced) in a chemically defined formulation with low-to-absent protein. As most clonal derivatives present, to one degree or another, their own metabolic phenotype, very large-scale producers face another issue: whether to use a commercially available medium or to invest in the optimization of a medium for a particular clone. Optimizing a medium to a particular clone and production format can easily increase production levels two- to fivefold. However, there are no guarantees, and in some cases, little or no improvement in yield over that provided by an off-the-shelf medium is obtained. The cost of such optimization ranges from moderate to quite high, depending on the facilities, expertise, and schedule available.

2.5.2.2. Media Selection

Commercial manufacturers offer a number of media for each of the above platforms that well support (given any required supplementation) the basic culture of both CHO and NS0. First, the required basic features, such as price or acceptable components of the medium, are established. This determines the platform of media to be employed. Acceptable vendors are determined by such factors as manufacturer validation, product quality, batch reproducibility, price, and customer service. Vendor and platform identification establish the number of media available for initial screening. It is common that multiple media (brands or platforms) are tested, even if the basic large-scale system has been established before, as variances in characteristics of individual clones can affect significant clone-specific media demands.

After adaptation of the clone of interest to the media selected, initial values for the performance of small-scale culture in each are developed. Performance factors examined include minimum successful seeding densities, cell division rate, peak cell density, peak viable density duration, product secretion kinetics, total product accumulated, product quality, raw product stability, performance in purification applications, and how well the culture performs physically (e.g., shear damage, foaming, precipitation, pH control, and metabolic waste accumulation).

Once one or more media are selected as candidates for production, further testing begins to establish one as an optimal material. Testing in a system most representative of the final production system provides the greatest assurance that values established in this process development will be predictive of true production runs. Features assessed in this stage include culture performance in scale-up, consistency of performance across multiple lots of medium, and the performance of the medium in large-scale specific procedures.

2.5.2.3. NS0

NS0, especially in SFM culture, demonstrates a particular sensitivity to nutrient timing and the effects of impeller, sparge, and cell separation technologies. Many of these issues have been solved, but some implementations of SFM, especially protein-free culture, remain problematic.

As NS0 cells cannot produce cholesterol, they must obtain 2 to 8 mg/L of it from medium supplementation. Because of the metabolic step they lack, functional supplements to the culture medium must be in the area of lathosterol to cholesterol. Most have chosen to employ cholesterol as the supplement, although some cholesterol precursors and other sterols (e.g., stigmastanol) do work. It has been observed that spontaneous cholesterol-independent revertants are quite readily generated, although it appears that these derivatives are generally reduced in their production capacity.

Glutamine supplementation is a requirement to varying degrees for the culture of animal cells. Glutamine may be available to cells in culture through absorption from the ambient medium or from the activity of GS on glutamate and ammonia or asparagine. One factor affecting the level of glutamine supplement required is a natural variability in the capacity of cells from various tissues to express GS. Another is the culture history of the cell line in that ambient glutamine levels can regulate endogenous GS expression. In the latter case, lines that appear to have a constitutive requirement can in fact be adapted to low or absent levels of ambient glutamine. In contrast, NS0 and derivatives have been shown to be auxotrophs for glutamine *(26)*. This means that culturists must either supply glutamine by supplementation or introduce an exogenous source of GS. It has been reported that, although NS0 cells do possess an endogenous GS structural gene, the generation of spontaneous glutamine-independent phenotype revertants is very rare, and those examined have not performed well.

Culture seeding densities, doubling times, and peak cell densities are very dependent on the (1) characteristics of the parent line employed; (2) individual clonally derived characteristics; (3) selection or copy number amplification system; (4) medium and feed or supplementation regime; and (5) specific culture mode. Therefore, accurate values for each must be determined for each case. However, in well-agitated suspension modes, seeding at 2 to 6×10^5, anticipating a doubling time from 22 to 35 hour and achieving peak cell densities of from 2 to 8×10^6 can normally be expected.

2.5.2.4. Chinese Hamster Ovary

Individual CHO lines display inherent characteristics that must be addressed in their handling and culture. For example, K1, DX B11, and DG44 are known to have an absolute requirement for proline resulting from an inability to convert glutamic acid to glutamic 5 γ-semialdehyde. This means that, although proline is normally a "nonessential" amino acid, media for these lines must be proline supplemented. DX B11 and DG44 are mono- and heterozygote for dhfr$^-$, respectively, and therefore, require proportional nucleosides or nucleotide precursors such as glycine, hypoxanthine, and thymidine in their media. K1 and DX B11 have been shown to have insulin receptors and to respond to insulin, or substitutes, in the media. DG44 has been shown not to present insulin receptors, and reports of their response to insulin supplementation are mixed.

The issues of carbon dioxide (CO_2), ammonia, and lactate buildup have been endemic problems in high-efficiency transgenic CHO culture. These buildups can negatively affect four distinct production parameters: peak culture density, cell-specific productivity, batch-specific productivity, and product quality. Ammonia and lactate can become limiting in any culture style, whereas CO_2 only becomes a significant issue in larger-scale culture. As with many culture variables, the derivative line, basal media, and culture mode greatly influence the appearance and effects of these issues. Modern serum-free formulations have addressed the ammonia and lactate issues by modulating the metabolic use of glutamine and glucose by the cell through a variety of means, including inducing the cells to employ alternative pathways. The effects of CO_2 and ammonia, as modulators of intercellular pH, are under examination in terms of ion channel and transporter manipulation.

CHO cells were originally cultured as attached monolayers in serum-containing medium. In scales that are large- to very-large, this mode of culture can be employed in such culture formats as roller bottles, multilayer stacked-plate monolayer systems, or microcarriers. The most popular approach is to convert them to suspension culture because most have found this easier to handle and more productive. CHO-S (Gibco, 1989) was the first suspension derivative generally available. Many popular lines are now available as suspension cultures, and most attached cultures may be adapted to suspension in 3 to 10 weeks using suspension-modified formulations that provide such features as reduced calcium and magnesium *(68)*. For those having trouble with cell aggregation, medium additives such as suramin and the polysulfate compounds dextran sulfate and polyvinyl sulfate have been reported to be very beneficial.

Intentional genetic modifications (and the associated clonal selection), such as the introduction of the genes for *de novo* transferrin and IGF-1 *(69)*, and incremental β1,4-galactosyltransferase *(59)* can significantly alter (1) specific media supplementation required; (2) the quality of the secreted product; and (3) the nutritional requirements of the new line. As with the various established lines and each recombinant producer clone, CHO lines genetically transformed to provide some added feature can display unique metabolic demands. Therefore, optimal performance, especially in large scale, can require clone-specific medium optimization.

As in other animal cell systems, the medium in which any particular derivative is cultured can itself influence the characteristics of the product. Production rate and glycoform complement from production in serum-containing media as a whole can be different from those from SFM *(70)*. The same has been reported for the levels of particular media components such as glucose and specific lipids and amino acids.

A number of metabolism and division rate-altering regimens have been employed to boost specific productivity. Of these, sodium butyrate is commonly reported to provide between 0.5 and 3 times the benefit in the quantity of product. However, butyrate use is also reported to increase the degree and extent of product molecular heterogeneity. As the extent and consistency of posttranslational processing required depends on both the structure of the original transcript and application of the final product, the effects of such production enhancers must be assessed for each case.

Viability reduction is observed in CHO suspension cultures as they reach peak cell densities after 4 to 6 days of culture. This loss of viability has been shown in studies of a number of lines to be caused by the onset of apoptosis. The DX B11 line in 2-L bioreactor culture was assessed through a variety of apoptosis-specific assays to contain apoptotic cells throughout culture, with significant levels beginning as the culture entered the stationary phase *(71)*. It was demonstrated that the vast majority of cells contributing to the reduction in overall viability of the culture were apoptotic, as opposed to necrotic. In that study, a correlation to glutamine depletion was not observed, but in other studies, the stress of nutrient deprivation and waste accumulation have been associated with the onset of apoptosis. Again, whether related to the aggregation or not, suramin, dextran sulfate, and polyvinyl sulfate have been reported to help reduce apoptosis up to 95%.

Extracellular insulin degradative activity and a neutral pH glycosidase activity have been observed in mature CHO cultures *(72)*. The former is implicated in culture viability reduction and the latter in the postsecretion generation of heterogeneous molecular weight and glycoforms in the product. Control of these phenomena through media supplementation was attempted with limited success. The uncharacterized insulin degradative activity can accumulate in prolonged culture to the extent of degrading even high levels of added insulin. Synthetic substrate assays demonstrated that the generation of heterogeneous glycoforms in the product was caused by ambient sialidase and galactosidase activity.

2.5.2.5. Adaptation to SFM

Moving a culture from its original serum-containing medium to a SFM can be a delicate procedure. It is not uncommon for one culturist to conclude that a line will not adapt to a particular medium, only to have another succeed. The following direct adaptation can sometimes move a cell line to SFM. The main advantage of this method is saving time; using this approach, the culture could be fully adapted to SFM in 4–8 passages. However, there are some cell lines and clones that are more sensitive to the physiochemical and nutritional changes; in some cases, it may take eight or more passages for full adaptation.

The following procedure is generally applicable for CHO or NS0, but expected peak cell densities and doubling times may have to be adjusted according to either the respective line or the SFM employed. If viabilities decrease to less than 50% or if the cultures are growing slowly (population doubling time is >40 hours) for more than two to three consecutive passages, then the use of a more laborious sequential adaptation is recommended. The cultures can be grown and maintained in conventional T-flasks, roller bottles, shaker flasks (approximately 110–120 rpm), and spinner flasks (approximately 65–75 rpm) in a CO_2 gassed incubator. It is recommended to start adaptation with higher cell densities. Use prewarmed medium for each of the following activities:

1. Split the cultures grown in the current SFM or serum-containing medium directly into the SFM at a seeding of about 0.5×10^6 viable cells/mL.
2. When the viable cell density reaches 1 to 3×10^6 viable cells/mL, subculture the cells to 0.5×10^6 viable cells/mL.
3. Subculture adapted cells one to two times per week when viable cell counts reach 2 to 4×10^6 viable cells/mL with at least 85% viability.
4. When the cells are fully adapted, viable cell counts of many constructs should routinely exceed 2×10^6 for CHO and 4×10^6 viable cells/mL for NSO after 3 to 6 days in culture (viabilities should be >85%). CHO cells may tend to aggregate in suspension SFM adaptation; in this case, counting may be accomplished by trypsinization. Dispersed cultures may be obtained by repeated removal of aggregates by mild centrifugation.
5. Fully adapted cultures should be cryopreserved as a master seed stock.

Should this procedure not provide success, a sequential adaptation is recommended by altering the above procedure to reduce the rate of introduction of new SFM to about 40% increment over the previous concentration in the first few passages. This means that, in each subsequent passage, the percentage concentration of SFM will be 40, 60, 80, and finally 100.

2.5.3. Culture Formats

Culture systems supporting the culture of attached cells include T-flasks, roller bottles, microcarriers, and new high surface area multilayer stacked-plate systems. As the cells reviewed here may all be maintained in suspension culture, only the T-flask system is introduced.

T-flasks are an acceptable means of maintaining most mammalian suspension cultures in very small scale, inexpensively, or when performing activities such as transfection. Suspension-type cells maintained in T-flasks will exist as a loosely adhered layer on the bottom face of the flask. Each line and derivative exhibits a slightly different characteristic in this mode. Some cells have a tendency to adhere slightly to the flask, but may be loosened by pipetting media across the face of the flask or tapping the flask gently on its side. Some suspension cultures will tend to form loose clumps in the absence of shaking. These are sometimes dispersible by pipetting, but may have to be accepted until the culture is moved into a physically agitated mode. A typical implementation uses a T-75 flask with 15 mL of media.

Suspension culture in many formats demands greater concern for the factors of culture surface-to-volume ratio, shear force effects, and medium foaming. In nonbioreactor suspension culture, diffusion of gases into and out of the culture occurs at the air-to-media interface. Even experienced attached cell culturists are often surprised by the degree of gas exchange required by suspension cultures. It is actually the same as for cells in monolayer culture, but becomes more apparent and problematic when the medium surface-to-volume ratio decreases, as in suspension culture. Shear forces are introduced as physical agitation of the culture is introduced to suspend the cells and increase gas exchange, and although generally they are not an issue in serum-supplemented media, they can be catastrophic in SFM. Foaming of the media usually begins to be an issue in sparged bioreactor culture. A number of procedural, chemical, and mechanical solutions to these issues exist *(73)*.

Shake flask culture is generally the best way of establishing and maintaining suspension cultures in small scale. Disposable shake flasks (modified Erlenmeyer) are recommended for eukaryotic cell culture and are available in a range of sizes to support culture volumes from 20 to 200 mL. The system is scalable through the choice of flask size and adjustment of the medium volume dispensed. Aeration, cell suspension energy, and shear force can be varied by adjusting the volume of medium dispensed and the speed or path of the agitation. Shake flasks are maintained in variable-speed orbital shakers designed for use in 37°C warm rooms or jacketed incubators.

Fernbach culture is essentially a larger scale of shake flask culture. As they are more easily cleaned and rinsed, glass flasks are acceptable, although disposables are available. Cultures from 200 to 2000 mL may be produced in Fernbach flasks in the same orbital shaker and incubator apparatus used in shake culture.

Spinner culture is a good way to produce larger quantities of raw product, and it supports the largest scale possible before moving to bioreactor culture. There are a few popular designs, but all rely on magnetic stirrers of special design (slow speed, high torque) to impel the culture and are maintained in 37°C warm rooms or jacketed incubators. Spinner flasks are very scalable and support cultures from 30 to 3000 mL. They may be the sole system used to adapt a line to suspension culture and expand to bioreactor scale. The main reasons to recommend other smaller-scale suspension approaches are spinner flask setup costs, flask maintenance, and cleaning.

Bioreactors are essentially chambers for cell culture that provide containment and agitation of the culture as well as temperature and gas exchange control. Applications include cultures that demand special gas ratios, greater oxygen delivery rate, or volume scale greater than batteries of the above suspension systems can deliver. Bioreactors are available to support culture volumes from 100 mL to 15,000 L. So many distinct types of bioreactors and bioreactor culture modes exist that even adequate definition of them is beyond the scope of this work. As sparged, marine impeller-driven, batch-style suspension culture is the most common means of larger-scale culture, it is introduced in Section 2.6.

2.5.4. Cryopreservation

2.5.4.1. Methods

Preserving cells in liquid nitrogen provides a source of seed cells with a defined passage history and any other criteria required by the user. Animal cells remain viable at cell freezer temperatures indefinitely. However, although the state of the frozen culture is for all intents and purposes fixed, the freezing-and-thawing process can be quite stressful to the cells. Only fully adapted cells in excellent condition from cultures in mid-log phase growth should be cryopreserved. Optima for the rates of ambient temperature gradients in the freezing-and-thawing process for individual cell types and media formats have been generally established. Automated control cell-freezing apparatuses are programmable for each application and provide the best results. The main advantages of these instruments are: (1) fine process control and programmability for process optimization; (2) extreme heat extraction rate and controllability to overcome the detrimental heat of crystallization effects on freezing; and (3) programmability for consistency in application. For manual methods, the general paradigm is to reduce the ambient temperature between 1 and 0.1°C per minute –70°C, followed by immediate immersion in liquid nitrogen. A basic cryopreservation medium consisting of 7.5% dimethyl sulfoxide in 50% fresh and 50% conditioned media is a recommended starting point. The addition of other cryoprotectants, such as BSA, glycerol, PVP, and sucrose can help with troublesome cultures.

2.5.4.2. Cell Banks or Stocks

The banking of cells and the expansion of cultures for large- to very large-scale production is a well-regulated procedure for production in the biopharmaceutical industry, and its basic principles are valuable even to small-scale producers in research. Establishing a master cell bank (MCB) from low passage number, well-characterized cultures can provide a source of known starting/reference material for years of work.

Essentially, an MCB is established by producing a batch of culture calculated to provide sufficient individual vials of cells to ensure availability for the duration of the project and beyond. For large-scale producers, 20 to 100 (1- to 2-mL) vials are sufficient because, under usual conditions, this stock is accessed infrequently, depending on the scale and regulatory status of the process. The MCB can be a source for purposes from product or process development to commercial transfer.

Following the production of the MCB, even research-grade producers should take every effort to ensure that the bank is excellent quality, perfectly represents the cell line established in all categories of definition, and is maintained in a way to ensure its permanence. The MCB becomes the definition of the cell line.

A working cell bank (WCB) is then usually established to provide vials of seed for each individual production. New WCBs may be established from the MCB, but for many producers, a judicious assessment of need allows all production to begin from just one WCB. As a fresh vial from the WCB is used for each culture expanded to production, and possibly in process development and failure analysis, the production of two to three times the anticipated production runs is recommended. For many, this is in the range of 100 to 300 (1- to 5-mL) vials. Accordingly, this stock is produced in 1- to 20-L batches.

2.6. Protein Production

Once the cell line, vector, and promoter have been chosen and the genetically transformed clone and medium format selected, an adapted stock is produced and frozen. Following the characterization of product secretion in small scale, the system is ready to be transferred to large scale.

2.6.1. Bioreactor Operation

This outline assumes a general knowledge of the setup procedure of the specific bioreactor in operation and of a step-by-step protocol for its operation. Steps presented here assume the most basic batch format (*see* Section 2.6.3). In-depth approaches to bioreactor operation are generally available *(74)*.

Cleaning: Toxic material (deteriorated cell mass, endotoxin caused by stored cell mass, or cleaning materials) left in the bioreactor vessel or head-plate ports after poor cleaning procedures can lead to poor cell growth or death. Except for the probes, all parts (including the vessel) can be cleaned with a sponge or brush and a detergent solution. Perform a sequential rinse of all the bioreactor parts with hot water (preferably deionized), 0.5*M* NaOH (optional), hot water, and a final rinse in deionized water. Allow to air-dry.

Reassembly: Care should be exercised to effect reproducibility in reassembly and tubing configuration to ensure batch-to-batch consistency in seeding, sampling, supplementation, and product removal. Refer to manufacturer's guides for proper pH electrode calibration, for autoclaving procedure, DO electrode calibration, and reassembly procedure. It is recommended to use fresh tubing for each run, especially if it is worn or discolored, if virus was employed in the last run, if contamination occurred, or if there is a change in cell lines or processes.

Seed Train: A seed train is a protocol established for use in each production run. It is the scheme for the generation of a production culture from a vial from the WCB. A typical seed train for large-scale producers begins with a vial recovered from the WCB, expanded in a shake flask to 20–200 mL, and expanded again in 1.5-L Fernbach or spinner flask to 500–2000 mL. This then becomes the seed for spinner batteries or bioreactor lots of from 5 to 20 L. The process may be shortened or extended one step to accommodate smaller or larger lots. However, such a defined process is recommended even if not required. Monitoring of passage number, culture split ratios, doubling times, and count at time of passage is imperative for high-yielding and consistent bioreactor runs, as is seeding from cultures of consistent cell count and day in their growth cycle. Seed is best produced from one source flask.

Reactor Inoculation: The dilution required to produce the seed cell concentration is calculated. For some cell facilities, cell lines, seed densities, or bioreactor volumes, it is necessary to pool multiple seed flasks or expand the culture in a reduced bioreactor charge prior to the final seed. Smaller reactors can be charged by configuring tubing or venting to 1-L bottles, but larger reactors require media purchased or hydrated in large-volume containers with integral porting. The reactor is first charged with the culture medium and equilibrated to initial operating settings, then the seed is introduced.

2.6.2. Performance Characterization and Optimization

The final word in "productivity" is based on the amount of purified, active product inventoried per the cost in time, equipment, and materials required. Efficiencies can be assessed and modulated in every step of the supporting systems, beginning with the selection and modification of the product's structural gene to the packaging of purified product. Even when narrowing the scope of production efficiency to the optimization of a particular transgenic line in a particular mode of culture, there are still multiple parameters that affect the net raw product collected, such as culture performance, product secretion efficiency, stability of the product postsecretion, and product collection or processing efficiency. The performance factors that involve the secretion of product from an established transgenic line in the production mode selected are addressed here.

2.6.2.1. Culture Optimization

The two basic ways to enhance product secretion from a given line in a given system are to raise the specific productivity on a cellular basis or to increase the number of cells producing per unit volume of culture. An increase in the net cell-specific production can be achieved by: (1) raising the rate of product secretion per time; (2) increasing the percentage of efficient producers in the culture; or (3) extending the elapsed production period. For example, in a culture of 3×10^5 cells/mL containing 40% of the cells secreting product at 1 pg/cell a d for 3 d, there are four ways of doubling the aggregate production: (1) doubling the peak rate of secretion to 2 pg/cell a day; (2) doubling the number of efficient secretors in the culture to 80%; (3) doubling the production period to 6 day; or (4) doubling the cell density to 6×10^5 cells/mL.

Means of approaching the two basic product secretion enhancements mentioned above include: (1) selection or engineering of vector, promoter, promoter-enhancing elements, and integration mechanisms or sites; (2) selection or engineering of the host cell; (3) amplification of the resident copy number in the host; (3) increasing the copy-specific transcription or translation rate; (4) selecting for efficiently secreting cells in the culture; (5) modulating the basal culture environment, including the medium nutrient composition or gas exchange rate; and (6) supplementing the culture with production-enhancing agents at the time of peak production. Many of these approaches affect more than one of the basic production enhancements; for example, medium nutrient composition optimization can increase the per cell secretion rate, extend the production period, or increase the peak cell density.

A complicating aspect of any of these approaches is that they are often interdependent, such that actions to drive one in a positive direction can inversely impact others. A commonly observed example of this is that nutrient supplementation that increases the culture density often reduces the cell-specific production rate. Only approaches 5 and 6 fall within the scope of this chapter and are further considered here.

Beyond optimization of the basal medium to the individual clone, the addition of concentrated supplements of those nutrients measured as highly consumed in midculture can result in significantly increased performance. Product expression, secretion, and net production have been shown to be very dependent on nutrient supplementation, alteration, timing, and depletion as well as on other culture environment perturbations *(22,50,51,75,76)*. Means commonly employed to assess the specific nutrients highly consumed in the clone or medium addressed include high-performance liquid chromatography for such components as amino acids, nucleosides, and some vitamins and lipids and gas chromatography for other lipids, such as cholesterol. A real asset in these assessments is the newer and automated instrumentation available, which provide a real-time panel of nutrient and waste product levels. For example, YSI Inc. (Yellow Springs, OH) distributes instruments providing values for lactose, sucrose, galactose, lactate, glutamine, glutamate, ethanol, methanol, choline, oxygen, carbon dioxide, ammonium, potassium, and hydrogen peroxide. They feature instruments that can provide six analyses at a time and even an aseptic monitoring and control system that pulls samples directly from a bioreactor, produces measurements, and immediately directs the replenishment of nutrients as it maintains the sterility of the process.

Reasons for not including additional quantities of these highly consumed nutrients in the basal medium prior to culture commencement include toxicity to low-density cultures, solubility and precipitation issues, component degradation in storage, and inefficient metabolism at extreme levels. Components found generally consumed at high rates are glutamine and glucose, some vitamins, and host line- or even clone-specific amino acids. Osmolality in the designed supplements can usually be maintained because most media components, such as inorganic salts, are not included in the supplement. Aspects of this approach anticipate the concept of the fed-batch method of culture outlined in Section 2.6.3.

Product secretion kinetics through the culture cycle is known to be a production issue. It is common to see cultures in their peak growth phases yield very reduced product, only to shift suddenly to a production mode at the point of reduction of division rate and viability. A related theme observed is that cultures either inhibited for (or at least introduced to a condition suboptimal for) nominal cellular proliferation or specifically arrested at particular stages of the cell cycle can be stimulated to higher overall product accumulation on a per cell, culture volume, or time basis. Some of these reports are contradictory within the scope of the parameters measured, and most appear to be derivative, medium, or culture configuration dependent. A variety of nutrient and culture environment perturbations applied to cultures at advanced stages of cell growth has been shown to induce increased production rates. These include the reduction in either essential or nonessential nutrients *(77)*; the addition of a bolus of such nutrients *(51,78,79)*; change in ambient pH *(75,80)*, tonicity *(50)*, ion complement *(22)*, or CO_2 tension *(51)*; and the addition of a variety of toxicants or cytostatic agents, including those that specifically inhibit cell division or DNA replication *(81,82)*.

2.6.2.2. Statistical Design and Modeling

Many mathematically based approaches to the optimization of complex problems, such as culture nutrient component design and bioreactor process engineering, have been developed. Basic strategies are referred to in terms such as borrowing, component swapping, and biological mimicry. A number of statistically driven methods of experimental design have been developed, including the partial factorial, Plackett–Burkman, and Box–Behnken. Advanced computational methods (neural networks and fuzzy logic) have been applied to deal with the types and volume of data returned from such systems. Mathematical model simulation of such general types as structured and unstructured can provide predictions of culture performance in unassayed conditions. Whether any of these techniques are employed, a familiarity with their principles can create insight to even less-rigorous experimental approaches and design, which aids in more rapidly designing or improving the conditions addressed. Many reviews of such design and modeling techniques are readily available *(83,84)*.

2.6.2.3. Bioreactor Optimization

Once the performance of a particular transgenic line is optimized in a particular medium in small scale, the scale-up to bioreactor production requires its own optimization. Environmental heterogeneities because of inherent consequences of bioreactor design have been well documented. Gradients in carbon dioxide, oxygen, pH, and nutrient substrate are established because of the style and extent of mixing established by the size and configuration of the bioreactor vessel, impeller, spurge dynamics, and intrusive porting and probes. Although this has been observed and measured in large scale, it becomes most evident in very large-scale culture.

The basic reason for the discontinuity in conditions during scale-up is that, in these complex and dynamic systems, maintaining a particular constant in scale-up can be difficult to impossible. For example, maintaining a constant speed of the impeller as the vessel size increases results in reduction in mean circulation time. Alternatively, maintaining a constant mean circulation time eventually results in unmanageable equipment requirements or harmful shear forces imposed by consequent fluid velocities. Similar scale-induced issues exist for other systems, such as gas exchange. Many of these issues can be simply predicted by the laws of Newtonian fluid mechanics. Others, although discovered empirically, can eventually be defined, albeit in sometimes complex explanations. Suffice it to say that the culture conditions present in a 100-mL shake flask culture cannot be uniformly established even in large scale. Nevertheless, by employing commercially engineered equipment and working within ranges of monitored values previously established for a general culture format, good, although not necessarily optimal, performance can be expected in a relatively short time.

2.6.2.4. Scale-Down Approach

For those knowledgeable in the general performance of their bioreactor with the media and cell line of interest, optimal performance can be more readily obtained in what is known as the scale-down approach. Here, the conditions possible in large scale are imitated in small scale to study and optimize or remedy more efficiently the results of culture production in those conditions. For example, if the oxygen tension possible for a particular culture at a particular density in the reactor is known, then experimentation in various nutrient conditions at that level can be accomplished to optimize the medium for the large-scale condition *(85)*.

2.6.3. Operating Formats

Batch: Batch culture is the simplest system to operate and scale. In batch culture, the reactor is seeded with a cell inoculum, medium, and required supplements, and from that point, nothing is added to, or removed from, the reactor vessel except gases and possibly alkali for pH control. The operator monitors dissolved gases, sparge and gas mixture rates, pH, temperature, and total and viable cell and sometimes product yields on a daily or bidaily basis. The cell line, vector, and culture medium determine the product secretion kinetics. A number of factors determine when the culture is harvested for product. These factors include the (1) nature of the expression system; (2) culture's secretion kinetics; (3) requirement for product molecular homogeneity; and (4) any protease activity associated with the particular culture.

Fed batch: Fed batch is becoming the most widely used approach to large- and very large-scale production. Here, cultures are supplemented one or more times with critical nutrients at rates reflecting how they have previously been observed to be metabolized. This allows the maintenance of optimal levels of these components, which is one parameter in allowing the culture to progress beyond levels obtained in batch mode. These nutrient supplements are added using either continuous or discontinuous (stepped) protocols. Nothing is removed from the culture during the run. Fed-batch processes generally focus only on components with high depletion rates. Typical components addressed are glucose, glutamine (or glutamic acid) and other selected amino acids, vitamins and trace elements, although each cell line and even sub-clone can present unique requirements. One example of this is NS0-derived clones, which can require additional cholesterol in the feed. The timing of nutrient addition in stepped fed batch and the rate of delivery in continuous methods can, in some cases, be as critical as the composition of the additives. The frequency of feed can be determined empirically or by following the rate of oxygen consumption, cell density, product accumulation, or indicator nutrient levels and it comparing to previously established values. As each cell line, clone, and medium combination defines a unique demand for any particular nutrient, it is difficult to generalize about specific nutrients or feed timing *(86)*. Fed-batch processing, although it supplies additional nutrients, does allow waste products to accumulate in the medium. However, with few exceptions, product levels will be higher in fed-batch approaches than in batch processes *(87)*.

Chemostat: In chemostat, a reactor is seeded, and culture is expanded normally. Once the cell density reaches a predetermined density, cells and medium are constantly removed, and fresh medium proportionally added to maintain a steady state. The concentrations of cells, nutrients, product, and waste are maintained at a constant density. Optimal density of the culture can be defined by such parameters as medium exhaustion, maximum product secreted, product quality, or cell metabolism. This optimal density can be sustained as long as the reactor systems can be maintained. Chemostat has not proven viable for large-scale production, but does provide valuable information for application to fed-batch or perfusion processes.

Perfusion: Perfusion processes allow the constant removal of secreted product and waste products as well as maintenance of nutrient levels. As spent medium is recovered, a normal complement of nutrients is fed into the system. This process relies on a cell separation device to retain cells in the system. Common cell retention devices are cell settlers, spin filters, and filtration membranes. Fresh culture medium is the principle nutrient supply used to feed cultures and maintain a constant volume by balancing the fresh medium input rate with the spent medium removal rate. Adjustment of the rate of perfusion in response to metabolic demands is the main process control under standard perfusion procedures. The exchange of one-half to one reactor volume of medium per day is a nominal rate. At operating cell densities, cells are exposed to a higher specific concentration of the full complement of nutrients and a lower concentration of waste than in chemostat or fed-batch systems. Therefore, higher peak cell densities are possible. Practically, the limit of cell density becomes one or both of two high cell density-determined events: the gas exchange rate to the medium and the rheology of the cell mass. This essentially means that, as cell densities increase beyond about 5×10^7 cells/mL (depending on cell line and other factors), the physical means of supplying oxygen cannot keep up with the culture demand, and the culture becomes too thick to operate in the distribution and filtration systems.

2.6.4. Example Run

This section describes the elementary batch-style production of a recombinant MAb from a proprietary, low-efficiency, GS NS0 transfectoma. Production is in a Collagen Plus® 20-L bioreactor (New Brunswick Scientific, Edison, NJ) using HyQ® SFM4MAb™ (HyClone) supplemented at seeding with HyQ LS1000™ lipid supplement (HyClone) and GS Supplement (JRH Inc.). Two- to threefold higher cell and MAb production may be obtained without gene copy amplification using a fed-batch mode and optimized supplementation. Depicted are the values obtained for viable cell counts and MAb production (Fig. 2).

2.6.4.1. Materials

1. Cell line: Proprietary GS NS0 producing MAb
2. Media: HyQ SFM4MAb without L-glutamine cat. no. SH30391
3. Supplements: HyQ LS-1000 SH30554 and JRH GS Supplement 58672-100M
4. Bioreactor: 20-L New Brunswick Celligen Plus
5. Impeller: Pitched-blade marine style
6. Tubing connection: Welded connections via Terumo 72721T tubing welder and C-Flex 1/8 id tubing
7. Filters: Pall 0.2-μm Acropac 300 and Pall 0.2-μm Acrovent
8. Enzyme-linked immunosorbent assay: Zeptometrix 0801182
9. Chemistry analysis: Nova Bio Medical BioProfile Analyzer 100

2.6.4.2. Setup

2.6.4.2.1. Bioreactor: Preautoclaving

1. Cleaning: Clean and rinse reactor vessel thoroughly.
2. Port Configuration:
   a. 15-cm tubing Luer fitted headspace port for supplementation during run.
   b. 1-M tubing Luer fitted headspace port for supplementation during run.
   c. 10-cm tubing length headspace port fitted with an Acrovent filter for headspace gas overlay.
   d. 10-cm tubing Luer fitted midvessel port for culture sampling.
   e. 1-M tubing welded closure midvessel port for culture and medium addition.
   f. 1-M tubing welded closure vessel-bottom port for culture removal.
   g. 10-cm tubing sparge ring port fitted with an Acrovent filter.

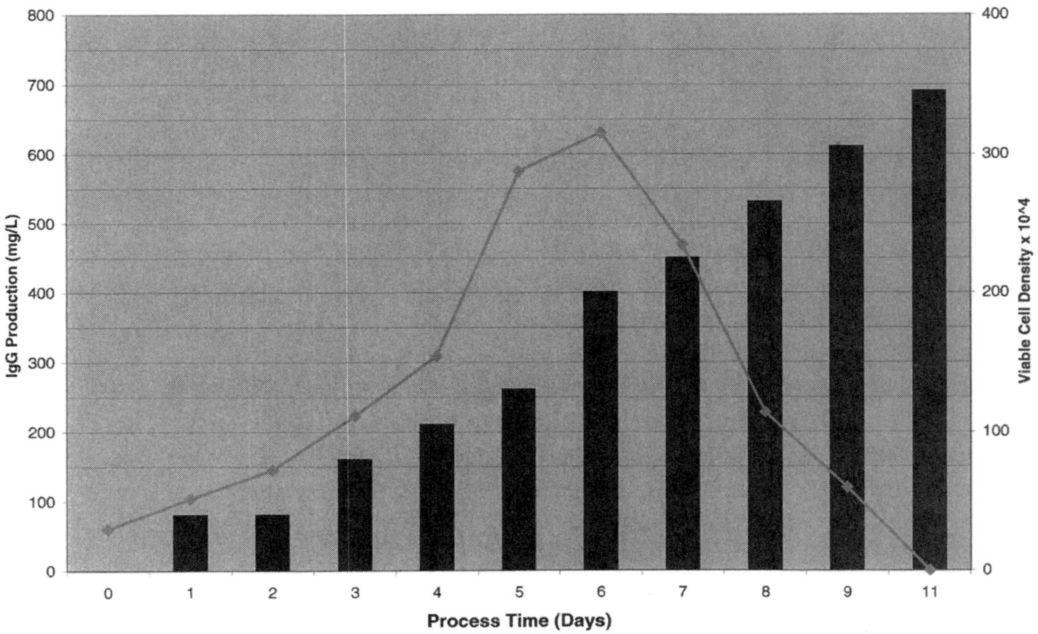

Fig. 2. Results of a bioreactor culture of a transgenic cell line. A 10-L Celligen Plus bioreactor was seeded at 3×10^6 cells/mL with an IgG-secreting GS NS0 in HyQ® SFM4MAb™ and supplemented with LS-1000™ high-cholesterol lipid supplement. The reactor was sampled daily for viable cell count and secreted product accumulation.

3. Probe configuration:
   a. P0727-5742 Broadley James pH probe
   b. P0720-5562 Broadley James DO probe
   c. Thermocouple probe for temperature measurement
4. Vessel: Add 4 L PBS to reactor vessel, seal welding closures, cap/wrap Luers.
5. Head plate: Secure, confirm impeller fit and rotation.
6. Dynamic testing: Install, connect, and operate impeller and sparge to ensure physical operation.
7. Probe calibration: Calibrate pH probe according to manufacturer's instructions.

2.6.4.2.2. Bioreactor: Postautoclaving

1. Vessel: Install and connect water jacket, condenser, gas inputs and outputs, and probe cables.
2. Control panel: Turn unit on and set DO calibration to "manual."
3. Incubation: Prime DO probe for 6 hours according to manufacturer's instructions.
4. Probe calibration: Calibrate DO probe according to manufacturer's instruction.

Inoculum

1. L culture in 2.8-L Fernbach flask seeded at 3 ¥ 10^6 cells/mL.
2. Incubated at 120 rpm at 37°C, and harveste on day 4 or 5 at 2.5 ¥ 10^6 cells/mL, 96% viable.

Seeding

1. Remove PBS from bioreactor using one vessel bottom port and a peristaltic pump.
2. Add 7 L fully supplemented medium via the same midvessel port and peristaltic pump.

3. Set instrument mode to "DO/pH" and "PID."
4. Set "RPM" to 50, "Temp" to 37, "DO" to 50, and "pH" to 7.2.
5. Add inoculum via a second midvessel port and peristaltic pump.

Operation

1. Initiate aspargation via the sparge ring and head gassing via headspace port. Set initial flow rates to 110 and 525 mL/minute, respectively.
2. Sample daily or bidaily via sample port by aseptically withdrawing 50 mL and discarding, followed by withdrawal of a second 50 mL for assay aliquots.
3. Assay for cell count, lactate, glucose, glutamic acid, and ammonia. Dispense or store samples for future product level assay.
4. Monitor gas saturation and adjust flow accordingly to maintain O_2 at 50% and pH at 7.2.
5. Add any supplements or feed postculture sampling by aseptically connecting syringe or bioprocess container and peristaltic pump to 15-cm or 1-M tubing on headspace port.

Harvest and Analysis

1. Terminate run at day of duration or cell viability predetermined in scale-down studies.
2. Collect raw culture mass into bioprocess container via vessel bottom port.
3. Examine metabolite consumption and waste buildup against culture progression and product accumulation.
4. Calculate cumulative and specific productivity. Design subsequent runs in light of accumulated results.

3. Insect Systems

3.1. Overview

The major method of producing recombinant proteins in insect cells is by using the baculovirus expression vector (BEV) system *(88–91)*. This helper-independent virus-based system has become one of the most widely used for those desiring to produce a significant amount of protein in small-to-large scale in a short period of time. It can often deliver gram-scale quantities of product much more quickly than recombinant CHO or myeloma. This is because constructing a recombinant baculovirus is considerably faster than generating a high-expressing recombinant mammalian line, and direct cloning methods of recombinant virus construction have further enhanced this feature. Although examples exist of its use in the very large scale and for therapeutic products, the BEV system has yet to become a major player in this area; however, recent significant developments might change this. These developments include transgenic cell lines that provide new cell culture and product-processing features, stably transformed lines that produce product continuously, and applications in the field of viral vectors.

The basic system exploits a natural lytic viral life cycle in cells derived from insect species that are natural hosts for baculoviruses. When insect cells are infected by a baculovirus, a natural protein produced by its most powerful promoter can accumulate by up to 50% of the total cellular complement. Replacing a native gene with that of the desired product causes the virus to produce the product in high levels. Derived from an animal, insect cells have the capacity for many required protein-processing and transport activities and are therefore a good candidate for expression of secreted proteins requiring folding, modification, or assembly. One disadvantage of the system is that, beyond the ubiquitously required steps of gene identification and cloning and cell culture and transfection, those who wish to employ the BEV system are required to become proficient in some theory and techniques in baculovirology.

The essential steps of the system are (1) selecting the insect cell, virus, or expression vector to be used; (2) producing a recombinant virus containing the gene for the desired product; and (3) infecting an insect cell culture to produce a lytic infection that results in a burst of product expression prior to cell death. Although over 500 baculoviruses have been identified, only 2 have been commonly accepted as expression vectors. Although both *Bombyx mori* nuclear polyhedrosis virus and *Autographa californica* nuclear polyhedrosis virus (AcMNPV) have provided features for large-scale producers, AcMNPV presents some practical advantages. AcMNPV has therefore become the dominant system and is presented here as a model. Even experienced mammalian cell culturists desiring to employ insect and BEV systems will have many new technologies for skilled development. A sufficient portrayal of them is beyond the scope of this chapter, but is available elsewhere *(90,92,93)*.

3.2. Baculovirus Basics

Baculoviruses, of the family Baculoviridae, are a diverse group of viruses known to be maintained only in arthropods; they are noninfectious in vertebrates, an advantage over many other viral systems for producers of therapeutic products. These enveloped viruses have rod-shaped capsids about 45–300 nm long. Their genome consists of one double-stranded, covalently closed, circular DNA strand, which for AcMNPV is approximately 130 kbp. The large size of the genome is an advantage to the system as it can accommodate quite large segments of exogenous DNA.

As with most animal viruses, their life cycle has been categorized into immediate early, early, late, and very late phases. The virion enters the cell by adsorptive endocytocysis and moves to the nucleus, where the DNA is released. DNA replication begins about 6 hours after infection and continues for approximately 70 hours, ending with the death of the host cell within an additional 2 to 4 days. Viral protein production begins at about 6 hours, and a cascade of various promoters is induced for the next 20 hours.

3.2.1. Native Baculovirology

The biology of the natural infection process underlies the power of the BEV system. A natural infection results in virions embedded in a crystalline protein matrix variously referred to as polyhedral occlusion bodies, polyhedria, occluded viruses, or polyhedral inclusion bodies (PIBs). The protein comprising the crystalline matrix in which the virion resides is known as *polyhedrin.* Insect larvae ingest these polyhedria, which are then dissolved in the gut to allow infection. Once this infection progresses from cell to cell throughout the larvae, it eventually ruptures and releases new PIBs into the environment to be consumed by other insect larvae, thus continuing the cycle.

3.2.2. In Vitro Baculovirology

The infection cycle in the laboratory begins with a culture of insect cells in log-phase growth. Nonoccluded (not entrapped in a PIB) baculovirus is introduced, and the culture becomes infected. In the succeeding few days, recombinant secreted product and new virus accumulate in the culture medium. Recovered medium is then a source of both product and virus to infect new cultures. When moving the baculovirus infection cycle to the laboratory, some proteins, such as polyhedrin, become unnecessary for the infection cycle, and their genes become candidates for replacement with the gene of interest.

Generation of recombinant virus through allelic replacement is accomplished either within a virally infected insect cell through homologous recombination with a transfected transfer plasmid or by direct cloning in *E. coli* using a shuttle vector. This process minimally takes from 4 days to 4 weeks, depending on the system used. Some additional time is required for optional in-depth screening of clones or product. Once a successful clone of virus is secured, a master stock of recombinant virus can be produced and titered in an additional 2 weeks. Stocks of virus are stable in refrigerated storage for months and in liquid nitrogen indefinitely.

3.3. Gene Vectors

Unlike production methods for stably transformed lines, the BEV system relies on the activity of a viral infection to drive the expression of the recombinant product. All means of establishing this productive viral infection are based on incorporating the structural gene of the product under the control of a viral promoter, such that the gene is expressed at a high rate during the infection. There are two major approaches to this end, as well as many variations of these approaches, which are selected according to such factors as (1) the speed with which production of each protein must commence; (2) the type and number of genes to be concurrently or sequentially expressed; and (3) the cost of establishing each system.

3.3.1. Transfer Vectors

The original system of recombinant virus generation was through allelic replacement via a transfer vector. It begins by the insertion of the gene of interest into the vector downstream of a promoter and between flanking regions homologous to an identified region in the viral genome. The vector is then mixed with baculoviral DNA and transfected into a culture of insect cells. Native enzymes within the cell mediate an occasional recombination event between the vector DNA within the homologous flanking regions and the viral DNA. This rare event within the culture results in some virus in the culture containing the gene of interest under control of the selected promoter. Plaque purification of the supernatant from this culture is required to separate successful recombinant virus from unsuccessful recombinants (e.g., single crossover) and the parent strain.

The essential principle in the plaque-based selection of a successfully recombined virus is the use of some marker system to allow visual distinction of successful double-crossover recombinant virus. A number of types of transfer plasmids have been developed, providing many options with features such as (1) the means of recombinant selection; (2) the number and type of promoters and enhancers employed; (3) the number of exogenous genes accepted, and (4) the cloning strategies accepted.

3.3.2. Shuttle Vectors

The construction and purification of a recombinant baculovirus for protein production using standard transfer plasmid and plaque assays can take as long as 4 to 6 weeks. This period can be reduced to several days, and multiple rounds of plaque purification can be eliminated, using a baculovirus shuttle vector (bacmid). The bacmid can replicate in *E. coli* as a plasmid and can infect common lepidopteran insect cells as a baculovirus. It is a recombinant virus genome that employs the mini-F replicon for bacterial replication and attTn7 as a site for the bacterial transposon Tn7-mediated cloning.

The system begins with insertion of the gene of interest into a cloning site within a donor plasmid. This donor plasmid contains an expression cassette consisting of a baculovirus promoter and a multiple cloning site, flanked by Tn7 transposon sequences. When cotransfected with the bacmid into *E. coli*, the expression cassette is transposed to the bacmid via the Tn7. Selection of successful events and isolation of recombinant bacmid DNA are accomplished using standard bacterial cloning techniques. Introduction of the bacmid DNA into insect cells results in a standard lytic infection. The progeny virus from this infection is then handled and employed as recombinant baculovirus *(94)*. A patented, commercially available bacmid system may be licensed from Invitrogen Corporation.

3.4. Insect Cell Lines

Of the many lepidopteran insect cells established in culture, only three, Sf-21, Tn-5B1-4 (Tn5 or High Five™), and Sf-9, have become dominant in the AcMNPV-based system. Cells may be secured from such sources as the ATCC, PharMingen, or Invitrogen. Sf-9 is a clone of IPLB-SF21, originally derived from *Spodeptera frugiperda*, the fall army worm, and is probably the most common for large-scale production. Nevertheless, many emphasize the higher production capability of Tn5, especially in low-passage stocks *(95)*. Factors to use in the selection of cell line include (1) growth rate; (2) support of virus production and plaque purification; (3) amenability to suspension and large-scale culture; (4) production and processing capabilities; and (5) pedigree for regulator and passage number use. As with mammalian expression systems, those desiring the very best rate, homogeneity, and authenticity in production have a formidable job in both general system and subsystem component selection. However, those with less-stringent requirements should approach the system that is most accessible to them from licensing, information, and materials points of view.

3.4.1. Genetically Transformed BEV Hosts

A major application of the BEV system is to produce human and other mammalian proteins. Proteins produced in insect cells display the results of many higher animal protein-processing pathways; however, significant differences have been noted for years between the glycoprotein-processing pathways of insects and higher eukaryotes *(96)*. Differences have been observed between many native *N*-glycan structures of glycoproteins and of those proteins produced in the BEV system *(91,97)*. *N*-Glycan processing in native lepidopterin cells is truncated and lacks both the sialyltransferase and the CMP–sialic acid required for full mammalian-type processing *(98)*. Isolated reports have been made on the ability of some established insect cell lines to produce some glycoproteins with complex, terminally sialylated *N*-glycans, and some sialylation activity has been provided, through various means, to the BEV system. However, until very recently, no reports showed large-scale production of properly sialylated mammalian proteins in the BEV system.

A transgenic Sf9 line has been developed (and is now commercially available) that has been genetically transformed to provide the mechanisms required to produce many mammalian proteins with authentic glycosylation. This was accomplished by producing a model of the pathways required for the observed *N*-glycosylation in both insect and mammalian products (Fig. 3) and introducing enzymes required for mammalian-

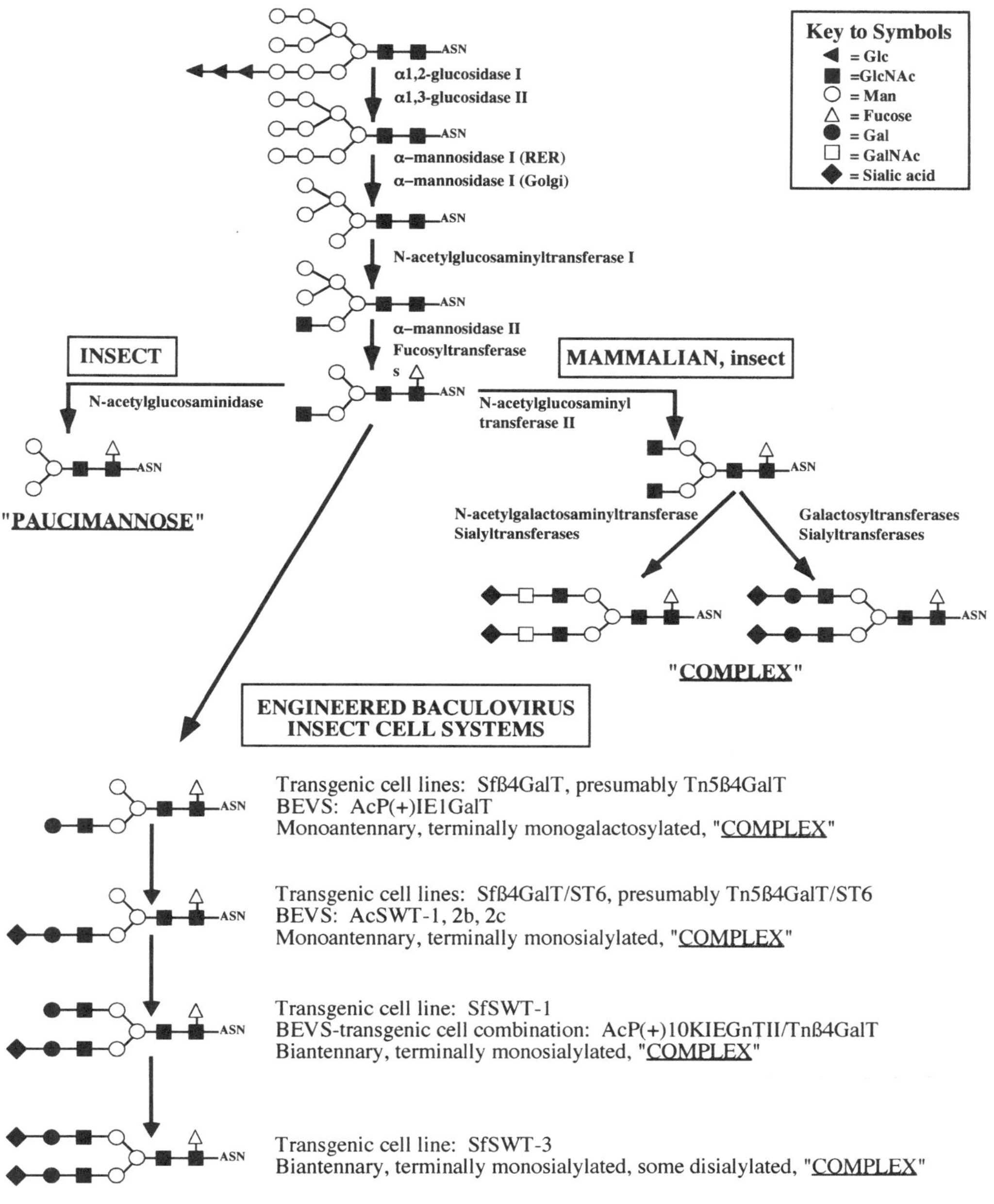

Fig. 3. Protein *N*-glycosylation pathways in insect and mammalian cells. Monosaccharides are indicated by their standard symbolic representations, as defined in the key. The structures of the *N*-glycans produced by new transgenic lepidopteran insect cell lines in the BEV system are shown as well.

type processing, but apparently absent, or silent, in the insect cell *(3)*. SfSWT-1 is an Sf9 cell that has been genetically transformed to constitutively express five mammalian glycosyltransferases *(99)*. Importantly, this line has retained normal growth properties and can support baculovirus infection and recombinant protein production.

Detailed structural analysis of recombinant glycoproteins produced in these cells using conventional vectors showed that they provide all of the *N*-glycan processing steps required for authentic mammalian glycoprotein production (Fig. 3). This means that the BEV system may now be employed in the production of recombinant protein with much more humanized *N*-glycans. It was demonstrated by specific analysis that product from these cells possesses terminally monosialylated glycans; however, it has also been demonstrated that they do not contain CMP–sialic acid, the donor of sialic acid to the sialotransferase. Initial study of this issue demonstrated that these cells required animal sera to support protein sialylation, and further work revealed that purified sialoglycoproteins would also work *(100)*. Apparently, these cells can salvage the sialic acid from extracellular proteins for use in processing the recombinant glycoprotein product.

As many employing the BEV system would like to avail themselves of the benefits afforded using SFM, work began on producing a cell line that could sialyate proteins without a serum protein donor. As it was known that Sf-9 cells could produce CMP–sialic acid when transiently transformed with CMP–sialic acid synthetase *(101)*, work was initiated to engineer SfSWT-1 cells further to provide sialylation activity in SFM.

A new line, designated SfSWT-3, has been reported to produce recombinant glycoproteins efficiently with complex, terminally monosialylated *N*-glycans in the absence of serum *(102)*. This line has had two more enzymes added to it that provide the CMP–sialic and sialic acid synthetase activity required for production of its own CMP–sialic acid from *N*-acetylmannosamine supplemented to the growth medium. Furthermore, it was discovered that these cells, grown in SFM, could process product more efficiently than SfSWT-1 grown in serum. Owing to its capacity to process complex glycoproteins authentically in SFM, this line, when available, has the promise of becoming a powerful tool in the production of human proteins for biomedical applications in the BEV system.

3.4.2. Stable Producing Lines

A second system for protein production employing insect cells is a cell-based system based on the genetic recombination of the cell to include the gene of interest under control of a suitable promoter. Successful implementations of such systems have been described for many years *(105)*, but have not gained much popularity because of their low levels of expression. However, this approach should be mentioned as it offers the advantages of continuous culture and the absence of degenerative cellular processes seen in the later stages of BEV system production. Promising improvements in the technology include the development of more powerful, and inducible, promoters and the construction of recombinant insect cell lines that provide new or improved features.

3.4.3. Insect Cell Culture

Many similarities are found between the culture of insect and of mammalian cells. They are of similar size, division rate, and essential nutritional requirements and have similar culture-handling techniques. The distinct differences that do exist require the attention of even experienced cell culturists.

3.4.3.1. Culture Basics

Sf-9 cells grow well in a suspension of loosely attached monolayer culture. They are unusual in that they are maintained at 27°C, and their media have no requirement for CO_2. Surprising to many cell culturists, they can even be maintained on a laboratory bench.

Under ideal conditions, populations will double every 18 to 24 hours. Standard cultures may be subcultured twice a week, but ideal culturing is accomplished by monitoring the cell count and passaging as the culture approaches the end of log phase. They are somewhat fragile and shear-sensitive cells, especially in SFM, but not unmanage-ably so.

As viral techniques require the use of very healthy cultures, it is important to establish a familiarity with the features of healthy cultures. Nicely rounded cells with distinct boundaries and discernible nuclei and without significant granulation or vacuolization indicate expanding cultures in a good environment. Viability may be assessed using trypan blue, similar to the way mammalian cells are assessed. Detailed procedures for insect cell culture are generally available *(90,104)*.

3.4.3.2. Culture Formats

T-flasks work well for maintaining insect cultures in small scale or when performing activities such as transfection. A typical implementation is the use of a T-75 flask with 15 mL of media and splitting as the culture becomes confluent. Monolayer cultures may be removed from their loose adherence to the flask bottom by gently, but firmly, shaking or tapping the nearly empty flask. Trypsinization is not required.

Shake flask culture is the best way of establishing and maintaining suspension cultures in small scale, and Fernbach or spinner flasks are best for larger-scale cultures. Section 2.5.3 provides a description of both. Suspension cultures are seeded at about 2×10^5 cells/mL and split at 2 to 5×10^6 cells/mL, depending on the medium used. An orbital shaker or flask assembly should be maintained in a 28 ± 0.5°C nonhumidified, nongas regulated environment. Aeration is accomplished by loosening the cap of the shake flask approximately one-quarter turn (within the intermediate closure position). In this condition, there is no oxygen limitation to the cells, and they proliferate to maximal rates. A 250-mL Erlenmeyer flask is inoculated with 100 mL of complete medium containing 2 to 3×10^5 viable cells/mL. The orbital shaker is set to 80 to 90 rpm for cultures maintained in medium supplemented with FBS. For cultures in SFM, the shaker should be maintained at 125 to 135 rpm. Serum-free cultures must be supplemented with a shear protectant, such as pluronic acid, and commercially available media are also supplemented.

Oxygen supply, as in mammalian cell culture, is an important consideration, especially when working with infected cultures. The cell-specific demand for oxygen increases considerably as a baculovirus infection proceeds. The two major parameters in ensuring sufficient supply are the means of exchange to the headspace in the flask (such as caps or ports) and the surface-to-volume ratio of the culture. Flasks are subcultured to 2 to 3×10^5 cells/mL twice weekly. Every 3 weeks, cultures may be gently centrifuged at 100*g* for 5 minutes and pellets resuspended in fresh medium to reduce accumulation of cell debris and toxic by-products.

3.4.3.3. Media

The first synthetic media formulations were developed in the 1960s, and some serum-dependent media developed then *(106)* are still in use today. The basic composition of insect cell culture medium is similar in many respects to the more commonly used mammalian cell media, with a major difference in the balance of essential nutrients in the formulations. A few insect-specific characteristics deserve mention as they can have an impact on performance optimization strategies. These media have a pH of about 6.2 and an osmolality of about 360, are buffered with sodium phosphate in addition to the standard bicarbonate, and employ sucrose among their carbon sources. Three basic serum-dependent formulations have evolved: Graces, IPL-41, and TC-100. Their formulations are publicly available in many references. As the nutritional requirements of insect cells do differ from mammalian cells, serum used in insect cell culture should be screened in insect cells prior to use.

A number of SFM formulations have been developed, mostly based on the IPL-41 formulation *(106)*. The current method to avoid the requirement of serum is essentially in the supplementation of an insect-balanced basal medium with a lipid emulsion, yeast extract, and Pluronic F-68. Much work has been done to remove the undefined hydrolysates, but to date this has not been accomplished. Many insect cell SFM are commercially available and have extensive application success in the published literature. For reasons including increased performance, serum availability, cost, and lot consistency; regulatory compliance; and product purification, many culturists prefer SFM formulations.

3.4.4. Baculovirus Techniques

3.4.4.1. Virus Identification and Purification

Regardless of the vector employed and the means of recombinant virus production, there are many good reasons to establish the clonal purity of the virus strain. Many report the ability to operate the BEV system without regularly clonally isolating their virus or titering their stocks. Although it is true that a desired amount of homogeneous product may be generated this way, it is also true that there are significant and demonstrated risks to this approach.

Plaque purification is the basic means of isolating a genetically uniform strain of virus. Heterogeneities in a stock of virus can arise through a number of means, including (1) the generation of multiple phenotypically "successful," but genetically diverse, events in the genetic engineering of a strain; (2) the natural mutation of progeny virus in passage; and (3) the generation of defective interfering particles through high multiplicities of infection (MOIs).

Plaque purification is basically accomplished by first seeding a lawn of insect cells on a culture dish. An appropriate dilution of the stock to be purified is then added, and some few individual cells on the dish become infected by a single infectious particle. Finally, a medium mixed with an immobilizing material (such as agarose) is overlaid, and the infection is allowed to continue. As an originally infected cell releases its progeny virus, only adjacent cells become infected. When fully developed, this produces spots, or plaques, on the plate that can be the result of an infection by a single infectious particle.

These plaques are identified by various means, and selected plaques can be harvested to obtain a stock of genetically homogeneous virus. The gene for β-galactosidase is often used as a marker in the construction process. When a colorimetric marker for the gene product (such as X-gal) is incorporated in the immobilizing overlay, either blue or white plaques (depending on how the marker is used) will indicate a successful clone.

3.4.4.2. Virus Titer

A plaquing system is also a way of determining the concentration of infectious particles, or titer, in an inoculum. By monitoring the dilution and final volume used to produce a particular number of plaques, the concentration of infectious particles in the original inoculum can be calculated. Significant dilutions are required because the number of infectious particles produced in a nominal infection is on the order of tens to hundreds of millions per milliliter, and it is only feasible to measure a concentration of less than 50 per milliliter. In certain circumstances, it can be important to distinguish the infectious titer of a virus stock from the number of virions or viral genomes present because these values can diverge by orders of magnitude. Means of such a determination, as well as other means of determining the infectious titer, are presented in commonly available BEV system manuals *(90)*.

3.4.4.3. Virus Production and Storage

Virus for use in producing stocks is a natural product of the infection cycle and accumulates in the culture medium. Supernatant from high MOI cultures, initiated in mid-log-phase growth and collected just as the culture loses viability, yields the best stocks. Such material generally contains 10^7 to 10^8 pfu/mL and is suitable for further culture infection or plaque purification. Cell-free harvested medium used for virus stocks should be protected from light and is stable for months when stored in the refrigerator. For all but the most casual purposes, stocks should be titered prior to use. Titering is accomplished through plaque-based or end-point dilution assays *(90)*. Cryopreservation in liquid nitrogen produces indefinitely stable stocks.

3.5. System Operation

Bioreactor applications of insect cell suspension cultures are very similar to those for mammalian cultures, with the main differences the temperature setting, lack of CO_2, specific nutrient depletion, production kinetics, and possibly the impeller speed. An overview is presented here, and many sources of more detailed procedures are available *(94,104,107)*.

3.5.1. Batch Culture

The large-scale batch culture of insect cells and its application in the BEV system requires particular attention to the optimization of the process parameters and quality of stock materials. Compromised procedures or ingredients of the system, which would only slightly diminish the efficiency of a mammalian cell run, can impair the quantity of product recovered significantly or totally. Elements of the process parameters that require particular attention include the culture kinetics, the concentration and timing of the viral inoculation, and the nutrient and gaseous environment of the culture during the productive stage of the infection. Essential process kinetics must be established in small scale before attempting to move to bioreactor-scale operation. Material qualities to be considered include the cleanliness of the reactor, the health of the cells at infection, and the quality and titer of the viral inoculum.

The first step in establishing a production run is the generation of a sufficient quantity of high-quality viral inoculum. MOIs of 2 to 8 are generally recommended for production infections. As inoculum viral titers are generally on the order of 10^8 pfu/mL and as cultures are infected at approx 2×10^6 cells/mL, a 20-L reactor will require on the order of 1 to 5 L of high-quality inoculum. The generation of the cell inoculum and reactor preparation is essentially the same as that for a mammalian culture (*see* Section 2.6.1). Antibiotics may be used in insect cell culture; however, they are discouraged, and the majority of culturists do not use them unless absolutely necessary.

3.5.2. Fed Batch

As with mammalian systems, the fed-batch approach has become a popular way of large-scale, high-efficiency production. Although in-process nutrient supplementation and waste buildup monitoring is not essential to the BEV system, much work has been done in pushing the BEV system to both higher volumetric and cell-specific yields. Regardless of the insect cell line used or the vector or promoter system chosen, the nutritional environment is a major component in the establishment of increased quality and quantity of production.

As in mammalian systems, fed-batch approaches have become the most popular implementation of this approach. Feeding design through metabolic demand analysis in serum-free systems has allowed Sf-9 cultures yielding nearly 10 times the peak total cell yield, 4 times the infectable density, and 4 times the total product yield of standard batch culture *(108)*. Factors allowing this include that Sf-9 cells do not readily build up toxic levels of ammonia or glutamine during culture or infection, and that modern SFM provide, qualitatively, the basis for very high-density culture. Another significant factor is that new bioreactor configurations and monitoring methods provide the information and control required for efficient feed solution design and delivery timing *(109)*.

4. Conclusion

Not that long ago, cells only produced their own endogenous cellular proteins. They did not produce proteins from other cells, nor did they produce polypeptides that had no biological function to the producing cells or other kinds of cells. Now proteins can be produced in an astonishing array of ways, including within cells to which they are not native. Almost any polypeptide can be produced in at least one cell system, and many polypeptides can be produced in a number of diverse systems. Many systems are only good at making particular classes of proteins or polypeptides. The optimal production of a protein in commercial quantities, in easily purified form, and with the right posttranslational modifications, and currently requires a thorough understand of the strengths and weaknesses of many production systems. This chapter described the state of affairs for producing commercial quantities of specific proteins, including appropriate posttranslational modifications, in higher eukaryote cell culture systems. Readers with greater interest are referred to the more specific and extensive information in the referenced works.

Acknowledgment

I gratefully acknowledge the assistance of Joyce Siler in production of the references section.

References

1. Goeddel, D. V. (1990) Systems for heterologous gene expression. *Methods Enzymol.* **185**, 3–7.
2. Gray, D. and Subramanian, S. (2000) Choice of cellular protein expression system. In *Current Protocols in Protein Science* (Coligan, J. E., Dunn, B. M., Speicher, D. W., and Wingfield, P.T., Eds.). John Wiley and Sons, New York, pp. 5.1.6.1–5.16.34.

3. Jarvis, D. L. (2003) Developing baculovirus-insect cell expression systems for humanized recombinant glycoprotein production. *Virology* **310**, 1–7.
4. Harrison, R. G. (1907) Observations on the living developing nerve fiber. *Proc. Soc. Exp. Biol.* **4**, 140.
5. Maitland, H. B. and Maitland, M. C. (1928) Cultivation of vaccinia virus without tissue culture. *Lancet* **215**, 596.
6. Eagle, H. (1945) Nutrition needs of mammalian cells in tissue culture. *Science* **122**, 501–504.
7. Griffiths, J. B. (2000) Animal cell products, overview. In *Encyclopedia of Cell Technology*, Vol. 1 (Spier, R. E., Ed.). John Wiley and Sons, New York, pp. 71–76.
8. Kohler, G. and Milstein, C. (1975) Continuous culture of fused cell secreting antibody of predefined specificity. *Nature* **256**, 495–497.
9. Gray, D. (1997) Overview of protein expression by mammalian cells. In *Current Protocols in Protein Science* (Coligan, J. E., Dunn, B. M., Speicher, D. W., and Wingfield, P. T., Eds.). John Wiley and Sons, New York, pp. 9.15–9.18.
10. Twyman, R. M., et al. (2000) Genetic engineering, animal cell technology. In *Encyclopedia of Cell Technology*, Vol. 2 (Spier, R., Ed.). John Wiley and Sons, New York, pp. 737–819.
11. Schmidt, E. V., Christoph, G., Zeller, R., and Leder, P. et al. (1990) The cytomegalovirus enhancer: a pan-active control element in transgenic mice. *Mol. Cell. Biol.* **10**, 4406–4411.
12. Senapathy, P., Shapiro, M. B., and Harris, N. L., et al. (1990) Splice junctions, branch point sites, and exons: sequence statistics, identification, and applications to genome project. *Methods Enzymol.* **183**, 252–278.
13. Palmiter, R. D., Sandgren, E. P., Avarbock, M. R., Allen, D. D., and Brinster, R. L., et al. (1991) Heterologous introns can enhance expression of transgenes in mice. *Proc. Natl. Acad. Sci.* **88**, 478–82.
14. Carswell, S. and Alwine, J.C. (1989) Efficiency of utilization of the simian virus 40 late polyadenylation site: effects of upstream sequences. *Mol. Cell. Biol.* **9**, 4248–4258.
15. Karreman, C. (1997) Fusion PCR, a one-step variant of the "mega-primer" method of mutagenesis. *Biotechniques* **24**, 736–742.
16. Morris, A. E., et al. (1997) Expression augmenting sequence element (EASE) isolated from Chinese hamster ovary cells. In *Animal Cell Technology* (Carondo, M. J. T., Ed.). Kluwer, Boston, MA, pp. 529–534.
17. Baer, A. and Bode, J. (2001) Coping with kinetic and thermodynamic barriers: RMCE, an efficient strategy for the targeted integration of transgenes. *Curr. Opin. Biotechnol.* **12**, 473–480.
18. Chu, L. and Robinson, D. K. (2001) Industrial choices for protein production by large-scale cell culture. *Curr. Opin. Biotechnol.* **12**, 180–187.
19. Jenkins, N. (2003) Analysis and manipulation of recombinant glycoproteins manufactured in mammalian cell culture. In *Handbook of Industrial Cell Culture: Mammalian, Microbial, and Plant Cells* (Vinci, V. A. and Parekh, S. R., Eds.). Humana, Totowa, NJ, pp. 3–20.
20. Dorner, A. J. and Kaufman, R. J. (1994) The levels of endoplasmic reticulum proteins and ATP affect folding and secretion of selective proteins. *Biologicals* **2**, 103–112.
21. Youakim, A. and Shur, D. B. (1994) Alteration of oligosaccharide biosynthesis by genetic manipulation of glycosyltransferases. *Ann. NY Acad. Sci.* **745**, 331–335.
22. Zhang, Y., Fong, W., and Yung, P.. et al. (1998) Optimization in hybridoma cell culture. *Chin. J. Biotechnol.* **14**, 187–193.
23. Puck, T. T., Cicciura, S. J., and Robinson, A., et al. (1958) Long-term cultivation of euploid cells from human and animal subjects. *J. Exp. Med.* **108**, 945–959.
24. Wirth, M. and Hauser, H. (1993) Genetic engineering of animal cells. In *Biotechnology: A Multi-volume Comprehensive Treatise*, Vol. 2. (Puhler, A., ed.) VCH Publishers, New York, pp. 663–744.
25. Birch, J. R. and Froud, S. J. (1994) Mammalian cell culture systems for recombinant protein production. *Biologicals* **2**, 127–133.

26. Bebbington, C. R., Renner, G., Thomson, S., King, D., Abrams, D., and Yarranton, G. T., et al. (1992) High-level expression of a recombinant antibody from myeloma cells using a glutamine synthetase gene as an amplifiable selectable marker. *Biotechnology* **10**, 169–175.
27. Dempsey, J., et al. (2003) Improved fermentation process for NS0 cell lines expressing human antibodies and glutamine synthetase. *Biotechnol. Prog.* **19**, 175–178.
28. Galfre, G. and Milstein, C. (1982) Chemical typing of human kappa light chain subgroups expressed by human hybrid myelomas. *Immunology* **45**, 125–128.
29. Chen, J. K., Okamoto, T., Sato, J. D., Sato, G. H., and McClure, D. B., et al. (1986) Biochemical characterization of the cholesterol-dependent growth of the NS-1 mouse myeloma cell line. *Exp. Cell. Res.* **163**, 117–126.
30. Sato, J. D., Kawamoto, T., and Okamoto, T., et al. (1987) Cholesterol requirement of P3-X63-Ag8 and X63-Ag8.653 mouse myeloma cells for growth in vitro. *J. Exp. Med.* **165**, 1761–1766.
31. Sauerwald, T. M. and Betenbaugh, M. J. (2002) Apoptosis in biotechnology: its role in mammalian cell culture and methods of inhibition. *Bioprocessing J.* **Summer**, 61–68.
32. Lasunskaia, E. B., Fridlianskaia, II., Darieve, Z. A., da Silva, M. S., Kanashiro, M. M., and Margulis, B. A., et al. (2003) Transfection of NSO myeloma fusion partner cells with HSP70 gene results in higher hybridoma yield by improving cellular resistance to apoptosis. *Biotechnol. Bioeng.* **81**, 496–504.
33. Vosastek, T. and Franek, F. (1993) Kinetics of development of spontaneous apoptosis in B cell hybridoma cultures. *Immunol. Lett.* **35**, 19–24.
34. Tinto, A., Gabernet, C., Vives, J., Prats, E., Cairo, J. J., and Godia, F., et al. (2002) The protection of hybridoma cells from apoptosis by caspase inhibition allows culture recovery when exposed to non-inducing conditions. *J. Biotechnol.* **95**, 205–214.
35. Lengwehasatit, I. and Dickson, A. J. (2002) Analysis of the role of GADD153 in the control of apoptosis in NSO myeloma cells. *Biotechnol. Bioeng.* **80**, 719–730.
36. Sauerwald, T. M. and Betenbaugh, M. (2002) Apoptosis in biotechnology: its role in mammalian cell culture and methods of inhibition. *Bioprocessing J. Summer*, 61–68.
37. Tey, B. T., Singh, R. P., Piredda, L., Piacentini, M., and Al-Rubeai, M., et al. (1999) Influence of Bcl-2 over-expression on NS0 and CHO culture viability and chimeric antibody productivity. In *Animal Cell Technology: Products From Cells, Cells as Products* (Bernard, A., et al., Eds.), Kluwer, Dordrecht, The Netherlands, pp. 59–61.
38. Tey, B. T., Sigh, R. P., Piredda, L., Piacentini, M., and Al-Rubeai, M., et al. (2000) Bcl-2 mediated suppression of apoptosis in myeloma NSO cultures. *J. Biotechnol.* **79**, 147–159.
39. Ibarra, N., Watanabe, S., Bi, J. X., Shuttleworth, J., and Al-Rubeai, M., et al. (2003) Modulation of cell cycle for enhancement of antibody productivity in perfusion culture of NS0 cells. *Biotechnol. Prog.* **1**, 224–228.
40. Frantisek, F. (2003) Antiapoptotic activity of synthetic and natural peptides. Paper presented at 18th Annual ESACT Meeting, May 11–14, Granada, Spain, in press.
41. Keane, J. T., Ryan, D., and Gray, P. P., et al. (2003) Effect of shear stress on expression of a recombinant protein by Chinese hamster ovary cells. *Biotechnol. Bioeng.* **81**, 211–220.
42. Mercille, S., Johnson, M., Lanthier, S., Kamen, A. A., and Massie, B., et al. (2000) Understanding factors that limit the productivity of suspension-based perfusion cultures operated at high medium renewal rates. *Biotechnol. Bioeng.* **67**, 435–450.
43. Al-Rubeai, M., Emery, A. N., Chalder, S., and Jan, D. C., et al. (1992) Specific monoclonal antibody productivity and the cell cycle comparisons of batch, continuous and perfusion cultures. *Cytotechnology* **9**, 85–97.
44. Barnes, L. M., Bentley, C. M., and Dickson, A., et al. (2000) Advances in animal cell recombinant protein production: GS-NSO expression system. *Cytotechnology* **32**, 109–123.

45. Nikolaenko, N. S., Tsupkina, N. V., and Pinaev, G. P., et al. (1992) The cultivation of mouse and human lymphoid cells on serum-free media. *Tsitologiia* **34**, 88–95.
46. Barnett, B., et al. (2003) NSO and NSO derived Hybridoma: MAb production in large-scale formats. Paper presented at 18th Annual ESACT Meeting, Granada, Spain, May 11–14, in press
47. Keen, M. J. and Hale, C. (1996) The use of serum-free medium for the production of functionally active humanized monoclonal antibody from NS0 mouse myeloma cells engineered using glutamine synthetase as a selectable marker. *Cytotechnology* **18**, 207–217.
48. Whitford, W., et al (2003) NS0 and NS0 derived Hybridoma: SFM Culture Applications. Paper presented at IBC Life Sciences 10th Annual Antibody Production and Downstream Processing Conference, La Jolla, CA, March 5–7.
49. Froud, S. J. (1999) The development, benefits and disadvantages of serum-free media. *Dev. Biol. Stand.* **99**, 157–166.
50. Lee, M. S., and Lee, G. M. (2000) Hyperosmotic pressure enhances immunoglobulin transcription rates and secretion rates of KR12H-2 transfectoma. *Biotechnol. Bioeng.* **68**, 260–268.
51. deZengotita, V. M., Schmelzer, A. E., and Miller, W. M., et al. (2002) Characterization of hybridoma cell responses to elevated pCO^2 and osmolality; intracellular pH, cell size, apoptosis, and metabolism. *Biotechnol. Bioeng.* **77**, 369–380.
52. Keen, M. J., and Steward, T. W. (1995) Adaptation of cholesterol-requiring NS0 mouse myeloma cells to high density growth in a fully defined protein-free and cholesterol-free culture medium. *Cyotechnology* **17**, 203–211.
53. Hartel, S., Diehl, H. A., and Ojeda, F., et al. (1998) Methyl-β cyclodextrins and liposomes as water-soluble carriers for cholesterol incorporation into membranes and its evaluation by a microenzymatic fluorescence assay and membrane fluidity-sensitive dyes. *Anal. Biochem.* **258**, 277–284.
54. Baker, K. N., Rendall, M. H., Hills, A. E., Hoare, M. Freedman, R. B., and James, D. C., et al. (2001) Metabolic control of recombinant protein *N*-glycan processing in NS0 and CHO cells. *Biotechnol. Bioeng.* **73**, 188–202.
55. Goochee, C. F., Gramers, M. J., Anderson, D. C., Bahr, J. B., and Rasmussen, J. R., et al. (1991) The oligosaccharides of glycoproteins: bioprocess factors affecting oligosaccharide structure and their effect on glycoprotein properties. *Biotechnology* **12**, 1347–1355.
56. Jenkins, S., Parekh, R. B., and James, D. C., et al. (1996) Getting the glycosylation right: Implications for the biotechnology industry. *Nat. Biotechnol.* **14**, 975–981.
57. Weikert, S., Papac, D., Briggs, J., et al. (1999) Engineering Chinese hamster ovary cells to maximize sialic acid content of recombinant glycoproteins. *Nat. Biotechnol.* **11**, 1116–1121.
58. Yoshikawa, T., Nakanishi, F., Ogura, Y., et al. (2000) Amplified gene location in chromosomal DNA affected recombinant protein production and stability of amplified genes. *Biotechnol. Prog.* **5**, 710–715.
59. Hammill, L., Welles, J., and Carson, G. R., et al. (2000) The gel microdrop secretion assay: Identification of a low productivity subpopulation arising during the production of human antibody in CHO cells. *Cytotechnology* **34**, 27–37.
60. Barnes, L. M., Bentley, C. M., and Dickson, A. J., et al. (2001) Characterization of the stability of recombinant protein production in the GS-NS0 expression system. *Biotechnol. Bioeng.* **4**, 261–270.
61. Omasa, T., et al. (2003) Selection and Stability for Recombinant CHO Cell Line Expressing Human GM-CSF in Gene Amplification. Available on-line at: www.bio.eng.osaka-u.ac.
62. Brown, M. E., Renner, G., Field, R. P., and Hassell, T., et al. (1992) Process development for the production of recombinant antibodies using the glutamine synthetase (GS) system. *Cytotechnology* **9**, 231–236.
63. Allison, D. S., et al. (2003) Rapid development of CHO cell lines for high-level production of recombinant antibodies. *Bioprocessing J.* **2**, 33–40.
64. Wurm, F. and Bernard, A. (1999) Large-scale transient expression in mammalian cells for recombinant protein production. *Curr. Opin. Biotechnol.* **2**, 156–159.
65. Mueller, P. P., et al (2003) Genetic approaches to recombinant protein production in mammalian cells. In *Handbook of Industrial Cell Culture: Mammalian, Microbial and Plant Cells* (Vinai, V. A. and Patekh, S. R., Eds.). Humana, Totowa, NJ, pp. 21–49.

66. Meissner, P., Pick, H., Kulangara, A., Chattelard, P., Friedrich, K., and Wurm, F. M., et al, (2001) Transient gene expression: recombinant protein production with suspension-adapted HEK293-EBNA cells. *Biotechnol. Bioeng.* **75**, 197–203.
67. Freshney, I. R. (Ed.). (1999) *The Culture of Animal Cells. A Manual of Basic Techniques.* Wiley-Liss, New York.
68. Mather, J. P. (1998) Laboratory scaleup of cell cultures. *Methods Cell Biol.* **57**, 219–227.
69. Pak, S. C. O., et al. (1996) Super-CHO: a cell line capable of autocrine growth under fully defined protein-free conditions. *Cytotechnology* **22**, 139–146.
70. Lifely, M. R., Hale, C., Boyce, S., Keen, M. J., and Phillips, J., et al. (1995) Glycosylation and biological activity of CAMPATH-1H expressed in different cell lines and growth under different culture conditions. *Glycobiology* **8**, 813–822.
71. Moore, A., et al. (1995) Apoptosis in CHO cell batch cultures: examination by flow cytometry. *Cytotechnology* **17**, 1–11.
72. Lao, M. S., et al. (1996) Degradative activities in a recombinant Chinese hamster ovary cell culture. *Cytotechnology* **22**, 43–52.
73. Griffiths, B. (2001) Scale-up of suspension and anchorage-dependent animal cells. *Mol. Biotechnol.* **3**, 225–238.
74. Gray, D. R. (2000) Bioreactor operations- preparation, sterilization, charging, culture, initiation and harvesting. In *Encyclopedia of Cell Technology*, Vol 1 (Spier, R. E., Ed.). John Wiley and Sons, New York, pp. 138–174.
75. Cherlet, M. and Marc, A. (1998) Intracellular pH monitoring as a tool for the study of hybridoma cell behavior in batch and continuous bioreactor cultures. *Biotechnol. Prog.* **4**, 626–638.
76. Osman, J. J., Birch, J., and Varley, J., et al. (2001) The response of GS-NS0 myeloma cells to pH shifts pH perturbations. *Biotechnol. Bioeng.* **1**, 63–73.
77. Hansen, H. A., Damgaard, B., and Emborg, C., et al. (1993) Enhanced antibody production associated with altered amino acid metabolism in a hybridoma high-density perfusion culture established by gravity separation. *Cytotechnology* **2**, 155–166.
78. Hencsey, Z., Fizil, A., Inzelt-Kovacs, M., Veszely, G., and Bankuti, I., et al. (1996) Effect of medium composition on hybridoma growth and antibody production. *Acta Microbiol. Immunol.* **4**, 359–370.
79. Sauer, P. W., Burky, J. E., Wesson, M. C., Sternard, H. D., and Qu, L., et al. (2000) A high-yielding, generic fed-batch cell culture process for production of recombinant antibodies. *Biotechnol. Bioeng.* **5**, 585–597.
80. Miller, W. M., Blanche, H. W., and Wilke, C. R. (2000) A kinetic analysis of hybridoma growth and metabolism in batch and continuous suspension culture: effect of nutrient concentration, dilution rate, and pH. *Biotechnol. Bioeng.* 67, 853–871. Reprinted from *Biotechnol. Bioeng.* 1988, **32**, 947–965.
81. Balcarcel, R. R. and Stephanophoulos, G. (2001) Rapamycin reduces hybridoma cell death and enhances monoclonal antibody production. *Biotechnol. Bioeng.* **1**, 1–10.
82. Watanabe, S., Shuttleworth, J., Al-Rubeai, M., et al. (2002) Regulation of cell cycle and productivity in NS0 cells by the over-expression of p21CIP1. *Biotechnol. Bioeng.* **1**, 1–7.
83. Kompala, D. S. Cell growth and protein expression kinetics. In *Encyclopedia of Cell Technology*, Vol. 1. (Spier, R. E., Ed.). John Wiley and Sons, New York, pp. 383–391.
84. Rose, S., et al. (2003) Mammalian cell culture. In *Handbook of Industrial Cell Culture: Mammalian, Microbial and Plant Cells* (Vinci, V. A. and Parekh, S. R., Eds.). Humana, Totowa, NJ, pp. 69–103.
85. Palomares, L. A. and Ramirez, O. T. (2000) Bioreactor scale-down. In *Encyclopedia of Cell Technology*, Vol. 1 (Griffiths, B. and Scragg, A. H., Eds.). John Wiley and Sons, New York, pp. 174–183.

86. Bibila, T. A., Ranucci, C. S., Glazomitsky, K., Buckland, B. C., and Aunins, J. G., et al. (1994) Monoclonal antibody process development using medium concentrates. *Biotechnol. Prog.* **1**, 87–96.
87. Bibila, T. A. and Robinson, D. K. (1995) In pursuit of the optimal fed-batch process for monoclonal antibody production. *Biotechnol. Prog.* **1**, 1–13.
88. Jarvis, D. L. (1997) Baculovirus expression vectors. In *The Baculoviruses* (Miller, L. K., Ed.). Plenum, New York, pp. 389–431.
89. Lukow, V. L. and Summers, M. D. (1988) Trends in the development of baculovirus expression vectors. *Biotechnology* **6**, 47–55.
90. O'Reilly, D. R., Miller, L. K., and Luckow, V. A. (Eds.). (1992) Baculovirus expression vectors. In *A Laboratory Manual.* W.H. Freeman, New York.
91. Marchal, I., et al. (2001) Glycoproteins from insect cells: sialylated or not? *Biol. Chem.* **2**, 151–159.
92. King, L. A. and Possee, R. D. (1992) *The Baculovirus Expression System. A Laboratory Guide.* Chapman and Hall, London.
93. Murphy, C. I., et al. (1995). Choice of cellular protein expression system. In *Current Protocols in Protein Science* (Coligan, J. E., Dunn, B. M., Speicher, D. W., and Wingfield, P. T., Eds.). John Wiley and Sons, New York, pp. 5.16.1–5.16.34.
94. Maiorella, B., Inlow, D., Shauger, A., and Harano, D., et al. (1988) Large-scale insect cell-culture for recombinant protein production. *Biotechnology (NY)* **6**, 1406–1410.
95. Li, G. (2003) Growth characteristics and expression of recombinant proteins by new cell clones derived from *Trichoplusia ni* (BTI Tn5B1-4) High Five™ cells. *Bioprocessing J.* **2(1)**, 35–40.
96. Marz, L., Altmann, F., Staudacher, E., and Kubelka, V., et al. (1995) Protein glycosylation in insects. In *Glycoproteins*, Vol. 29a (Montreuil, J., et al., Eds.). Elsevier, Amsterdam, The Netherlands, pp. 543–563.
97. Altmann, F., Staudacher, E., Wilson, I. B., and Marz, L., et al. (1999) Insect cells as hosts for the expression of recombinant glycoproteins. *Glycoconj. J.* **2**, 109–123.
98. Hooker, A. D., Green, N. H., Baines, A. J., et al. (1999) Constraints on the transport and glycosylation of recombinant IFN-gamma in Chinese hamster ovary and insect cells. *Biotechnol. Bioeng.* **63**, 559–572.
99. Hollister, J. Grabenhurst, E., Nimtz, M., Conradt, H., and Jarvis, D. L., et al. (2002) Engineering the protein *N*-glycosylation pathway in insect cells for production of biantennary, complex *N*-glycans. *Biochemistry* **41**, 15,093–15,104.
100. Hollister, J., Conradt, H., and Jarvis, D. L., et al. (2003) Evidence for a sialic acid salvaging pathway in lepidopteran insect cells. *Glycobiology* **13**, 487–495.
101. Lawrence, S. M., Huddleston, K. A., Tomiya, N., et al. (2001) Cloning and expression of human sialic acid pathway genes to generate CMP-sialic acids in insect cells. *Glycoconj. J.* **18**, 205–213.
102. Aumiller, J. J., Hollister, J. R., and Jarvis, D. L., et al. (2003) A transgenic insect cell line engineered to produce CMP-sialic acid and sialylated glycoproteins. *Glycobiology* **13**, 497–507.
103. Jarvis, D. L., Fleming, J. A., Kovacs, G. R., Summers, M. D., and Guarino, L. A., et al. (1990) Use of early baculovirus promoters for continuous expression and efficient processing of foreign gene products in stably transformed lpidopteran cells. *Biotechnology* **8**, 950–955.
104. Bernard, A., et al. (1995) Protein expression in the baculovirus system. In *Current Protocols in Protein Science* (Coligan, J. E., Dunn, B. M., Speicher, D. W., and Wingfield, P. T., Eds.). John Wiley and Sons, New York, pp. 5.5.1–5.5.18.
105. Grace, T. D. C. (1962) Establishment of four strains of cells from insect tissues grown in vitro. *Nature* **195**, 788–789.

106. Weiss, S. A., Smith, G. C., Kalter, S. S., Vaughn, J. L., and Dougherty, E., et al. (1981) Improved replication of autographa californica nuclear polyhedrosis virus in roller bottles: characterization of the progeny virus. *Intervirology* **15**, 213–222.
107. Chen, S., et al. (1998) Production of recombinant proteins in mammalian cells. In *Current Protocols in Protein Science* (Coligan, J. E., Dunn, B. M., Speicher, D. W., and Wingfield, P. T., Eds.). John Wiley and Sons, New York, pp. 5.10.1–5.10.41.
108. Elias, C. B., et al. (2003) Advances in high cell density culture technology using the Sf-9 insect cell/baculovirus expression system—the fed batch approach. *Bioprocessing J.* 2(1) 22–28.
109. Kamen, A. A., Bidard, C., Tom, R., Perret, S., and Jardin, B., et al. (1996) On-line monitoring of respiration in recombinant-baculovirus-infected and uninfected insect cell bioreactor cultures. *Biotechnol. Bioeng.* **50**, 36–48.

6

Biosafety, Ethics, and Regulation of Transgenic Animals

Raymond Anthony and Paul B. Thompson

1. Introduction

Transgenic animals—animals with genes added to their deoxyribonucleic acid (DNA) (either from organisms or eventually synthesized genes never before present in living organisms)—will no longer be limited by the gene pool of their parents. Such animals are slated to be created expressly to provide vital and novel benefits for human beings. These animals can have desirable characteristics or traits from virtually any gene pool and may also possess properties not present in nature or available through conventional breeding. They will be created for the production of new medical and pharmaceutical products and to enhance meat, dairy, and fiber production efficiency.

Transgenic animals such as antimalaria mosquitoes and cows that can produce desired pharmaceuticals in their milk have the potential to curtail or treat human diseases, respectively. It is likely that transgenic animals will also be created to produce tissues, living cells, and organs (with likely lower chance of immune rejection) as viable alternatives for transplant patients. Also, through cloning of herds of "elite transgenic livestock," animals can be engineered to provide more nutritious and efficiently produced foods, thus promising also to lower the cost of food for consumers. Other genetically modified animals such as the *Enviropig* of the University of Guelph in Canada may also help reduce agricultural waste and the number of animals needed for food and fiber production (for additional information, *see* refs. *1* and *2*). Similar animals may also serve to flush out agricultural pests, thus reducing the dependency on toxic pesticides.

Although animals with genomes that will be modified through manipulation of recombinant DNA, such as fish, sheep, cattle, pigs, goats, and insects, hold promise to improve our future, their existence also raises important ethical and public policy questions and concerns. Are products that have undergone genetic reconstitution safe? How are they substantially different from conventional products? What are the environmental impacts and risks of undesired gene transfer? Do transgenic animals present novel hazards? Will they create new pathways for animal disease to become hazardous to future generations of humans and animals? What are the ethical implications for the health and welfare of animals used in agricultural and biomedical research? How adequate are the national and international regulatory frameworks, respectively, and

From: *The GMO Handbook: Genetically Modified Animals, Microbes, and Plants in Biotechnology*
Edited by: S. R. Parekh © Humana Press Inc., Totowa, NJ

legal edifice in meeting the challenges raised by animal biotechnology? Is there international uniformity regarding standards and regulations? Are the government and intergovernmental agencies sufficiently prepared to respond to a biotechnological calamity should it happen? Are there sufficient guidelines to promote responsible research and application among private industries and researchers? Does the ability to modify mammals raise questions about the potential applications of genetic modification to human beings?

This chapter begins by providing an overview of some of the central concerns identified for the development of transgenic animals. Some have argued that those responding to the environmental risks and animal welfare and health issues should be paying greater attention to ethics and normative values, both in the process of risk assessment and management and in risk communication. Others have argued that there should be stricter regulation of biotechnology. Central issues in both the ethical and regulatory debates are reviewed. We conclude with a discussion of the role that working scientists should be expected to play in attending to these issues and the need to be attentive to public perception of environmental and animal welfare impacts from transgenic techniques. In limiting the focus of the discussion to transgenic animals developed for agricultural, industrial, and therapeutic applications, we do not consider issues associated with animals such as "knockout" mice altered solely for the purpose of studying gene functions.

2. Ethics, Public Policy, and the Regulation of Biotechnology

Generally, philosophical ethics is the study of a community's vision of how they ought to live responsibly and well. It concerns how individual and collective action should have an impact on others; which actions are morally permissible, impermissible, and insignificant; and embodies the values that a community sanctions as legitimate for promoting the good life. Philosophical ethics is an articulation and critical analysis of the norms, values, and framing assumptions that help determine the path a community should take and directs them on how to meet their responsibilities.

Central ingredients to living responsibly include having consideration of and respect for others, both human and nonhuman beings who matter from the moral viewpoint. Here, living responsibly enjoins recognizing vulnerability and dependency in others and promoting an environment of trust among members of the moral community. Living responsibly entails curbing specious profit-seeking behaviors and placing limits on institutions, practices, and technology that conspire against maintaining trust and respect for others. Finally, living responsibly also entails minimizing harms and maximizing benefits for all those with moral significance and encouraging equitable access to material goods that promote better quality of life. To ensure that an equitable distribution of the benefits and harms are shared among the members of the moral community in a trustworthy and respectful way, open communication between members is important so the interests of others as well as their values and concerns may be understood and considered in moral deliberations and during the formation of public policies. Thus, vibrant discussion and debate are central in establishing a community's shared vision of the ethical life.

A community's shared ethical vision is often mirrored by regulation and through public policy. Public policy is influenced by the moral arguments associated with the development of certain technologies and practices. Regulations and public policy concerning the public good, the role and limits of government, and equitable distribution of goods also serve to promote the value assumptions and political consensus of a community's vision for living well and responsibly. However, public policy has a prescriptive dimension of its own and may serve to bind a community together despite the presence or absence of ethical consensus. When ethical consensus is absent or moral stalemates or division persist, regulation and public policy serve to establish, administer, and enforce practical compromises and political solutions that can be adopted by all constituents. When collective and univocal ethical judgments are present, regulation and public policy may serve as a positive guide to allow for effective and efficient actualization of these moral ends.

Animal biotechnology, although promising to improve quality of life, poses significant risks as well. Even if the prospects are exciting, mistakes have terrifying consequences. Not only can mistakes cause harm, but also the reaction to even minor mistakes can substantially undermine public support for biotechnology. Because the stakes are high if something goes awry, both the scientific community and the nonscientific public must come together to clarify their responsibilities to those who will be impacted by its applications, and to set limits so that what is potentially promising is not also inimical.

This idea of limits connotes some form of regulation. Animal biotechnology is regulated at three interlocking levels that affect research, product development, and use of and commerce in transgenic animals. *Institutional regulation* is conducted by the organizations themselves, although often according to legal mandates handed down from governmental authorities. *Governmental regulation* is conducted by specific agencies of local or national governments with specific areas of authority created by legislative or judicial actions. Finally, *international* organizations such as the World Trade Organization (WTO), the United Nations (UN), and certain multinational treaty and covenant bodies such as the North Atlantic Free Trade Association (NAFTA) have authority to coordinate and resolve conflicts that may arise as a result of diverse governmental regulatory regimes.

Institutional regulation represents the first line of regulation for animal biotechnology. Organizations (e.g., nonprofit scientific research institutes, hospitals, or universities) or for-profit private corporations have specific committees or institutional officers to establish internal policies for biosafety and animal use and oversee compliance with food safety and environmental or other regulations. Although these institutional regulatory bodies operate in accordance with minimum standards dictated by the legal requirements of local and national governments, most institutions conducting work with transgenic animals have adopted internal policies that exceed these minimums. For example, the US Department of Agriculture (USDA) requires organizations conducting animal research to establish provisions of internal oversight through an institutional animal care and use committee (IACUC). The IACUC has responsibility for ensuring that basic requirements of animal welfare are met, but the USDA specifically excludes birds, rats, mice, and farm animals from required oversight. Nevertheless, most US organizations conducting animal research have adopted internal regulations requiring IACUC supervision of these excluded species *(3–7)*.

Governmental regulation includes legal requirements enacted and enforced by local, regional, and national governments. The organization of governmental regulation varies considerably from one country to the next, for example, with environmental regulations exclusively administered by local authorities in some instances, by national agencies in others, and by a combination of local, regional, and national authority in the majority of cases. In most countries, it is typical for different agencies to regulate food safety, environmental, animal welfare, and commercial activity, and in many countries, each subclass of regulatory action is subject to judicial review.

The authorizing legislation for each of these distinct regulatory activities specifies the scope and aims of the regulatory activity, and detailed discussion of regulatory activity is quite technical and varies considerably from one locale to another. However, there is a remarkable consistency in the general aims of regulatory authorities across the globe, and the discussion that ensues substantively addresses issues generally (but not in every case) subject to governmental or institutional regulation *(8–13)*.

Increasingly, however, governments are not the final authority for regulatory decision making because governments participate in international forums that harmonize regulatory activity. The Codex Alimentarius, for example, is a body within the Food and Agricultural Organization (FAO) of the UN that has long had the task of ensuring consistency among global standards for food identity and food safety. The WTO has undertaken review of national regulatory decisions on genetically engineered plants. The basis for this are components of the treaty that established WTO that are intended to limit the possibility that member states will erect spurious food safety or environmental regulations as *de facto* trade barriers. Similar actions within the WTO framework could certainly affect the fate of transgenic animals *(14)*.

The direction that animal biotechnology takes should be determined on the basis of ethics, that is, on the basis of a shared vision of how people ought to live, and on actions collectively sanctioned out of respect for the members of the moral community. Regulations represent the institutionalization of this vision. The definition of ethics provided here is consistent with a philosophical viewpoint that is sometimes stated by opposing the importance of ethics. Some commentators believe that regulations should be based solely on scientifically demonstrable risks and benefits to human beings and presume that the word *ethics* implies something more. But, the belief that regulations should be based solely on scientifically demonstrable risks and benefits is itself a vision of how communities should live, and as such it is an ethical viewpoint. Ethics should not be understood to indicate a single or dominant philosophical vision. The term *ethics* serves to call attention to the norms, goals, and shared assumptions that guide actions and policies and should include debate over differences in these norms, goals, and assumptions, which itself is a characteristic of a community's ethical life. As discussed here, the scientific community has a leadership role to play that entails deliberating with the public and relevant government agencies on the ethical and social implications and risks associated with animal biotechnology in advance before policy is formulated.

3. Ethics and Science-Based Concerns Associated With Animal Biotechnology

In August 2002, the National Research Council's (NRC) *ad hoc* committee on Agricultural Biotechnology, Health, and Environment published a report that identified central science-based risk issues associated with animal biotechnology and its products.

The committee determined risks in terms of criteria that assessed the immediacy and severity of impact of animal biotechnology and the rapid fluidity of technological change. The committee evaluated risk in terms of the nature of the technology, possible undesired and unanticipated effects of application and misapplication, novel ethical and social questions raised by animal biotechnology, and whether the various government agencies and the present-day statutory infrastructure and technological expertise were sufficiently ready to meet these challenges.

The committee applied these criteria to four major areas of concern associated with agricultural biotechnology and genetic modification of animals, including cloning. These areas are environmental concerns, animal welfare issues, food safety, and policy matters and institutional concerns *(15)*. These areas have ethical significance because they concern our vision for how best to advance the good life at the expense of and for others and in the face of different kinds and levels of risks. We have expanded this list of concerns to include issues related to biomedical ethics and social consequences. Furthermore, because animal biotechnology raises questions regarding the roles and responsibilities of scientists, we consider public trust in science another important area of concern that must also be explored.

3.1. Environmental Concerns

The biosafety of animal biotechnology represents a technically complex and emerging area of science. As applications of animal biotechnology are in their infancy, so are scientific approaches to characterize and measure the risks that these applications may pose to the larger environment. The diversity of animal species and the complexity of their respective types of interaction that both managed and unmanaged or natural ecosystems entails make it impossible to offer more than the most general discussion of issues relevant to biosafety in the present context. There is already extensive literature emerging for assessing the environmental risks of transgenic insects *(16,17)* and fish *(18,19)*. The aim here is to provide a general and conceptually oriented overview of the problems of biosafety as related to transgenic animals. Any attempt to assess or manage environmental risks from transgenic animals will require substantially more detailed scientific study of the species and environments involved.

Environmental risks are a function of hazard and exposure. An organism of any sort poses an environmental *hazard* when the presence of that organism in an environment can be interpreted as the possible source or cause of adverse events. The identification of hazards involves both a general and often speculative basis for linking the triggering event to subsequent outcomes and the normative judgment that these outcomes are unwanted, undesirable, or in some sense worse than other alternatives. However, many (if not most) hazards do not actually result in any harm. As such, an analysis of environmental risk also involves an account of *exposure*, the mechanisms that would produce the unwanted outcomes, and quantification of the likelihood for each stage or sequence of events comprising these mechanisms.

Characterizations of hazard and exposure for environmental risk, on the one hand, may be fairly broad and conceptual heuristic devices for thinking about the possible environmental consequences that might follow a triggering event; on the other hand, they might be technically specific and carefully determined measurements that reflect a high degree of empirical investigation and statistical sophistication. In either case, a characterization of environmental risk may be distinguished from *risk management*, which indicates the principles, policy, and general plan that will be undertaken in deciding whether to mitigate, insure against, or simply accept the risk in question.

The intentional or accidental release of transgenic animals into the environment represents the triggering event for characterizing environmental risk from animal biotechnology. The NRC committee presumes that this release poses a hazard that could result in unwanted changes in the composition of plant and animal species comprising an ecosystem. The primary basis for this presumption is the recognition that some nontransgenic species have become invasive when introduced into new ecosystems, resulting in extensive changes in those environments that have disrupted both human use of the environment and the suitability of the environment as habitat for native species of plants and animals. The general approach that the NRC committee *(15)* recommends for biosafety is to draw on and model experiences with nontransgenic invasive species as a theoretical framework for anticipating risks from transgenic species. Thus, they argued that transgenic animals do not constitute a novel class of hazards when compared to their conspecifics.

Given this general approach to hazards, the presence of nontransgenic conspecifics in both wild and managed ecosystems provides an empirical basis for estimating exposure. Based on prior studies conducted for transgenic plants, the estimation of exposure involves two questions. First, do the transformations confer phenotypic characteristics on transgenic animals that could be expected to result in significantly different environmental effects from those observed for nontransgenic conspecifics? Second, is there a potential for transgenes themselves to migrate to other species, resulting in phenotypic effects on nontarget organisms that could, in turn, result in environmental impacts *(20)*?

Prevailing assumptions among biologists dictate that the only mechanism for gene migration in animals is through interbreeding with interfertile populations (wild or domesticated) extant in ecosystems. If this is correct, the probability of cross-species gene migration among animals is vanishingly small, suggesting that there is little need to worry about the second question in animal biotechnology risk assessment. However, it should be noted that experimental studies of environmental risk from transgenic plants resulted in a significant revision of prevailing assumptions about the potential of controlling environmental risks from transgenic plants through isolation strategies. Experimental risk analysis demonstrated significant potential for cross-species gene migration among plants *(21)*. These results testify to the need for experimental validation of critical assumptions.

If the prospects of cross-species gene migration can be discounted, estimating exposure from transgenic animals becomes a problem of first characterizing how transgenes will confer different phenotypic characteristics on transgenic animals and then estimating how these different characteristics will in turn lead to adverse environmental outcomes when compared to the behavior of nontransgenic conspecifics. This problem can be analyzed in terms of the likelihood that transgenic species will become established as breeding populations and the subsequent impact that established populations possessing the transgene might have on predator–prey relationships. The details of both reproductive fitness and predator–prey relationships involve considerable empirical knowledge that will be specific to the animal species and ecosystem involved. Any more detailed discussion of approaches to the assessment of biosafety risks thus involves considerable empirical and technical specification *(22)*.

The above discussion suggests that it is, in principle, possible to characterize the environmental risks of transgenic animals, although such characterizations may be difficult, especially in light of existing gaps in knowledge. However, the extent to which such gaps qualify the ability to understand environmental risks from transgenic organisms lies at the heart of hotly contested debates over the future of genetically engineered organisms of all kinds.

These debates have taken many forms, but the best known involve specification and application of the precautionary principle or the precautionary approach to environmental risks. The debate over precautionary approaches to genetically engineered organisms has often been subsumed into the politics of international trade because advocates of agricultural biotechnology have accused those who deploy the terminology of precaution of allowing protectionist aims to override scientific principles *(23,24)*. Nevertheless, there is a serious issue to be faced by anyone who considers the environmental risks of transgenic organisms in deciding how to use the characterization of risk that is developed by systematically analyzing hazard and exposure.

The possible responses to this question can be simplified for the purposes of exposition into two diametrically opposed alternatives. One approach was articulated in Bentham's statement of utilitarian ethics over 200 years ago. Bentham advocated an approach to quantifying the likelihood and value of consequences of an action that anticipates the general approach to estimation of hazard and exposure described above and argued that this approach allows determination of the risk-based elements of the expected value associated with that action. These elements can be weighed against expected benefits to determine the overall expected value, and the utilitarian approach dictates taking the course of action with the greatest overall expected value *(25)*.

Several analysts of the debate over genetic engineering have argued that mistrust in the ability to adequately anticipate the consequences of recombinant DNA techniques is closely tied to the rejection of the utilitarian approach in general. In place of an approach that accepts weighing costs and benefits, they see people advocating norms of respect for nature. This approach is far more prejudicial with respect to the ethical acceptability of biotechnology in general and dictates that transgenic animals would be acceptable only if we could assure ourselves that developing them was consistent with largely qualitative characterizations of human responsibilities toward the natural world *(26,27)*. This kind of argument has indeed been made by at least some advocates of precautionary approaches in environmental affairs *(28)*.

It is difficult to say how a path might be charted between these two extremes, and authors who have attacked precautionary approaches would almost certainly argue in favor of simply taking the utilitarian approach. Nevertheless, others have argued that precaution can be understood in terms of giving additional weight to catastrophic hazards without regard to their likelihood. *Catastrophic hazards* are adverse outcomes that have geographically widespread or extremely damaging effects, especially when these effects are irreversible *(22)*. Others argue that the high degree of uncertainty and ignorance that pervades ecological assessments provides a basis for extreme caution in releasing organisms that would be likely to survive and interbreed with their conspecifics *(19)*. If the precautionary approach is understood in this way, it does not involve an abandonment of risk assessment so much as it recognizes circumstances in which a norm of minimizing the chance of worst-case outcomes should be substituted for the more typical utilitarian norm of seeking the greatest expected value.

3.2. Animal Well-Being and Health Issues

Animals typically used in agricultural or biomedical research (excluding insects) are considered sentient creatures that have a well-being or a good of their own. In contrast to the view of them as mere resources (as in traditional human-centered ethics) is the belief that these animals have a life that can go either better or worse for them. This last view is held by animal protection movements such as animal rights and animal welfare.* The well-being of these animals can be understood to have three major components. They include the animal's capacity to feel well both psychologically and physiologically, to function well, and to engage in species-specific natural behaviors *(39)*. These components of well-being are brought to bear on the question of permissible modification of animals by genetic engineering.

Genetic engineering has stimulated interest in the moral permissibility of animal use in research and challenges both the scientific community and the public sector to reexamine basic attitudes toward the moral status of animals and what is owed to them commensurate with their status and needs. Genetic engineering also raises questions about appropriate standards of well-being for research animals and spotlights the need for setting appropriate limits of modification and manipulation of animals.

Ethical concern regarding how modern biotechnology will affect animal well-being and health can take five general forms *(40)*:

1. That animals may suffer directly as a result of the effects of modification and manipulation.
2. That animals may suffer indirectly as a result of the effects of modification and manipulation.
3. That animals may suffer from consumption of or treatment with genetically modified products.
4. That, by using genetic transfer, the natures of animals are changed in substantial ways not for the benefit of the animals themselves, but for ours.
5. Procedural concerns related to the governance of animal use in general.

---

*Although there is disagreement over the philosophical underpinning for taking animal interests seriously, four positions stand out. These include a sentientist view made popular by Singer *(29–31)*, which endeavors to optimize the total balance of sentient experience in a species-neutral way; a strong rights-based approach that is synonymous with Regan *(32,33)*, which holds that certain animals are "subjects-of-a-life" and hence have noninstrumental value, and by this view, any experimentation on animals is prohibited if they are not also direct beneficiaries of research. There is an ethics of care view imputed to Midgley *(34)* that considers our kinship and interspecies connectedness with some species as sufficient for establishing acquired duties to care for their well-being; and there is an integrity or "natures" view attributed to (among others) Rollin *(35,36)*, Rutgers and Heeger *(37)*, and Fox *(38)* and basically states that animals have an intrinsic nature or unique species-specific purposes that underscore the content of our responsibilities toward them, that is, they have "a nature, a function, a set of activities, intrinsic to [them] evolutionarily determined and genetically imprinted" *(35,36)* that is morally obligating in character.

3.2.1. Direct Effects

There is significant public concern that modifying the genetic constitution of animals will lead to increased physical pain and psychological suffering, whether inadvertent, unwanted, unexpected, or intentional. This concern is amplified given (1) widespread belief that we have special responsibilities, which include minimizing or not inflicting unnecessary harm, to care for animals in our charge; (2) the present underdeveloped state of the technology; and (3) the impossibility of anticipating the impact of modification on an animal's well-being, especially if the animal's constitution departs greatly from its evolutionarily determined genome.

Because the science and technology of genetic engineering are still in their inception, it is feared that genetic engineering will contribute to animal suffering by producing dysfunctional animals that must endure physical, physiological, and psychological harm, behavioral abnormalities, or health maladies. Although unhealthy transgenic animals will almost certainly be euthanized, an increase in the rate of euthanasia is not without ethical significance. Concern over the present inefficiency of production techniques is joined by questions over the utility value of animals in general and the morality of creating animals with pathological conditions to serve as research models for human beings. Concern for transgenic animals in this way is not without precedent.

The 1985 Beltsville pigs were among the first transgenic animals produced by the USDA Agricultural Research Service. Scientists microinserted the gene for human growth hormone into pig embryos in one of the early experiments that applied bioengineering to food animals so that they would grow faster, use less feed, and produce leaner meat. Nineteen animals made it to maturity, but they experienced painful arthritic conditions and endured physical deformities, ulcers, and decreased immune resistance. These crippled pigs were euthanized *(36)*.

Dolly, the famous cloned sheep from Scotland, was euthanized on February 14, 2003, after experiencing premature aging and virus-induced lung cancer. Dolly was 6 years old, approximately half the life expectancy of her breed. Her premature death, as did circumstances surrounding her conception, raises questions about the ethics and practicality of copying life *(41,42)* In the case of events leading up to her conception, Wilmut's team struggled approximately 300 times to fuse nuclei from adult cells with denucleated blastocysts. Of the 29 successful transfers to host wombs, only one clone, Dolly, was produced. Many fetuses were used and destroyed as the team also applied the same technique using nuclei from fetal and embryonic cells *(43,44)*.

The NRC committee cited a few examples highlighting direct deleterious effects of novel techniques on animals. They indicated that knockout and cloned mice showed increased levels of aggression and suffered impaired learning and motor skills in certain trials. A number of hoofed animals produced by either in vitro culture or nuclear cell transfer tended to have higher birth weights and longer gestation periods than conspecifics produced by artificial insemination. Of these animals, some experienced difficulty during birth and required specialized procedures like caesarean section and respiratory assistance and therapy *(15)*.

Another direct concern warranting attention includes the potential transmission of newly acquired diseases or traits from transgenic animals (such as those used in xenotransplantation) to conspecifics with no immunity against the disease.

This short litany of problems is an echo of a general concern that it is difficult to predict just how the psychological, behavioral, and physiological well-being of animals will be affected as a result of genetic modification of the very constitution of animals. Such incidents raise public concerns about the governance of science and accountability of scientists. In the mid-1980s, the same questions regarding responsibility, professional ethics, and accountability were sparked in lieu of *Silver Springs vs Dr. Taub*, and *University of Pennsylvania Head Injury Clinic baboons vs Dr. Genarelli (45)*. Scientists need to be sensitive to nonutility views of animals as well as practice good husbandry and care for research animals. It is thus important that the scientific community remain vigilant in their assessment of compatibility between the animal's well-being and its adaptability to its environment when pursuing research.

3.2.2. Indirect Effects

New applications of animal biotechnology, like xenotransplantation, raise secondary concerns related to creation of animals that deviate substantially from their traditional roles. In xenotransplantation of livestock animals, concerns related to management and housing of highly sophisticated and social source animals such as pigs and nonhuman primates used as research subjects raise eyebrows. To prevent or minimize transmission of diseases to potential human organ, cell, and tissue recipients, these animals must be housed in isolated and sterile living quarters. This form of housing may involve low stimulation and poorly enriched environments and may cause these animals to exhibit abnormal behaviors such as fear, anxiety, aggression, and patterns of stereotypical behaviors or boredom. Similar questions may be posed for transgenic animals developed to secrete pharmaceutical or industrial products in their milk. Whether subjecting animals to impoverished or frustrating environments can ever be justified remains a contentious issue. In any case, it demands serious attention by the scientific community, especially because this form of neglect or impairment of an animal's well-being is within the sight of many in the public.

3.2.3. Biotechnology Product Application

Genetic engineering may also be applied not directly to manipulate animals' genomes, but to produce drugs, therapies, and feed for animals. Issues associated with the approval of these products are, in one sense, no different from those for any other drug or additive. However, the use of advanced life science techniques may heighten controversy. For example, genetic engineering was used to produce recombinant bovine somatotrophin (rBST) to boost milk production of dairy cattle. Animal protection groups protested when rBST use was linked to increased incidence of mastitis and lameness and lower productivity *(46–48)*. The US Food and Drug Administration (FDA) concluded that these health problems were typical of high-production animals, and therefore that rBST should not be identified as a cause. However, regulatory agencies in other countries, including Europe countries and Canada, have cited animal health issues in refusing to approve rBST. The lesson of rBST is that researchers should be prepared for increased scrutiny (and possibly higher standards) when a biotechnology product has equivocal impact on animal health.

3.2.4. Changing the Nature of Animals

Although genome manipulation may produce animals that are able to transform feed with greater efficiency or animals better suited to their environments, the moral permissibility of altering or infringing the genetically encoded set of physical and psychological capacities that give rise to the basic interests of an animal (i.e., the "pigness of pigs" or "horseness of horses") remains a contentious subject. Moral harm is "perceived" to be committed when an animal is prevented from performing behaviors commensurate with the way it has evolved *(49)* or if the animal's genetically predetermined "set of functional needs" are thwarted *(38)*. Proposals for genetic modification that create duller or decerebrate animals so that they will be more conducive to conditions of intensive farming or sterile laboratory housing have been especially controversial among the public. Creating insentient beings purposefully for human ends and preventing them from living in accordance with their natural ends in life is perceived as a perversion of the sanctity of nature or natural boundaries.

Rollin *(36)* has suggested the principle of welfare conservation to help mitigate inhumane procedures related to genetic engineering. The principle states that genetic engineering is prohibited if it would make animals worse off than nongenetic animals in comparable circumstances; that is, it is unethical to create animals worse off with respect to suffering and deprivation comparable to conspecifics begotten through conventional breeding. Rollin's principle leaves two important implications:

1. A more palatable one: It is permissible to alter an animal's genetic constitution and biological function if it leads to less suffering or improved well-being.
2. A highly contentious one: It is permissible to modify an animal's experiential capacity if it relieves suffering, even if that suffering is caused by less-optimal living conditions.

In lieu of the perceived integrity of animals, Thompson has argued (with respect to implication 2) that, if it is wrong to alter a human being so the person would no longer be characteristic of human species, then without offering relevant differences, doing the same to an animal (i.e., depriving the pig of its pigness or the horse of what makes it the thing that it is) is equally wrong *(50)*. It is wrong to "estrange" animals from the functional needs characteristic of their species not only because it jeopardizes their well-being, but also because their perceived intactness is connected with their species identity. Genetic engineering that detracts from the animal's own good reduces them to mere means to human ends *(37,49)*.

3.2.5. Procedural Concerns Related to Animal Use

Besides these substantive issues regarding moral status is the concern that animal biotechnology is moving ahead in the absence of public discussion and consensus. The subjects of species integrity and whether it is permissible to modify animals if no benefits are conferred to the animals themselves have been of particular interest to animal protection groups seeking a voice in planning the research agenda. Furthermore, it would appear that public concern over the lack of clarity over what counts as adequate provisions for promoting health and normal development for research animals in our charge has been ignored by some sectors of science.

In the United States and Canada, IACUCs have stipulated norms for research since the mid-1970s. Although the institutional structure varies, similar committee approaches to animal ethics are now found across the globe. When functioning well, such committees deliberate carefully on the morality and prudence of research projects based on the principles of the 3Rs. Briefly, the 3Rs (reduction, replacement, and refinement) proposed by Russell and Burch in 1959 are three general principles for the governance of humane animal-based science and experimentation) *(51)*.

There is some doubt whether these committees will be able to offer appropriate guidance with respect to transgenic technologies without adequate revision or supplementation with other normative principles *(52,53)*. In particular, the 3Rs do not cohere with deeply held intuitions about animal and species integrity that constrain what is acceptable to do to sentient beings or animals in our charge. Furthermore, review of guidelines and regulations from government oversight agencies to keep up with contemporary standards is necessary and will go a long way to ensure that experiments are sufficiently controlled and offer adequate protections for animal research subjects (do not suffer unnecessarily; have appropriate standards of well-being; have adequate living conditions, good husbandry, and appropriate veterinary care; and are privy to humane end points). Reviewing outmoded guidelines may also help to anticipate and establish standards of "good welfare" instead of ameliorating or reacting to current conditions.

3.3. Food Safety and Consumer Autonomy and Sovereignty Issues

Animal products created through genetic engineering or cloning may pose unique disease and health risks when consumed and challenge existing aesthetic and cultural notions of food purity and standards of food quality. Here, the NRC *ad hoc* committee noted that the entry of genetically modified and genomically reprogrammed nonfood animals into the food supply was the most serious risk issue based on a strict expected value analysis of risk *(15)*.

Animals genetically modified to produce pharmaceuticals or other chemical or biologic properties in their eggs or milk may inadvertently find their way into the food supply. Strict monitoring procedures, regulations, and customized procedures meant to detect or anticipate implications of these new biotechnologies in the food supply may be necessary. The committee was also concerned that unused animals (such as male chicks and bull calves from dairy operations) engendered by new biotechnologies or that come into contact with biologically engineered products (i.e., conventionally bred and genetically modified animals fed with unapproved genetically modified foods) may inadvertently find their way onto grocery shelves in the absence of forward-looking measures or policies to ensure that they do not. Transgenic animals meant for food, like transgenic swine, fish, poultry, beef, dairy cattle, and sheep, will be screened using the principle of substantial equivalence. This requires that proteins not previously found in human diets will be subjected to extensive clinical trials for safety and quality before approval.

Other hazards associated with transgenic animals meant for consumption are as follows *(15)*:

1. They may induce allergens that could pose health risks.
2. Exposure to bioactive constitutive parts on consumption could give rise to illness.
3. There is the potential for toxicity from transgenically derived organisms, especially if the toxins manage to elude detection surreptitiously under conventional assessment methods.
4. Inappropriate gene expression may occur.
5. Activation of quiescent viruses is possible.
6. Nutritionally deficient substitutions that pose human health risks may be made.
7. Application of cloning on a large scale may result in monocultures that may be less resistant to disease, and thus communities of people may be susceptible to risk of famine or financial ruin *(54,55)*.

The committee noted that strict measures can be taken to mitigate the risks associated with items 1–5. They include monitoring the method of gene transfer and vigilance when it comes to how the genes are recombined or resequenced.

Not unlike risk analysis of environmental concerns, food safety issues are also typically assessed as a function of probability of unwanted outcomes occurring and their expected severity and immediacy. Again, experts are delegated the task of optimizing the ratio of bad consequences to good outcomes. Interpreting the problem of risk management solely as an optimization problem, however, bypasses concerns related to consumer sovereignty and autonomy in the food system.

Issues related to informed consent on the part of those who will be exposed to food-borne risks (whether real or perceived) cannot be treated as "costs" in an optimization problem. They raise questions about market, political, and social mechanisms to protect consumer autonomy and liberty of conscience *(56,57)*. Consumers may have concern for purity of food and may have aesthetic or religious reasons or moral arguments (such as wanting to support forms of farming as a lifestyle) that must also be considered. The lack of viable alternatives to genetically modified products impedes autonomous decision making and liberty of conscience (and is a form of covert coercion). Another concern is the absence of mechanisms of informed consent (such as standardized labeling) to help consumers decide on their own to avoid foods they deem incompatible with their moral, health, or religious values.

Apart from these infrastructural-minded matters, the dearth of public data on the subject of the safety of meat and milk and other products produced from genetically modified animals and somatic cell cloned organisms does not inspire confidence. Longitudinal studies and vigilant monitoring of the effects of new products and products fed to commercially bred animals are encouraged by private industry and government. Industry has been reluctant to make data on animal health public because of competitiveness concerns, yet with respect to food safety, it will almost certainly be critical to have published data available.

3.4. Policy and Institutional Concerns

How will animal biotechnology serve the public good? Might it perpetuate discrimination or be available only to well-capitalized outfits and people? Have government agents and scientists communicated likely applications and risks to the public, taken the time to help inform the public, and listened to public concerns before moving forward?

While pondering these questions, the NRC committee also questioned whether the current regulatory and legal framework supported the unique concerns raised by animal biotechnology and whether appropriate federal agencies had the technical capacity and resources to review the technology and address potential hazards. They also saw a need to clarify the responsibilities of individual scientists, academic institutions, private companies, and various government agencies associated with the development and application of animal biotechnology by clearly delineating the regulatory jurisdiction of the USDA, FDA, and the Environmental Protection Agency and reforming outmoded policies if necessary *(15)*. Of special concern was the lack of public engagement mechanisms to generate meaningful debate and improve public understanding of risk factors in a trustworthy and thorough manner at both national and international levels.

The NRC *(15)* indicated the need for broad public discussion with ethicists, scientists, policymakers, commercial agents, animal advocates, lawyers, biopharmaceutical representatives, physicians, citizen representatives, and other stakeholders on the ethical and social implications of developing and applying animal biotechnology in conjunction with the development of institutional guidelines, safeguards, and regulations (both nationally and internationally) to steer the course of the technology. At this time, broad public discourse to ensure that science and societal values remain aligned is a "serious political deficiency" *(48)*.

Institutional concerns also interface with the issues of food safety, the environment, animal well-being concerns, and biomedical concerns. Governments, with the aid of scientists and acting on behalf of their citizens, should stay on top of the technological advancements (so that they can report to their citizens). Governments should help improve the knowledge base so that citizens can weigh in on the risks and benefits for themselves, provide mechanisms by which citizens can act in ways commensurate with their convictions (i.e., labeling of altered foods), and monitor partnerships between publicly funded academic institutions and commercial industries (to forestall improprieties and conflicts of interest).

3.5. Biomedical Concerns

Animal biotechnology may also be applied to improve human disease resistance, as treatment alternatives, and to help offset the shortfall of human tissues and organs. However, biomedical uses of transgenic animals may pose trade-offs between benefits for individual patients and potential deleterious effects for society at large. At present, most projects involving animal biotechnology for biomedical purposes fall into one of three main categories:

1. The use of organs, live cells, and tissues for cross-species transfer or xenotransplantation
2. The production of biopharmaceuticals for human beings and animals
3. The creation and use of raw genetic materials for engineering other products

The relationship between biomedical concerns and increasing biotechnology to study gene function is not discussed in this chapter. The focus primarily is on xenotransplantation, but the issues discussed may be germane to categories 2 and 3 as well.

Xenotransplantation involves the transfer of tissues, living cells, and organs from one animal species to another. The potential for animal-to-human transplantation promises to increase the supply of viable organs, including lungs, kidneys, livers, pancreases, and whole hearts for human recipients. Tissue research promises bone transplants, skin grafts, and corneal transplants for accident, burn, and optical patients, respectively. Patients with diabetes, Parkinson's disease, Alzheimer's disease, or other diseases may have added hope of viable treatments through the fruitful xenotransfer of living cells from animals. Although promising, xenotransplantation has been questioned with respect to the merits and proficiency of the science and technology, the depth of the ethical discourse, and the absence of much needed national and global guidelines and regulations.

Some central concerns and questions associated with xenotransplantation include the following:

1. Personal health risks caused by immune rejection and infection despite the use of immunosuppressive drugs.
2. Potential spread of novel infectious diseases or viruses (i.e., xenozoonosis) from source animals to organ and tissue recipients and eventually to contact persons and to the public at large *(58,59)*. That is, unlike human-to-human transplants, using nonhuman donors leaves human recipients and their contacts susceptible to novel infections.
3. Legal issues. Individuals who currently participate in medical or clinical trials, must give their informed consent. Because of the threat of xenozoonoses, animal-to-human organ transplant recipients may be subject to invasion of privacy and be obligated to disclose personal information once protected under conventional patient–physician confidentiality statutes. These recipients may also be subjected to state-imposed restrictions on their right to self-determination, including life-long surveillance, quarantine, restricted travel (including prohibition to enter countries that forbid animal-to-human transplants), prohibition against procreation, or ban against blood, plasma, and organ donation. Recipients may also have to disclose their sexual partners and frequent social contacts and agree to mandatory postmortem examination. They may not, as most research subjects can, opt out of clinical trials *(60,61)*.
4. Business ethical issues. Animal biotechnology has the potential to be very lucrative if the prospects can be actualized. Should companies driven by profit and answerable to shareholders be mandated to exchange sensitive information and data to help the public weigh in on the costs and potential conflicts of interest? Are there appropriate international regulations and guidelines to regulate the business environment? Should governments allow clinical trials before or despite social consensus about the risks and how to manage them *(59,60)*?
5. Public health cost issues. In the event that transspecies transplants become accepted as standard medical practice, should public funds be committed for preclinical trials, clinical trials, postclinical screening and monitoring, animal care, and slaughter and disposal of the carcasses *(60,61)*? Who is responsible for the financial outlay for large-scale and long-term surveillance of organ and tissue recipients, close contacts, and their possible quarantine *(60,61)*? What are the obligations of national funding agencies, health care financing institutions, insurance companies and health maintenance organizations and pharmaceutical and biotechnology companies *(61,62)*?

Although xenotransplantation promises many things, undue attention to it as the ultimate medical elixir (or without substantial guarantees) may divert funding from equally viable alternatives that may also be less controversial. Alternatives that are less divisive or risky, such as making human donation more attractive and efficient through a more concerted effort to seek out and distribute organs; adoption of a "presumed consent" policy of organ donation on death; investment in preventive measures that encourage healthy diets, appropriate exercise, and lower consumption of "vices" like alcohol and tobacco; the plausibility of human stem cell research; and mechanical and artificial gadgetry, for example, should not be ignored.

3.6. Other Social Issues

There are other ethical and policy issues to consider that do not fall neatly into the above categories. They include concerns over (1) distributive justice; (2) implications of animal biotechnology on what it means to be human; and (3) colliding sensibilities about animals and our responsibilities toward them. Although the decision point for addressing these issues is far from the working scientist's laboratory, researchers should have a basic understanding of them nonetheless.

3.6.1. Animal Biotechnology and Distributive Justice

Critics have alleged that application of novel animal biotechnology and precipitating changes to intellectual property rights may (absent regulation and public consensus) support the economic interests of industry giants such as pharmaceutical companies, agribusiness companies, and well-capitalized businesses to the detriment of smaller producers and rural communities (here and in the developing world), as well as have a negative effect on consumer choice. In the case of agriculture, for example, novel animal biotechnology that is too expensive for all levels of producers to adopt would give larger, well-capitalized agribusinesses a greater economic advantage over poorer ones (especially during the short run of transformation to the new technology). The development and marketing of this technology will likely be targeted toward producers in the former group; as a result, smaller producers who cannot compete or move quickly enough to adopt these new innovations may end up victims of bankruptcy. It is likely that the communities in which smaller producers are located may face irrevocable structural changes to their futures and way of life *(63)*.

3.6.2. Implications for Human Cultural Identity

Is animal biotechnology a good way to advance the quality of life? Even if the technology turns out to be safe, should it be pursued, especially if crossing the species barrier may have an adverse impact on traditionally cherished values of what it means to be human? What are the limits to interfering with "natural" species boundaries?

On November 13, 2002, under the auspices of Rockefeller University and the New York Academy of Sciences, a panel of North American experts convened to discuss the morality and science of injecting human embryonic stem cells into an early mouse embryo (a blastocyst) to test the potential of special stem cells to help fight disease. By producing this "embryonic chimera," scientists hoped to learn whether human stem cells (the kind that have the ability to grow into just about every tissue type) can contribute to the development of tissues within the mouse embryo *(64)*. Such research on human embryos is presently taboo.

In this particular instance, the panel focused primarily on questions related to the sanctity of the human genome and on how best to approach the subject of weighting benefits to individual patients against traditional conceptions of humanity. For example, the panel debated the nature of this chimera. That is, to what extent is this creature still a mouse if it produces human sperm or if its brain is made up mainly of human cells?

They also debated whether and just how this sort of experiment varied from previous experiments involving the insertion of human genes into other nonhuman animals, for example, into pigs to reduce organ rejection during xenotransplantation. They were also concerned whether ethically charged experiments like the present one would create public backlash to more mainstream forms of stem cell research.

3.6.3. Sensibilities About Animals

The perception of the moral status of animals is not uniform, and the instrumental view of animals as research subjects, genetic commodities to be manipulated and modified for human purposes held by most working in animal biotechnology collides with the view of animals as beings deserving of respect and sympathy and requiring good care and appropriate husbandry. The research community, entrenched in a utility view of animals, should be sensitive to contrasting views of animals as having value and integrity in their own right. Keener attention to the emotional and cultural significance of animals to the public as well as acknowledgment of the religious aspects of animal use and consumption are encouraged. In the latter case, scientists and regulatory bodies and funding agencies should be sensitive to just what forms of animal uses are taboo, socially sanctioned today, or sacrilegious and ensure that these animals and their products or relevant genetic material are not present in the food supply or in commercial products without informing the public.

So science and policymakers do not run with the ethically charged technology ungoverned, public discussion of the above issues is needed. Through public education, positive regulation and legislative prohibitions may be generated that do not stifle truly beneficial research and product development. As discussed in the next section, the social consequences of novel technology can be minimized by allowing the public (including producers) to debate on how technology should serve the public interests and on issues of just desert and fair commerce and resource distribution. The need for a well-functioning political mechanism for deliberating these issues is urgent. Regulation on the basis of social issues borne out of animal biotechnology should also be considered to ensure that policies are in step with the values and concerns of the nonscientific community.

4. Animal Biotechnology and the Role-Defined Responsibilities of Science

If the many promised societal benefits of animal biotechnology are to come to fruition, then there must be public acceptance both in the direction in which it is progressing and of products of animal biotechnology. Although better public understanding of the risks and benefits associated with animal biotechnology will go a long way in engendering such public acceptance, this is but one facet of what must be undertaken by the scientific community to promote animal biotechnology. The other facet (which is often taken for granted with technological innovation) concerns public perception of the trustworthiness and credibility of scientists or proponents of the new technology *(57,65)*. If animal biotechnology is to gain a foothold and flourish in the future, scientists should also be prepared to devote their time and energy to securing the confidence of the public in the scientific community itself as well as in their initiatives.

As the public becomes more scientifically literate, conscientious with respect to the concerns raised in Section 3, and suspicious of conflicts of interest that may arise from research bankrolled by industry, the burden of proof rests with scientists. They must demonstrate that the increasing industry-driven science has not compromised the integrity and objectivity of the trusted professional *(66)* and prove that their technological innovation is safe and socially acceptable (or at least morally indifferent) and that these advances are undertaken in the interest of the public good.

In the case of animal biotechnology, rapid expansion and increasing specialization encourage knowledge gaps between the scientific community and the nonscientific public, which in turn places the public in a position of vulnerability. This vulnerability is amplified because of the inability to foresee dangerous consequences and unwanted outcomes associated with animal biotechnology. Relative to the lay public, scientists as "experts" or specialists are perceived as powerful authorities with decision-making capacity. As such, the responsibility of managing knowledge on behalf of the public good has been deferred to the technically skilled scientific community. But, as history has demonstrated many times, power inequilibrium can often lead to the exclusion and disenfranchisement of vulnerable parties by those in positions of privilege. The public's trust in the leadership of the scientific community should not be taken lightly by the scientific community.

In cultivating the public's trust and avoiding public suspicion, it is thus important that scientists recognize this social inequilibrium and their positions of privilege and make every effort not to let their positions of expertise detract from their public responsibilities. Instead, scientists as experts have a professional responsibility to reassure the public that animal biotechnology will be developed and applied in morally justifiable and socially responsible ways. This responsibility enjoins scientists to recognize the dignity and autonomy of others and to take measures to neutralize feelings of vulnerability by the public. Scientists can do this by being cognizant of the interests and values of others and by being aware that they are one part of a trust relationship.

The responsibilities associated with cultivating public trust encompass both personal and professional ethics. In general terms, they include establishing a climate of "participatory science," that is, of transparency and open communication, accountability, restraint, and leadership. Participatory science establishes a social component for scientists and serves to neutralize inequality among the various stakeholders by encouraging bridge building and partnership and by aspiring toward shared goals.

Presently, science, apart from being a field of study valuable in itself or aspiring to improve the quality of human life through problem solving and technological innovation, has also acquired advisory and regulatory roles in society *(67)*. In this last capacity, science and scientists are looked on to help people make good decisions as they aspire to live well. The voice of science has been an integral part in steering humankind along the path of a good life. Hence, the pursuit of knowledge and technological advancement is not a disinterested matter; it is not simply about excursions into curiosity for their own sake. Instead, knowledge and technological advancement are imbued with social, ethical, political, economic, or religions meaning. As such, scientists should take it on themselves to be cognizant of the ethical and social implications of their research and strive to understand better the various dimensions of risks and harms associated with the development and application of novel technology.

When it comes to questions of risk, science typically focuses on unwanted consequences, the probability of the actualization of various harmful scenarios, a valuation of their respective levels of harm, and how to mitigate the harm *(65,68)*. For the public, risk is inextricably tied to trust and is not limited to consequences and statistical probability that harm will result from a given practice or technology or from the possibility of mistakes. The public views risk in terms of anxiety, vulnerability, and feelings of security and well-being *(65,69–71)*.

For the public, the element of risk also encompasses how they feel about their lack of control over the speed and direction of biotechnological development, unfamiliarity with the nature of the science and technology, fear and doubt in the ability of scientists and governments to respond adequately to a possible biotechnological calamity, and a lack of confidence in and suspicion of scientific authority. Disinterestedness by scientists in addressing larger ethical and social implications of their work, a nonchalant scientific attitude about changing the natures of animals, professional arrogance, undisclosed affiliations and commercial bankrolling, and a lack of identification with broad public values are also factors that account for how the public perceives it is at risk with respect to animal biotechnology *(70)*.

Hence, effective risk management includes sensitivity to how the public feels about and assesses risk and according responses to the public's vulnerability. A scientific community that engages the public in genuine reflection, commitment, and open communication will be seen as trustworthy and credible advisors. By including the public in open and conscientious deliberation (i.e., addressing their questions and fears thoroughly about the purposes and justification for applying and developing these new technologies and their attendant problems and prospects), scientists demonstrate their willingness to take seriously the public's ethical and social concerns *(48)*. Furthermore, forming these partnerships will help ensure that discussion related to the concerns raised in the Section 3 will be robustly discussed and debated to promote responsible development and application of animal biotechnology.

As mentioned, scientists working in the development of animal biotechnology are in prime position to anticipate unwanted consequences and to calculate degrees and probability of harm. This special expertise gives rise to leadership responsibilities that include educating the public so that they are reasonably science literate and well informed about the substantial issues. Scientists should provide the public with relevant information and alternatives or contingencies so the public may make considered decisions commensurate with their values and interests *(72)*. Leadership and accountability also involve self-reflection and cognizance of how scientists as a group understand themselves and their responsibilities and respond to challenges and public criticisms *(73)*. Thompson *(74)* suggested a few things that scientists can do to cultivate good habits and meet the goals of transparency, better public communication, restraint, and leadership. We delineate this list with some additions. Scientists should do the following:

1. Participate in citizen conferences regularly (both formally and informally), that is, be concerned about the issues about which the public is also concerned, consider the interests of other stakeholders at the table, and engage the public on the ethical impacts of scientific advances.
2. Engage critics with the same care as if writing journal articles and encourage peer-reviewed articles on ethical and social implications by scientists on their work and be prepared to show compelling reasons for the procedures that are adopted and for wanting to move forward.

3. Make explicit how animals are impacted by the different forms of genetic engineering and how this technology will influence the shape of human genetic engineering and cloning *(54)*.
4. Develop teaching curricula, including courses or degrees in bioethics in universities and high schools, which have ethics components and raise questions about the social roles and values assumptions of the scientific community, especially highlighting conflicts of interest and professional responsibility.
5. Encourage graduate students to take classes in ethics and public policy and to discuss the social implications of their work.
6. Make use of ethics centers and other public assets and mechanisms for discourse and critical exchange or include ethicists in research groups.
7. Make academic and commercial affiliations transparent, that is, disclose research sponsorships and industry collaborations, especially if partially supported by public monies *(66)*
8. Show self-criticism and reflectivity by demonstrating how professional ethical issues are attended, debated, reviewed, and awarded by the scientific community.
9. Ensure that no professional backlash befalls those in the scientific community who: (1) are willing to engage in ethical self-review of the community; (2) are critical of the direction that science may be progressing; and (3) are willing to expose the values and deficiencies of the scientific method.

5. Conclusion

Human beings have long depended on animals for help and as resources for different purposes and have a notable, albeit erratic at times, history of responsible use and care of these animals. Today, modern genetically based animal biotechnology offers new opportunities to employ the services of animals, but it also challenges us to revisit our responsibilities to both human beings and animals alike.

Animal biotechnology, for all its prospects, is beset with moral concerns that may or may not be surmountable. Because it involves intervening in the lives of others and may have unforeseeable and radical consequences, it is therefore urgent that the scientific community engage the public to be forthright about their responsibilities and to determine the risks and limitations of the applications in virtue of respect for the relevant stakeholders so that what is beneficial may not also be detrimental. Scientists alone should not be left to decide the direction and means of technological progress. Instead, both the scientific and nonscientific communities should come together to build a public science agenda as a way to anticipate and preclude corporate control of science or unscrupulous and socially unfettered individual research ambition.

Animal biotechnology has important implications for the nature of human relationship with animals, the environment, food safety, biomedical safety, distributive justice, and what it means to be human. Serious and thorough debate and dialogue among scientists, government and commercial agents, and the general public to lay the ethical foundation for the development and use of animal biotechnology are urgently needed before genetically modified organisms or cloned animals make their appearance on farms, in grocery stores, or in the environment. Such discourse is necessary to forestall unnecessary societal suspicion or prejudice that has befallen entry of genetically modified crops and foods of plant origin into the food supply. A vibrant public discourse will ensure that animal biotechnology will be developed and used consistent with the shared vision of responsible living and for the benefit of advancing the quality of life for both human beings and animals and not simply for economic or discriminatory gain of a few.

With any new technology, the public's attitude toward the research community is the key to its success. Responsible animal biotechnology development and application will be sanctioned by the public when scientists improve their image, have direct relationships with the public, and are keenly aware of their own assumptions and values. Responsible animal biotechnology enjoins scientists to consider the interests of the public, be aware of their leadership and advisory roles, and understand the implications of their work.

References

1. Guelph transgenic pig research program. Available on-line at: http:// www.uoguelph.ca/enviropig/. Accessed October 1, 2002.
2. Communications and Public Affairs. (2001) U of G's "Enviropig" a Success, New Study Reveals. July 31. Available on-line at: http://www.uoguelph.ca/mediarel/archives/000404.html. Accessed October 1, 2002.
3. Shapiro, H. T. (2000) Federal policy making for biotechnology, Executive Branch, National Bioethics Advisory Commission. In *Encyclopedia of Ethical, Legal, and Policy Issues in Biotechnology* . (Murray, T. H. and Mehlman, M. J., eds.). John Wiley and Sons, New York, pp. 241–257.
4. Wagner, W. E. (2000) Federal policy making for biotechnology, Congress, EPA. In *Encyclopedia of Ethical, Legal, and Policy Issues in Biotechnology*. (Murray, T. H. and Mehlman, M. J., eds.). John Wiley and Sons, New York, pp. 227–234.
5. Angelo, M. J. (2000) Agricultural biotechnology, law and EPA regulation. In *Encyclopedia of Ethical, Legal, and Policy Issues in Biotechnology* .(Murray, T. H. and Mehlman, M. J., Eds.). John Wiley and Sons, New York, pp. 26–37.
6. Miller, H. I. (2000) Agricultural biotechnology, law, and food biotechnology regulation. In *Encyclopedia of Ethical, Legal, and Policy Issues in Biotechnology*. (Murray, T. H. and Mehlman, M. J., Eds.). John Wiley and Sons, New York, pp. 37–46.
7. Browne, W. P. (2000) Agricultural biotechnology, law, and social impacts of agricultural biotechnology. In *Encyclopedia of Ethical, Legal, and Policy Issues in Biotechnology* .(Murray, T. H. and Mehlman, M. J., eds.). John Wiley and Sons, New York, pp. 46–56.
8. Tourine-Moulin, F., Vicari, M., and Ruano, M. (2000) International aspects, national profiles, France. In *Encyclopedia of Ethical, Legal, and Policy Issues in Biotechnology* . (Murray, T. H. and Mehlman, M. J., eds.). John Wiley and Sons, New York, pp. 703–711.
9. Kettner, M. (2000) International aspects, national profiles, Germany. In *Encyclopedia of Ethical, Legal, and Policy Issues in Biotechnology* . (Murray, T. H. and Mehlman, M. J., eds.). John Wiley and Sons, New York, pp. 711–722.
10. Macer, D. R. J. (2000) International aspects, national profiles, Japan. In *Encyclopedia of Ethical, Legal, and Policy Issues in Biotechnology* . (Murray, T. H. and Mehlman, M. J., eds.). John Wiley and Sons, New York, pp. 722–731.
11. Hasson, M. (2000) International aspects, national profiles, Scandinavia. In *Encyclopedia of Ethical, Legal, and Policy Issues in Biotechnology* . (Murray, T. H. and Mehlman, M. J., eds.). John Wiley and Sons, New York, pp. 731–739.
12. Bieri, F., Ghisalba, O., Kappeli, O., and Reutimann, H. (2000) International aspects, national profiles, Switzerland. In *Encyclopedia of Ethical, Legal, and Policy Issues in Biotechnology* . (Murray, T. H. and Mehlman, M. J., eds.). John Wiley and Sons, New York, pp. 739–747.
13. Ashroft, R.E., Capps, B., and Huxtable, R. (2000) International aspects, national profiles, United Kingdom. In *Encyclopedia of Ethical, Legal, and Policy Issues in Biotechnology* . (Murray, T. H. and Mehlman, M. J., eds.). John Wiley and Sons, New York, pp. 747–761.

14. National Research Council. (2000) *Incorporating Science, Economics, and Sociology in Developing Sanitary and Phytosanitary Standards in International Trade. Report by the Board on Agriculture and Natural Resources.* National Academy Press, Washington, DC, pp. 238–245.
15. National Research Council. (2002) *Animals Biotechnology: Identifying Science-Based Concerns. Report Prepared by the Committee on Defining Science-Based Concerns Associated With Products of Animal Biotechnology.* National Academy Press, Washington, DC.
16. Hoy, M. A. (2001) Transgenic arthropods for pest management programs: risks and realities. *Exp. Appl. Acarol.* **24**, 463–495.
17. Braig, H. R. and G. Yan. (2001) The spread of genetic constructs in natural insect populations. In *Genetically Engineered Organisms: Assessing Environmental and Human Health Effects* . (Letourneau, D. K. and Burrows, B. E., eds.). CRC, Boca Raton, FL, pp. 251–314.
18. Hallerman, E. M. and Kapuscinski, A. R. (1995) Incorporating risk assessment and risk management into public policies on genetically modified finfish and shellfish. *Aquaculture* **137**, 9–17.
19. Muir, W. M. and Howard, R. D. (2001) Methods to assess ecological risks of transgenic fish releases. In *Genetically Engineered Organisms: Assessing Environmental and Human Health Effects* . (Letourneau, D. K. and Burrows, B. E., eds.). CRC, Boca Raton, FL, pp. 355–383.
20. National Research Council. (2002) *The Environmental Effects of Transgenic Plants.* National Academy Press, Washington, DC.
21. Ellstrand, N. C., Prentice, H. C., and Hancock, J. F. (1999) Gene flow and introgression from domesticated plants into their wild relatives. *Annu. Rev. Ecol. Syst.* **30**, 539–563.
22. Kapuscinski, A. R. (2002) Controversies in designing useful ecological assessments of genetically engineered organisms. In *Genetically Engineered Organisms: Assessing Environmental and Human Health Effects* . (Letourneau, D. K. and Burrows, B. E., eds.). CRC, Boca Raton, FL, pp. 385–416.
23. Anonymous. (2000) Swapping science for consensus in Montreal. *Nat. Biotechnol.* **18**, 239.
24. Miller, H. I. and Conko, G. (2001) Precaution without principle. *Nat. Biotechnol.* **19**, 202–203.
25. Bentham, J. (1789/1948) *An Introduction to the Principles of Morals and Legislation.* Hafner, New York.
26. Saner, M. A. (2000) Ethics as problem and ethics as a solution. *Int. J. Biotechnol.* **2**, 219–256.
27. Magnus, D. and Caplan, A. (2002) Food for thought: the primacy of the moral in the GMO debate. In *Genetically Modified Foods: Debating Biotechnology* . (Ruse, M. and Castle, D., eds.). Prometheus Books, Amherst, NY, pp. 80–87.
28. M'Gonigle, R. M. (1998) The political economy of precaution. In *Protecting Public Health and the Environment: Implementing the Precautionary Principle* . (Raffensparger, C. and Ticknor, J., eds.). Island, Washington, DC, pp. 123–147.
29. Singer, P. (1990) *Animal Liberation*, rev. ed. Avon Books, New York.
30. Singer, P. (1979) *Practical Ethics.* Cambridge University Press, Cambridge, UK.
31. Singer, P. (1975) *Animal Liberation.* Avon Books, New York.
32. Regan, T. (2001) *Defending Animal Rights.* University of Illinois Press, Champaign-Urbana.
33. Regan, T. (1983) *The Case for Animal Rights.* University of California Press, Berkeley.
34. Midgley, M. (1983) *Animals and Why They Matter.* University of Georgia Press, Athens.
35. Rollin, B. E. (1998) On Telos and genetic engineering. In *Animal Biotechnology and Ethics* . (Holland, A. and Johnson, A., es.). Chapman and Hall, London, pp. 156–171.
36. Rollin, B. E. (1995) *The Frankenstein Syndrome: Ethical and Social Issues in Genetic Engineering of Animals.* Cambridge University Press, New York.
37. Rutgers, B. and Heeger, R. (1999) Inherent worth and respect for animal integrity. In *Recognizing the Intrinsic Value of Animals, Beyond Animal Welfare* . (Dol, M., Fentener van Vlissingen, M., Kasanmoentalib, S., Visser, T., and Zwart, H., eds.). Van Gorcum, Assen, The Netherlands, pp. 41–51.

38. Fox, M. W. (1990) Transgenic animals: ethical and animal welfare concerns. In *The Bio-Revolution: Cornucopia or Pandora's Box.* (Wheale, P. and McNally, R., eds.). Pluto, London, pp. 31–54.
39. Fraser, D., Weary, D., Pajor, E. A., and Milligan, B. N. (1997) Scientific conceptions of animal welfare that reflect ethical concerns. *Animal Welfare* **6**, 187–205.
40. Orlans, B. (2000) Research on animal, law, legislative, and welfare issues in the use of animals for genetic engineering and xenotransplantation. In *Encyclopedia of Ethical, Legal, and Policy Issues in Biotechnology.* (Murray, T. H. and Mehlman, M. J., eds.). John Wiley and Sons, New York, pp. 1020–1030.
41. Griffin, H. D. (2003) Dolly's Death Update. February 17. Available on-line at: http://www.roslin.ac.uk/news/press/articles/175.html. Accessed April 1, 2003.
42. Griffin, H. D. (2003) We are sorry to report that Dolly the sheep is dead. February 14. Available on-line at: http://www.roslin.ac.uk/news/press/articles/174.html. Accessed April 1, 2003.
43. Wilmut, I., Schnieke, A. E., McWhir, J., Kind, A. J., and Campbell, K. H. S. (1997) Viable offspring derived from foetal and adult mammalian cells. *Nature* **385**, 810–813.
44. Varner, G. (1999) Should you clone your dog? An animal rights perspective on somacloning. *Animal Welfare* **8**, 407–420.
45. Sideris, L., McCarthy, C., and Smith, D. H. (1999) Roots of concern with nonhuman animals in biomedical ethics. *Inst. Lab. Anim. Res. J.* **40**, 1–14.
46. Comstock, G. (1988) The case against BST. *Agric. Hum. Values* **5**, 36–52.
47. Krimsky, S. and Wrubel, R. (1996) *Agricultural Biotechnology and the Environment: Science, Policy and Social Issues.* University of Illinois Press, Urbana.
48. Thompson, P. B. (1998) Biotechnology policy: four ethical problem and three political solutions. In *Animal Biotechnology and Ethics.* (Holland, A. and Johnson, A., eds.). Chapman and Hall, London, pp. 243–262.
49. Bovenkerk, B., Brom, F. W. A., and Van Den Bergh, D. J. (2002) Brave new birds: the use of animal integrity in animal ethics. *Hastings Cent. Rep.* **32**, 16–22.
50. Thompson, P. B. (1997) Ethics and the genetic engineering of food animals. *J. Agric. Environ. Ethics* **10**, 1–23.
51. Russell, W. M. S. and Burch, R. L. (1959) *The Principles of Humane Experimental Technique.* Metheun, London.
52. De Cock Buning, T. (2000) Genetic engineering: is there a moral issue? In *Progress in the Reduction, Refinement and Replacement of Animal Experimentation.* (Balls, M., van Zeller, A.-M., and Halder, M. E., eds.). Elsevier Science, Amsterdam, The Netherlands, pp. 1457–1464.
53. Costa, P. (2000) Welfare aspects of transgenesis. In *Progress in the Reduction, Refinement and Replacement of Animal Experimentation.* (Balls, M., van Zeller, A.-M., and Halder, M. E., eds.). Elsevier Science, Amsterdam, The Netherlands, pp. 1719–1723.
54. Thompson, P. B. (1999) Ethical issues in livestock cloning. *J. Agric. Environ. Ethics* **11**, 197–217.
55. Rollin, B. E. (1997) Send in the clones ... Don't bother, they're here! *J. Agric. Environ. Ethics* **10**, 25–40.
56. Jackson, D. (2000) Labeling products of biotechnology: towards communication and consent. *J. Agric. Environ. Ethics* **12**, 319–330.
57. Thompson, P. B. (1997) Food biotecnology's challenge to cultural integrity and individual consent. *Hastings Cent. Rep.* **27**(4), 34–38.
58. American Medical Association Council on Ethical and Judicial Affairs. (1994) *Financial Incentives for Organ Procurement: Ethical Aspects of Future Contracts for Cadaveric Donors.* American Medical Association, Chicago.

59. Daar, A. S. (1997) Ethics of xenotransplantations: animal issues, consent and likely transformation of transplant ethics. *World J. Surg.* **21**, 975–982.
60. Nuffield Council on Bioethics. (1996) *Animal to Human Transplants: The Ethics of Xenotransplantation*. Nuffiled Council on Bioethics, London.
61. Bach, F. H., Fishman, J. A., Daniels, N., et al. (1998) Uncertainty in xenotransplantation: individual benefit vs collective risk. *Nat. Med.* **4**, 141–144.
62. Morris, P. J. (1997) Pig transplants postponed [editorial]. *BMJ* **314**, 242.
63. Comstock, G. (2000) *Vexing Nature? On the Ethical Case Against Agricultural Biotechnology*. Kluwer, Boston, MA.
64. Abraham, C. (2002) Scientists look at creating a human-mouse embryo. *Globe and Mail* November 28. Available on-line at: http://www.globetechnology.com/servlet/ArticleNews/tech/RTGAM/20021128/wxmous1128/Technology/techBN/. Accessed November 28, 2002.
65. Thompson, P. B. (1997) *Food Biotechnology in Ethical Perspective*. Chapman and Hall, London.
66. Sharpe, V. A. (2002) Science, bioethics and the public interest: on the need for transparency. *Hastings Cent. Rep.* **32**, 23–26.
67. National Science Foundation. (1994) *Science, Technology and Democracy: Research Issues on Governance and Change*. National Science Foundation Workshop, Washington DC.
68. Stich, S. (1978) The recombinant DNA debate. Philosophy and Public Affairs. **7(3)**, 187–205. Reprinted in Ruse, M., ed. (1989) *Philosophy of Biology*. Macmillan, New York, pp. 229–243.
69. Stern, P. C. and Fineberg, H. V. (Eds.). (1996) *Understanding Risk: Informing Decisions in a Democratic Society*. Washington, DC.
70. Thompson, PB. (1995) Risk and responsibility in modern agriculture. In *Issues on Agricultural Biotechnology*. (Mepham, T. B, Tucker, G. A., and Wiseman, J., eds.). Nottingham University Press, Nottingham, UK, pp. 31–45.
71. Thompson, P. B. and Dean, W. (1996) Competing conceptions of risk. *Risk Health Safety Environ.* **7**, 361–384.
72. Ziman, J. (1992) Not knowing, needing to know, and wanting to know. In *When Science Meets the Public*. (Lewenstein, B. E., ed.). American Association for the Advancement of Science, Washington, DC, pp. 13–20.
73. Kunkel, H. O. and Hagevoort, G. R. (1994) Construction of science for animal agriculture. *J. Anim. Sci.* **72**, 247–253.
74. Thompson, P. B. (1999) From a philosopher's perspective, how should animal scientists meet the challenge of contentious issues? *J. Anim. Sci.* **77**, 372–377.

7

Transgenic Aquatic Animals

Sarad R. Parekh

1. Introduction

Aquatic organisms are valuable sources of food and other useful materials; they have been exploited by humans for a long time *(1,2)*. However, recently more focus has been placed on (1) farming useful species through aquaculture and (2) screening marine invertebrates as potential resources of complex bioactive compounds for application in therapeutics and agriculture *(3)*. Although in the mid-1980s the introduction of recombinant DNA fueled rapid advances in biotechnology, interest in applying biotechnology to opportunities in marine organisms has been slow. This may be because of the scientific community's limited knowledge about marine molecular biology and its knowledge about gene expression, reproduction, growth regulation, and defenses against diseases. In addition, there are available only a few reliable invertebrate models for biomedical applications in marine organisms.

Aquatic genetically modified organisms (GMOs) are those with stable modification of their deoxyribonucleic acid (DNA), either because of the insertion of exogenous genes or because of the manipulation of an endogenous DNA sequence. For these modifications to be inherited by the offspring, the DNA of interest must be introduced as a stage of the germline *(3)*. In plants, heterologous DNA (transgene) has been artificially introduced and integrated into the cells by infection with *Agrobacterium tumefacieance* or by physical means such as ballistic bombardment.

To bioengineer aquatic invertebrates, the first step is to choose the appropriate aquatic species *(4)*. Second, a specific transgene must be prepared so that it has the right orientation of structural genes for temporal, spatial development *(2)*. The gene segment is introduced into the fertilized egg by injection, and the injected embryo is incubated in vitro, implanted into embryonic tissue before the germ cells segregate during embryogenesis, or inserted into the germline itself. In most cases, with these techniques, multiple copies of transgenes are integrated at random sites in the genome of the transgene individuals. If the transgene is linked with the functional promoters, expression of the transgene as well as display of the phenotype is expected in some of the transgene individuals. The transgene events are transmitted further through the germlines into subsequent generations.

From: *The GMO Handbook: Genetically Modified Animals, Microbes, and Plants in Biotechnology*
Edited by: S. R. Parekh © Humana Press Inc., Totowa, NJ

The success of these methods typically depends on the particular features of the aquatic invertebrates, the method of isolation, and screening of the individual with transgene *(3)*. In applied biotechnology, aquatic GMOs offer unique opportunities for improving the genetic brood stock for aquaculture and then employing them as "engines" for producing valuable proteins for industrial and pharmaceutical applications. In this chapter, selected technologies and exploitation of aquatic GMOs for biotechnology applications are addressed.

Global harvest of aquatic and marine-derived fishery products depends on the surrounding populations of shellfish, crustaceans, and finfish and the availability of freshwater. In recent times, the worldwide harvest of fish products has surpassed more than 150 million tons *(5)*. To cope with the global demand for fish and aquaculture products and with the escalating price of fish-derived products, several countries have turned to aquaculture and are breeding fish in contained environments. Since the 1990s, the worldwide production of maroalgea, shellfish, and finfish has increasing steadily at a rate of 8% *(6)*.

To meet the future market demands, use of transgenic technology will be essential. To sustain continued growth in the aquaculture industry, newly developed technologies in molecular biology and transgenics need to be integrated and increasingly applied. These technologies need to be targeted toward enhancing growth rates, controlling reproductive cycles, producing new vaccines, developing better feed composition, resisting diseases, developing elite brood stocks, and manipulating growth hormones (GHs) and growth factor genes. These points are reviewed in this chapter *(5)*.

2. Technologies and Methods for Deriving GM Invertebrates

A transgene (a gene used for modifying the organism) employed in producing transgenic aquatic invertebrates should be a recombinant gene construct that expresses the gene product at an appropriate level in the desired time and in the specific tissue *(5,6)*. The prototypical transgene is constructed in a plasmid and harbors appropriate promoter-enhancement segments of DNA and a structural gene sequence.

Depending on the GMO application, the transgene is grouped into one of three types of use: (1) improvement of gene function; (2) identification and reporter function; or (3) loss of gene activity. The gain-of-function transgene is designed to have an additive effect and enhance functions in the transgenic individual or assist in identification of transgene events when the genes are expressed properly in GMOs. Transgenes containing the structural gene derived from a foreign source or aquatic growth hormone (GH) or their DNA fused to functional promoters, such as fish β-actin gene promoters, are examples of gain-of-function transgene constructs. Expression of GH in transgenetic invertebrates increases production of GHs and ultimately the growth of the GMO *(7,8)*. Bacterial β-galactosidase or luciferase gene fused with a functional promoter is a typical example of a transgene with reporter or identification function for successful modification. A more important function of the reporter gene is to assess and identify the strength of a promoter–enhancer DNA segment. An example is the structural gene of β-galactosidase or luciferase fused to the promoter element *(7)*. Following the successful construction of GMO and gene transfer, the final outcome of the reporter gene activity in the invertebrates is employed to estimate the transcription regulatory sequence of the gene or the potency of an inserted promoter *(7)*.

The loss-of-function transgene is constructed and applied in GMOs to disrupt expression of the host gene. These genes may interfere with the posttranscriptional process or translations of endogenous messenger ribonucleic acid (RNA). In a few cases, transgenes are inserted that may encode a ribozyme (a catalytic RNA) that can prematurely cleave specific mRNA and thereby arrest production of normal gene output. Such a transgene may find utility in producing disease-resistant transgenic brood stocks for invertebrates and aquatic GMO models for basic research *(9)*.

3. Selection of Target Species

Transgenetic studies have been executed successfully in several invertebrates and aquatic species, including goldfish, catfish, tilapia, crabs, lobsters, clams, and shrimp *(5)*. Depending on the target transgene species, some species are more easily suited to genetic manipulation studies than others. For example, Japanese medaka and zebra fish have short life cycles, taking but a couple of months to go from hatchlings to mature adults that can produce hundreds of eggs. Because the Japanese medaka and zebra fish do not require a critical seasonal breeding cycle, they can be genetically manipulated in the laboratory *(10)*. Eggs from these species are relatively large, have thin semitransparent chorions, and offer an easy means for microinjection of DNA into unfertilized eggs. However, a big drawback is their relatively small body size, making them unsuitable in certain instances for some biochemical and endocrinological analyses.

Rainbow trout, salmon, and lobsters are commonly used for transgenic studies requiring models with a large body size *(3)*. In these species, the basic physiology, comparative reproductive biology, and application of an aquatic transgene are well worked out. However, the trade-off for their size is their long maturation time and the potential for only a single spawning cycle per year; both of which limit the speed of experimentation.

Goldfish, tilapia, and shrimp are the other group of suitable model aquatic invertebrates with body sizes and maturation cycles that allow easy transgenetic manipulation and assessment of progeny *(5)*. The one major drawback is the lack of well-defined physiology, genetic background, and known behavior patterns for these species.

4. Procedures and Protocols

Techniques such as direct microinjections, lipofections, electroporations, calcium phosphate precipitation, and viral infections have been used successfully to introduce foreign DNA into invertebrates. Among the various methods (*see* Fig. 1), direct microinjections and electroporations into freshly fertilized eggs offer reliable methods for aquatic invertebrates *(11)*.

4.1. Microinjections

Microinjection into eggs or embryos has been successfully used since 1985 for developing transgenic fish and other aquatic organisms *(9)*. The gene constructs employed for transgenic fisheries include human GH, *Escherichia coli* hygromycine resistance marker fused with a β-galactosidase gene. Described below is a more general procedure about how a transgenic aquatic invertebrate (fish) is derived through microinjection technology (Fig. 1).

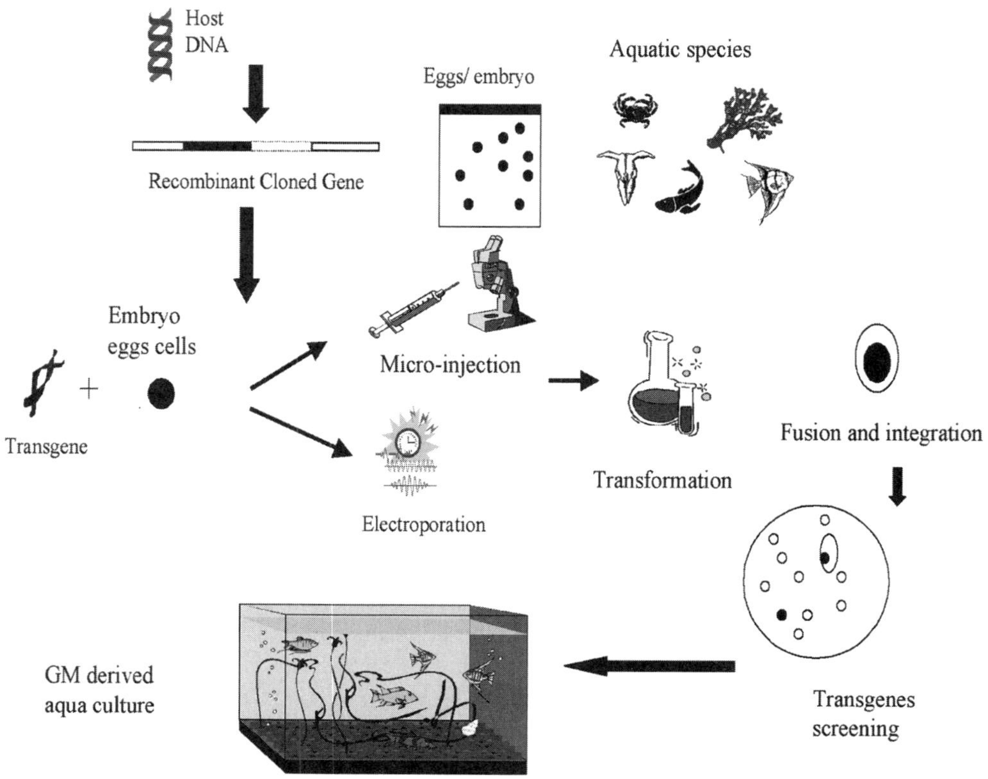

Fig. 1. Recombinant technology procedures for deriving aquatic GMOs.

1. Eggs and sperms are collected in separate containers.
2. Appropriate amounts of water and sperm are added to the eggs and then mixed to fertilize the eggs.
3. Fertilized eggs are microinjected postfertilization.
4. A microscopic needle and micropipette are used to inject a linearized transgene DNA (with or without plasmid) into the cytoplasm, maintaining careful sterility and avoiding puncturing the cells.

The embryos are incubated in water until they hatch. (For each species-specific strain, temperature and light are artificially induced.)

1. Success, viability, and recovery of transgenic events are species dependent, for example, fish embryos range from 30 to 80% percent; the rate of DNA integration ranges between 10 and 70%.
2. Tough chorions (outer membranes) of fertilized embryos in some species make microinjection difficult and are the root cause of low success rates.

In special instances, to facilitate and improve the frequency of transgene insertion via microinjections, microsurgery is used, a micro-opening is created through a micropyle, or a digestive enzyme is used to create pores on the chorion before injecting directly into the unfertilized eggs and further inhibiting the hardening of fertilized eggs treated with glutathione *(11)*.

4.2. Electroporation

The electroporation method has become popular *(10)* and has been more reliable for transferring transgene in invertebrates (fish). The electroporation principle relies in delivering a series of short electrical pulses to the cells to permeate the cell membrane and then facilitate the entry of external DNA molecules into the cells. Depending upon the species and methods the rate of DNA insertion and integration in electroporated embryos in certain cases is 20% or higher than other transformation procedures *(10)*.

Compared to microinjections, the rate of success may be equal and comparable. However, electroporation is far more rapid than microinjection in the time required to handle and manipulate the embryos. In addition, in some instances transgene and foreign gene can be successfully electroporated in the fish sperm instead of the embryos. It seems that electroporation offers a more versatile technology for transgene transfer in fish *(10)*.

4.3. Use of Retroviral Vectors for Transgene Transfer

In certain invertebrates, electroporation and microinjection have limitations for achieving successful transgene events. These limitations may be because of event sorting of individuals from mosaic germlines or possible delay in integration of the transgene. A gene transfer vector that is a defective pantropic retroviral vector has been employed to overcome these limitations. This vector contains a unique long terminal repeat sequence of the leukemia virus, and the transgene is typically packaged in the viral envelope, which then is successfully integrated into the phospholipid component of the host cell *(4,9)*. This retroviral vector has a broad range of hosts and has been used successfully as a vehicle to transfer transgenes at high efficiency into zebra fish or such marine invertebrates as dwarf surf clams *(5)*.

5. Transgene Characterization

5.1. Screening and Identification

Postinsertion of a foreign gene, the next step is the identification of the individuals from the population. Conventionally, high-throughput dot-blot or Southern blot hybridization of genomic DNA isolated from the presumed transgenic individuals identifies them. To handle a large sample volume and to assay it rapidly, polymerase chain reaction (PCR) amplification and assay of genomic DNA from tissue is adopted. Although this method does not illustrate whether the transgene is integrated stably into the host genome, it serves as a rapid and sensitive screening method for selecting transgenic individuals *(5,6)*.

5.2. Expression

The next step is to detect the expression of the transgene event. The success of this is depends on the level of expression of the transgenic product in the individuals. Efforts here focus on developing more sensitive and precise analytical procedures to raise the level of reliability. Several procedures are employed depending on the expression level, including RNA-Northern blot or dot-blot method, reverse transcriptase PCR, or immunoblotting assay *(8)*. Among these, reverse transcriptase PCR is the most sensitive and needs the least amount of sample.

5.3. Integration and Inheritance of Transgene

Following the success of the selection and expression, the integration of the transgene needs to be confirmed, especially the presence of multiple copies of the transgene; whether the orientation of the transgene is in the form of head to head, head to tail, or tail to head; and the site of the integration *(11)*.

Finally, it is critical to confirm that the transgene is stably integrated, that it is vertically transferred successfully to subsequent generations, and that an elite line has been established *(9)*. This final validation is typically done by backcrossing the first generation of the transgenic line with the nontransgenic individuals; then, the progeny are assayed for the loss or the continued presence of the transgene by the standard PCR process described above. A firm confirmation of the stable integration of transgene evaluates several generations of successive lines and the broad-range germline for the consistent expression and presence of transgene, thus ensuring that no delayed or dilution effect has occurred.

6. Biotechnological Application of Aquatic GMOs

Tremendous opportunities exist for the application of transgenics to vastly improve aquaculture production of marine invertebrates. However, as with other GMOs (microbes, plant, and animals), significant obstacles need to be overcome, especially concerns pertaining to containment of transgenic marine invertebrates *(5)*.

It is very difficult to ensure that the freshwater or seawater used to culture the transgenic aquaculture will be free of transgenic larvae or gametes when returned to the environment, and that these waters will not accidentally release transgenic organisms to the environment; thus, it is hard to predict that no event will occur that could be detrimental to the marine ecosystem. No governmental agencies openly support the release of transgenic metazoans into the environment; therefore, most transgenic breeding of aquatic species is performed in captivity.

When political and regulatory opposition is overcome, then subsequent crossbreeding beyond the boundaries of research and contained environment may be possible *(3)*. In addition, more detailed knowledge about the marine ecosystem and invertebrate biology (sea urchins, crabs, sea algae, and related phyla) needs to be incorporated into the impact of transgenes in aquaculture technology *(12)*. This will warrant further assessment and clearance from government agencies before commercial practice can commence.

6.1. Marine Invertebrates

Oysters, sponges, snails, corals, clams, crabs, and other aquatic invertebrates have been isolated and used for recovery of specialty compounds such as dyes, pearls, and building materials *(5,9)*. Marine invertebrates have been used for a long time in the development of novel products such as glues from mussels or chitin, chitosan, or chitotriose from shells of crabs, lobsters, and shrimp because of their unique properties for application in medicine and agriculture *(13)*.

To derive contamination free endotoxin clinical specimens from marine invertebrates GMO based technology has been adopted in the horseshoe crab amoebocyte lysate. For example a specific lysate preparation from the *Limulus polyphemus* crab is specifically employed by the US FDA for diagnosis and tests associated with releasing pyrogen-free pharmaceutical preparations and contamination-free endotoxin clinical specimens *(5)*. Because of subtle variations in lots of lumulus lysate pools derived from culturing different populations of stocks of these crabs and because of seasonal and environmental changes, application of cloning and molecular techniques to standardize the production of recombinant and characterized products of horseshoe crab lysate is gaining acceptance *(11)*.

In addition to useful materials, marine populations are a diverse source for procuring natural compounds that affect growth, metabolism, reproduction, and therapy toward other organisms, referred to here as *bioactive molecules*. These include effective diagnostic compounds with antiviral, parasitic, immunosuppressive, cytotoxic, antitumor, antibacterial, and antifungal properties. These compounds are derived from *Byrozoa*, *Mollusca*, *Cnidaria*, *Porifera*, and *Urochordata* species *(2,5,9)*. Bioactive molecules from these invertebrates offer information on the mode of action, potential design for drug discovery, and development and molecular modeling for higher expression via cloning and large-scale manufacturing systems for therapeutics *(5)*.

In addition, marine invertebrates are rich sources for procuring carbohydrate-binding proteins (lectins). Lectins have utility in histochemistry, tumor diagnosis, insulin-like activity, and cytokine-like activity *(12)*. Some sulfonated polysaccharide ligands from marine invertebrates have antiviral properties and have been exploited as therapeutic agents.

Last, certain sponges, such as *Halichondria okadai*, produce halichondrins or polyether macrolides. Halichondrin B is a potent antimitotic agent that was selected as a chemotype for an anticancer drug *(14)*.

6.2. Transgenic Aquaculture and Fisheries

As the global demand for seafood continues to increase substantially, as it is projected to do into the next decades, there are intensified efforts to meet the supply of fisheries and aquatic-derived food sources *(2,4)*. Aquaculture (farming and harvesting of aquatic cultures in a controlled environment) is becoming a major industry to satisfy the growing demands and appetites for seafood. However, the culture of marine organisms such as shellfish, prawns, shrimp, lobsters, mollusks, oysters, and clam species has a variety of problems related to reproduction, larva settlement, growth regulation, disease diagnosis and prevention, and public acceptance regarding palatable taste. Here, innovative manipulation via molecular and recombinant DNA techniques and application of transgenics to aquaculture stocks is anticipated to play a major role in addressing and circumventing these issues.

7. Opportunities in Growth Regulation and Reproduction

Marine invertebrates and fish offer new challenges because of the lack of knowledge of genetics and growth regulation *(15–17)*. However, a few models exist that suggest the growth of marine invertebrates can be influenced and improved using GHs and similar factors. The cloning of GH and GF genes and the derivation of transgenes in aquaculture has seen some ongoing fast track efforts to accelerate development and supply of oysters to fisheries.

Recently, GH and cyclic DNAs have been isolated and characterized so that when genetically manipulated they resulted in enhanced growth in several fish species (trout, tilapia, striped bass, salmon) and oysters compared to the control species *(5,9)*.

Osmoregulation, the physiological status of species, salinity, and temperature activity have been shown to influence GH activity. Cloning of antifreeze proteins and subsequent development of the transgenes has shown a substantial impact on enhanced growth in invertebrates.

7.1. Reproduction Issues

Components that will drive the success of the culture of crustaceans are to understand their life cycle in captivity, understand their reproduction cycle, and to have a reliable master seed stock supply. Recombinant transgenetic technology is important for characterization of the molecular spawning mechanics, molecular aspects of marine endocrinology, ecdisteroids maturation, and a hormonal approach to correlate biotechnological approaches that control and manipulate the spawning species in captivity *(16,17)*. Such efforts also allow establishment of genetically improved and robust brood stock desensitized to the risk of diseases. In addition, metamorphosis from embyrogenesis of larvae and juveniles, especially the biochemical systems, is also important for successful breeding of mollusks and shellfish. Transgenics with modified and manipulated morphogenic pathways and novel functional regulations have shed light on various environmental factors that control the timing, survivorship, and aquaculture setting for successful development of master seed stock *(11)*.

7.2. Resistance to Diseases: Diagnosis and Control

Control and diagnosis of pathogenic bacterial, viral, and protozoal diseases are recognized as critical factors for the success of a wild or transgenic marine invertebrates cultivation industry as a potential source of edible food for human consumption. Massive mortalities are affected in certain coastal regions by pollution, protozoal infection, and microbial infection, thus causing catastrophic reduction of natural and commercial populations of oysters, marine shellfish, mollusks, and fish and loss of productivity *(7,14,12)*.

One of the main difficulties in addressing disease controls either by virus or microbes is the lack of invertebrate cell lines. Therefore, isolation, genetic characterization of viruses, and identity fail to constitute the diagnosis. This also slackens the pace of developing appropriate strategies and technologies for rapid detection of parasites and correct therapeutic targets for controlling the disease *(9)*.

Thus, from disease control and diagnosis perspectives, there is urgent need to integrate recombinant DNA technologies that will allow in vitro transformation and development of continuous immortal stable invertebrate cell lines and brood stocks *(6,12,14)*. It seems that there is some success achieved in this area through GMO technologies, but it is far from the accomplishments in mammalian, microbial, and plant cells.

8. Conclusion

When the fact that more than two-thirds of the earth's surface is covered with water is combined with the novel breakthroughs in recombinant DNA technologies, it is easy to realize that tremendous opportunities exist for the use of transgenics in aquaculture production of marine invertebrates. Specifically, explosive growth in improved productivity and technology has occurred for shellfish and transgenic species. By introducing desirable genetic traits into finfish or shellfish, superior transgenic varieties have been achieved. These traits may include faster-growing or disease-resistant varieties, improved food conversion efficiency, tolerance to low oxygen concentrations, and survival in and tolerance to subzero temperatures. In addition, the use of GMO aquatic cultures is anticipated to accelerate identification of genes relevant to reproduction and growth in marine invertebrates, characterization of their regulatory regions, or selectable markers that direct the expression of transgenes at optimal level. Furthermore, it is anticipated that such data will assist in determining physiological, nutritional, and environmental factors that maximize transgenic aquatic performance.

Because immunoglobulin antibodies are not synthesized in shellfish, established methods for disease control for selected varieties are difficult. Rational vaccination programs therefore cannot be applied for species that are economically relevant, such as oysters, calms, shrimp, lobsters, and crabs. Assessment of the mode of their internal defense mechanism and identification of defensive molecules will warrant equivalent investigation. Such study through GMOs is anticipated to shed light on the characterization and provide detailed understanding of the effect that genes have that enables disease resistance in transgenetic invertebrates. Research efforts in diagnostics, employing molecular tools for detection of potential pathogens, and immunology toward improving disease control perhaps will help expand and manage the aquaculture industry.

References

1. Jaenisch, R. (1990) Transgenics. *Science* **240**, 1468–1477.
2. Gordon, J. (1989) Transgenic animals. *Int. Rev. Cytol.* **155**, 171–229.
3. Chen, T. and Powers, D. (1990) Transgenic fish. *Trends Biotechnol.* **8**, 209–215.
4. Fletcher, G. and Davis, P. (1991) Transgenic fish for aquaculture. *Genet. Eng.* **13**, 331–370.
5. Gerado, R. and Adam, M. (1998) Biotechnology of marine invertebrates: current approaches and future direction. In *Agricultural Biotechnology* . (Altman, A., ed.). Marcel Dekker, New York, pp. 563–583.
6. Chen, T., Jenn, K. L., and Fahs, R. (1998) Transgenic fish technology and its application in fish production. In *Agricultural Biotechnology*. (Altman, A., ed.). Marcel Dekker, New York, pp. 527–547.
7. Scambott, M. and Chen, T. (1992) Identification of a second insulin-like growth factors in a fish species. *Proc. Natl. Acad. Sci. USA* **89**, 8913–8917.
8. Moav, B., Liu, Z., Groll, Y., and Hackett, P. (1992) Selection of promoters for gene transfers into fish. *Mol. Mar. Biol. Biotechnol.* **1**, 338–345.
9. Chen, T., Vrolijk, N., Lu, J., Lin, C., Reinschuessel, R., and Dunham, A. (1996) Transgenic fish and its application in basic and applied research. *Biotechnol. Annu. Rev.* **2**, 205–236.
10. Buono, R. and Linser, P. (1992). Transient expression of RSVCAT in transgenic zebra fish made by electroporation. *Mol. Mar. Biol. Biotechnol.* **1**, 271–275.
11. Hackett, P. (1993) The molecular biology of transgenic fish. In *Biochemistry and Molecular Biology of Fish*. (Hochachka, P. and Mommsen, T., eds.). Elsevier Science, Amsterdam, The Netherlands, pp. 230–240.
12. Hansen, P., Von Westernhagen, H., and Rosenthal, H. (1985) Chlorinated hydrocarbon and hatching success in Baltic herring spring spawners. *Mar. Environ. Res.* **15**, 59–76.
13. Drickamer, K. and Taylor, M. (1993). Biology of animal lectins. *Annu. Rev. Cell. Biol.* **9**, 237.

14. Hirata, Y. and Uemura, D. (1986) Halichondrions—anti-tumor macrolides from a marine sponge. *Pure Appl. Chem.* **58**, 701.
15. Chang, S. (1989). Endocrine regulation of molting crustaceans. *Crit. Rev. Aquatic Sci.* **1**, 131.
16. Morse, D. (1984) Biochemical and genetic engineering for improved production of abalones and other valuable molluscs. *Aquaculture* **39**, 263.
17. Coon, S., Fit, W., and Bonar, D. (1990) Competence and delay of metamorphosis in the Pacific oyster *Crassostrea gigas* (Thunberg). *Mar. Biol.* **106**, 379.

IV

Plant GMOs

8

Development of Genetically Modified Agronomic Crops

Manju Gupta and Raghav Ram

1. Introduction

The development and introduction of genetically modified (GM) crops are important steps toward improving crop productivity and food quality around the world. The initial hurdle in developing GM crops was the availability of a routine transformation technology for agronomic crops. This was partly because of the lack of efficient tissue culture methodologies to enable transformation and regeneration and efficient means of delivering deoxyribonucleic acid (DNA) into cells. Crops such as rice, maize, and wheat, which belong to the monocotyledonous group, were once considered recalcitrant in tissue culture, inaccessible to the *Agrobacterium* method of gene transfer, and thus dependent on the development of alternative methods *(1–4)*. Since the 1980s, a surge in successful transformation of maize, rice, wheat, and other cereal crops has been reported as a result of efficient tissue culture and transformation strategies *(5)*. This has made genetic modification of agronomic crops possible, although it came much later than for other crops, such as tomatoes.

The second hurdle for advancing GM crops has been in the area of public acceptance. This has resulted in isolation of GM crops from certain geographies of the world, thus creating new challenges for global organizations that export grain and food products to different countries. The cost of physical isolation and segregation of GM grain from non-GM grain could be significant to producers of GM products. Despite these hurdles, several classes of GM products have been commercially introduced over the years (Table 1); details can be found on the AGBIOS Web site (http://www.agbios.com).

As the benefits of herbicide and insect resistance genes expressed in crops became evident, a wave of products such as Round-up Ready soybeans, *Bacillus thuringiensis* (*Bt*) corn, and *Bt* cotton was introduced into the marketplace *(6)*. Thus, advances in tissue culture and transformation methods *(7)*, availability of efficient molecular tools such as selectable marker genes and other components of a vector *(8)*, sound strategies for maintaining the quality and integrity of products throughout the developmental pipeline *(9)*, and development of analytical tools for detection and traceability of GM products *(10,11)* throughout the value chain have all been essential components for the

From: *The GMO Handbook: Genetically Modified Animals, Microbes, and Plants in Biotechnology*
Edited by: S. R. Parekh © Humana Press Inc., Totowa, NJ

Table 1
Genetically Modified Crops and Traits

Trait	Crops
Herbicide resistance	Maize, rice, cotton, canola, chicory, carnation, soybean, flax, linseed, tobacco
Insect resistance	Cotton, tomato, potato, maize
Oil modification	Canola, soybean
Male sterility	Canola
Fertility restoration	Canola, chicory, maize
Delayed ripening	Melon, tomato
Viral resistance	Papaya, squash, potato

success of GM products. The entire process of developing GM products and introducing them to the marketplace takes several years and involves many steps (Fig. 1). We introduce some of these key steps in this chapter. Some of these steps can be more specific to maize, but will adapt to other agronomic crops.

2. Production of Primary Transgenics

Primary transgenics are produced using one of many different transformation methods. Depending on which transformation method is used, appropriate DNA vectors containing the transgenes are assembled. Essential components and the construction of transformation constructs have been extensively reviewed by Hinchee et al. *(4)* and Merlo *(8)*.

A basic transformation construct consists of DNA encoding the gene of interest and regulatory sequences controlling the expression of the gene. Associated with this DNA are a selectable marker gene and its regulatory components, which enable the selection of transgenic tissues during the tissue culture process. In addition, use of other functional DNA sequences such as matrix attachment region for the stability of transgene expression may be an option *(12)*.

2.1. Transformation Methods

Several transformation methods have routinely been used for the introduction of target DNA into plant cells and its subsequent integration into the genetic makeup of the cell. Subsequently, plants are regenerated from these transformed cells, seed is produced, and thus the introduced DNA is transmitted through seed generations. Fundamentally, there are two methods for introducing DNA into plant cells: biological and physical.

2.1.1. Biological Method

Members of a symbiotic soil bacterium, *Agrobacterium*, have naturally developed the capability to transfer DNA from their cells to plant cells during infection. Of particular interest is *Agrobacterium tumefaciens*, which in nature transfers DNA (T-DNA) present between two border sequences into cells of dicotyledonous plants and causes the host cells to divide and produce crown galls *(13)*. Scientists have used this basic ability of *Agrobacterium* to transfer synthetic DNA sequences containing genes of interest to plant cells. The tumor-inducing DNA sequences (that cause crown gall forma-

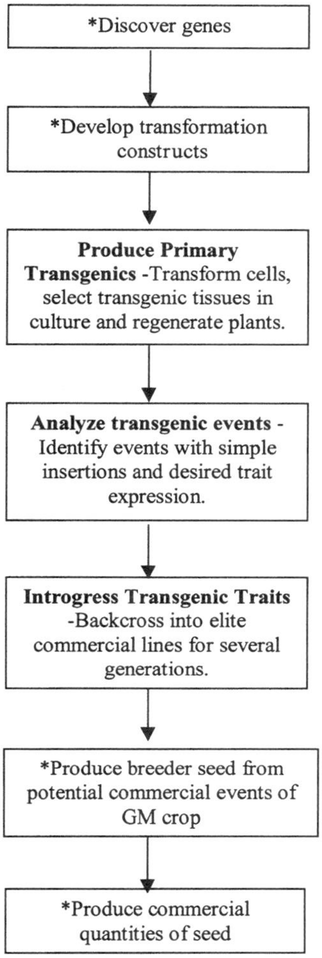

Fig. 1. A flow diagram showing multiple steps involved in developing a GM product. Regulatory requirements vary depending on the stage of development. *Not discussed in detail in this chapter.

tion) within the T-DNA were replaced with the DNA of interest for transfer into the plant cells *(14)*. Considerable advances have been made in improving the ability of the *Agrobacterium* to infect a wider variety of plant species and in extending their ability to transfer DNA to monocotyledonous species such as maize *(15)* and rice *(16)*.

2.1.2. Physical Methods

Cloned and isolated DNA containing all of the essential functional components for the expression of a trait of interest can be delivered into plant cells via a variety of physical or chemical methods. Particle bombardment, microinjection, sonication, silicon carbide (SiC) whisker treatment, and electric current pulse are examples of physical means applied for this purpose *(17)*. Chemical-based methods, such as transfection using liposomes and polyethylene glycol, have also been utilized to facilitate DNA entry into plant protoplast *(4)*. Some of these methods require treating cells with enzymes to remove the cell wall partially or completely and expose the cell membrane to facilitate DNA uptake. Difficulty in regenerating plants from protoplasts and low transformation frequency have made some of these methods less popular over the years.

Significant advances were made in crop transformation with the advent of DNA delivery methods that rely on particles propelled into cells *(4,7)*. Although the initial models of particle guns were damaging to the cells, subsequent improvements in the particle size, force used to deliver particles, and tissue culture methods have considerably improved the efficiency of producing transgenic events, facilitating the production of transgenic crops *(17)*. Another method that has been proven successful uses SiC whisker treatment. DNA and cell clumps in suspension are vigorously shaken to allow penetration of cells by the SiC whiskers and passage of DNA into the cell *(18)*. In most cases, successful regeneration of transgenic plants has been possible. However, application of both the methods has been limited to nonelite genotypes in maize.

2.2. Transgenic Seed Production

It is important to produce seeds from the primary transgenic plants (T0) to ensure that the introduced DNA is stably incorporated into the genetic makeup of a plant and that it can be propagated through meiosis to the next generation. Transgenic plants produced after undergoing the transformation procedures, selection pressures, and long culturing period during plant regeneration often show symptoms of stress. They require careful acclimation to the greenhouse and soil conditions. First-generation transgenic plants such as maize frequently do not produce adequate pollen in synchronization with silks (female florets) to enable self-pollination. In those cases, crosses are made with an elite genotype or the original parent genotype (sib pollination) to produce T1 seeds.

3. Transgene Analysis

Much of the early transgene analysis takes place prior to plant regeneration to reduce the number of cultures that enter the regeneration phase *(18)*. T0 plants are also a good place to test for the nature of the insert, transgene expression, and initial trait analysis. More detailed analysis begins at the T1 generation on confirmation that the transgene is stably incorporated into the nuclear genome and when discrete viable events are obtained. An *event* refers to a heritable integration of transgene into a unique location of the nuclear genome.

Large numbers (50–500) of events are typically created from a given construct. Not all of the events are suitable for a commercial product *(9)*. These events are sorted based on several criteria, including event complexity, transgene expression, and plant phenotype. Some of the criteria can be evaluated in one plant generation; others take multiple generations. Overall, the goal is to select the best candidate events during successive plant generations.

Understanding transgene structure and trait performance during event sorting requires sequential application of many analytical techniques. The type of technique used is normally driven by the assay reliability, speed, and cost. Some of the molecular tools and technologies that can effectively be applied at various stages of event sorting are reviewed next and summarized in Table 2.

Table 2
Event Sorting and Trait Introgression Via Stepwise Application of Various Analytical Techniques

Generation	Analysis Purpose	Technology Approaches	Results
T0	Transgene presence	Qualitative PCR	Select plants containing transgene
	Absence of antibiotic gene and plasmid DNA	Qualitative PCR	Eliminate plants containing antibiotic gene or plasmid DNA
	Transgene expression analysis of the targeted tissue	ELISA, RT-PCR, or biochemical analysis	Select events showing desirable expression
T1	Copy number estimation	Real-time PCR or Invader	Select simple (one- to two-copy) non-segregating events
	Transgene expression	ELISA, RT-PCR, or biochemical analysis	Select events with desirable expression
	Insertion analysis on simple events showing desirable expression	Southern analysis	Select unique stably inserted events
	Plant transcription unit analysis	Southern analysis	Select events with complete transformation unit
	Absence of antibiotic gene and plasmid DNA	Southern analysis	Eliminate plants containing antibiotic gene or plasmid DNA
T2	Transgene expression	ELISA or biochemical analysis	Select events showing stable expression
	Zygosity testing	TaqMan® quantitative PCR or Invader®	Select events showing no agronomic problems in the homozygous plants
	DNA methylation or siRNA analysis	Southern or siRNA on homozygous and hemizygous plants	Select stable expressing events
T3	Transgene expression	Elisa, RT-PCR, or biochemical analysis	Select stable expressing events
	DNA methylation or siRNA analysis	Southern or siRNA analysis	Select stable expressing events
BC_1, BC_2, and BC_3	Marker-assisted introgression	SSR, RFLP, AFLP, and SNP markers	Select plants with shortest linkage drag and most converted to contain recurrent parent genome

3.1. Transgene Structure Analysis

The level and stability of transgene expression to a large extent are correlated to transgene structure and its location in the genome. Complex events often show gene expression instability because of gene-silencing phenomenon, which is reviewed in Chapter 9. Prior knowledge of transgene structure instills more confidence in event sorting because eliminating or advancing events based on inappropriate understanding could be detrimental to GM crop development.

3.1.1. Transgene Presence and Copy Number Estimation

Transgenic traits are frequently associated with herbicide markers, which allows rapid confirmation of transgene presence in the transformed plant *(8)*. Nevertheless, verification of the presence of all of the genes in a transgenic construct is still crucial because genes can become dissociated during plant transformation or transgene integration. Detection based on polymerase chain reaction (PCR) is a simple method for rapid confirmation of transgene presence in the early phases of event sorting. Typically, it is accomplished by amplification of target genes, followed by gel electrophoresis. Identification of the correct size DNA fragment confirms the transgene presence. However, this method may not be suitable for large-scale screening. A 5' nuclease-based technology or TaqMan® technology (Applied Biosystems, Foster City, CA) offers a high-throughput system.

The TaqMan technology relies on the use of a fluorogenic oligonucleotide probe between the two PCR primers, such that when the PCR reaction proceeds in the presence of the target DNA, fluorescence is released because of the degradation of the probe resulting from the 5'-to-3' exonuclease activity of the Taq DNA polymerase *(19)*. The released fluorescence is quantitatively read by a plate reader. The reactions are run generally as biplex assays *(20)*. One assay amplifies the target gene, and another corresponds to an endogenous gene native to the transformed plant species. The probe molecules for the two targets are labeled with different fluorophores, thereby allowing both amplifications to be read simultaneously by the plate reader. The biplex format and the use of 96- or 384-well plate make the TaqMan assays more robust and ensures correct data interpretation when the transgene is absent.

Screening events for low copy number (1–3 copies) are critical for maintaining high trait performance. Low copy number events are also preferred for product registration because of fewer efforts involved in event characterization. In addition, complex events often lead to transgene silencing *(21–23)*.

There are several technologies and methods to filter events for low copy number. Of these, TaqMan and invasive cleavage (*see* Section 3.2.2) technologies are preferred methods. TaqMan in conjunction with real-time PCR permits quantitation of transgene copy number with high fidelity *(18,20,24)*. The target genes are amplified and concomitantly read in instruments, such as the ABI Prism™ 7700 (Applied Biosystems) or equivalent, as the reactions are progressing. Details of the technology are described elsewhere *(25,26)*. Only a small amount of DNA (5–10 ng) is needed for copy number estimation. The results can be obtained within a few hours.

Because the amplification target of the TaqMan technology is approximately 100 bp, there is a strong likelihood of not detecting partial or rearranged transgenic DNA fragments present either in *cis-* or *trans-* configuration to the full-length gene. Therefore, 1–3 copy events are further analyzed on Southern blots to verify the number of copies (*see* Section 3.1.2).

In addition to transgene presence, verification for the absence of undesirable vector DNA or "backbone DNA" is also critical because it may incorporate into genomic DNA during transgene insertion *(27)*. The presence of vector DNA has practical implication during product registration and must be avoided. Early detection of the backbone DNA is accomplished with PCR-based assays, and the events carrying such DNA are eliminated. However, there is a strong likelihood that partial copies or fragments of backbone DNA may escape such detection. Therefore, Southern blot analysis is required for final confirmation (*see* Section 3.1.2).

3.1.2. Stable Insertion and Structural Integrity of Transgenes

A 1:1 segregation ratio of the transgene at the T1 generation (when T0 plants have been crossed with elite lines) confirms its stable incorporation into the nuclear genome. Deviation from this ratio might reflect insertions at more than one locus. Therefore, each insertion needs to be identified uniquely to differentiate it from other insertions within and between events and to study their structural stability over several generations. Southern blot is a preferred method for identifying unique insertion of events. Transgene integration analysis in this technique requires the use of a restriction enzyme that cleaves only once within the transgenic DNA, with the other sites present in the flanking genomic DNA close to the junction *(28)*. On probing the blot with an appropriate transgene DNA fragment, each event displays a unique band position on the blot (Fig. 2). If two events show the same band position, then they should be further characterized by digesting with different restriction enzymes to establish the distinction. The chance that a transgene will insert at the same genomic location in two different events is highly unlikely.

It should be noted that one band observed on a Southern blot does not necessarily imply the presence of a single transgene copy *(28)*. Depending on the band size and restriction enzyme used, there can be other full or partial transgene copies in the same band. Multiple bands on a Southern blot denote insertions either at one locus or at multiple loci. Multiple-loci insertions can be confirmed by analyzing several T1 plants. Any segregation in the banding patterns is indicative of more than one locus insertion.

Analysis of the entire plant transcription unit (PTU), which includes promoter, coding sequence, polyadenylation sequence, and any other associated sequences, ascertains the insertion of a full-length transgene into the genomic DNA *(28)*. A Southern blot strategy for such analysis employs a restriction enzyme that cleaves at the termini of a PTU. A typical Southern blot result shows one band of the same size across all events originating from the same construct (Fig. 3). Presence of more than one transgene insertion in a given event will still produce a single PTU band as long as both transgenes are full length. Bands of unexpected size would indicate rearrangements in the inserted DNA fragments. Presence of partial copies will result in additional hybridizing bands, which can be easily identified. Full and partial copies for all of the PTU components can be identified by sequential hybridization with corresponding probes to the PTU blot. It is critical to identify all of the partial fragments because they can lead to gene silencing *(22,23,29)*.

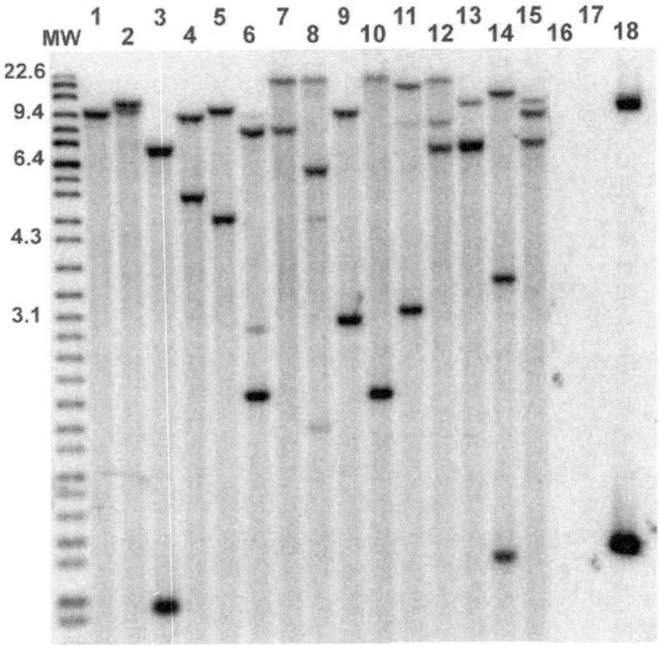

Fig. 2. An example of transgene integration analysis using Southern blotting technique. Note the unique banding patterns of each event. Some events show more than one band, which means they contain more than one copy of transgene. *Lane MW,*: DNA molecular weight standard; fragment sizes are in kilobases. *Lanes 1–15*: different events of a transgenic construct. *Lanes 16* and *17*: nontransformed plant genomic DNA used as negative controls. *Lane 18*: nontransformed plant genomic DNA spiked with the plasmid DNA carrying the transgene used in transformation.

All three sources of information, copy number (TaqMan), integration, and PTU analysis, must be evaluated together to estimate the number of full-length and partial transgene copies in a given event. Nevertheless, DNA sequencing will be the ultimate tool to understand completely the structure and orientation of transgene components and their effect on trait expression.

Southern blots made for integration and PTU analyses can also be used for backbone DNA screening and the associated bacterial antibiotic genes. Events containing such sequences must be eliminated.

3.2. Transgene Zygosity Analysis

Depending on the event, the homozygous state of the transgene can be disadvantageous and lead to counterselection of homozygous plants across generations *(30)*. This can happen because of either higher additive expression or lower silenced state expression of transgenes. Either situation may result in inadvertent elimination of homozygous plants. Therefore, it is critical to identify the homozygous plants early to ensure proper Mendelian segregation, desirable trait expression, and normal plant phenotypes.

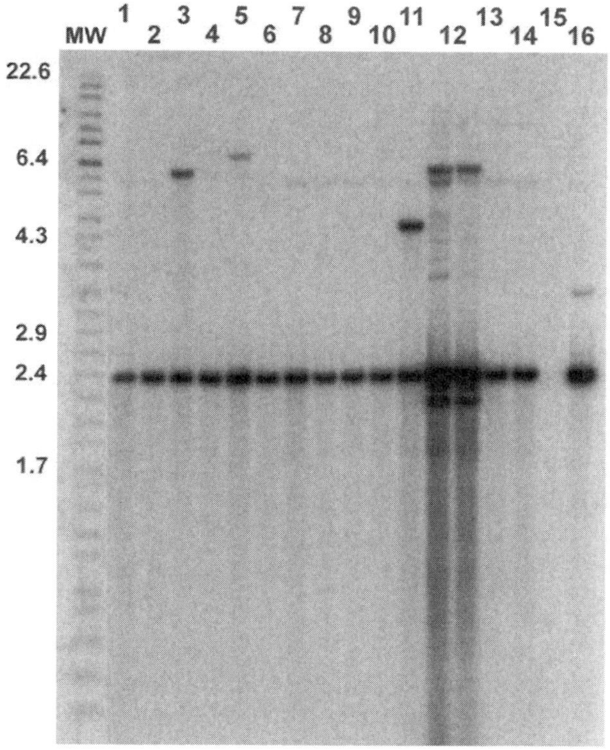

Fig. 3. Example of transgene PTU analysis using Southern blotting technique. Note that all of the events contain the same PTU band at the 2.4-kb molecular weight position. This band position also matches that of the plasmid control shown in lane 16. *Lane MW*: DNA molecular weight standard; the fragment sizes are in kilobases. *Lanes 1–14*: different events of a transgenic construct. *Lane 15*: nontransformed plant genomic DNA. Lane16: nontransformed plant genomic DNA spiked with the same transgenic plasmid DNA used in transformation.

3.2.1. 5' Nuclease-Based Assay

The real-time TaqMan platform, described in Section 3.1.1 for copy number estimation, has been successfully used for accurate identification of homozygous and hemizygous (zygosity) plants *(20)*. In a biplex assay, the reference endogenous gene signal remains constant whether the transgenic plants are hemizygous or homozygous; the transgene signal varies depending on zygosity. When using the same quantity of DNA, the ratio of the signals from transgene compared to the endogenous gene determines the transgene zygosity.

Because the real-time TaqMan technology requires expensive instrumentation, alternative non-PCR-based invasive cleavage assays can be used for zygosity determination as described next.

3.2.2. Invasive Cleavage Assay

Invasive cleavage or Invader® (Third Wave Technology Inc., Madison, WI) assay is the first gene detection technology capable of quantifying low copy sequences from

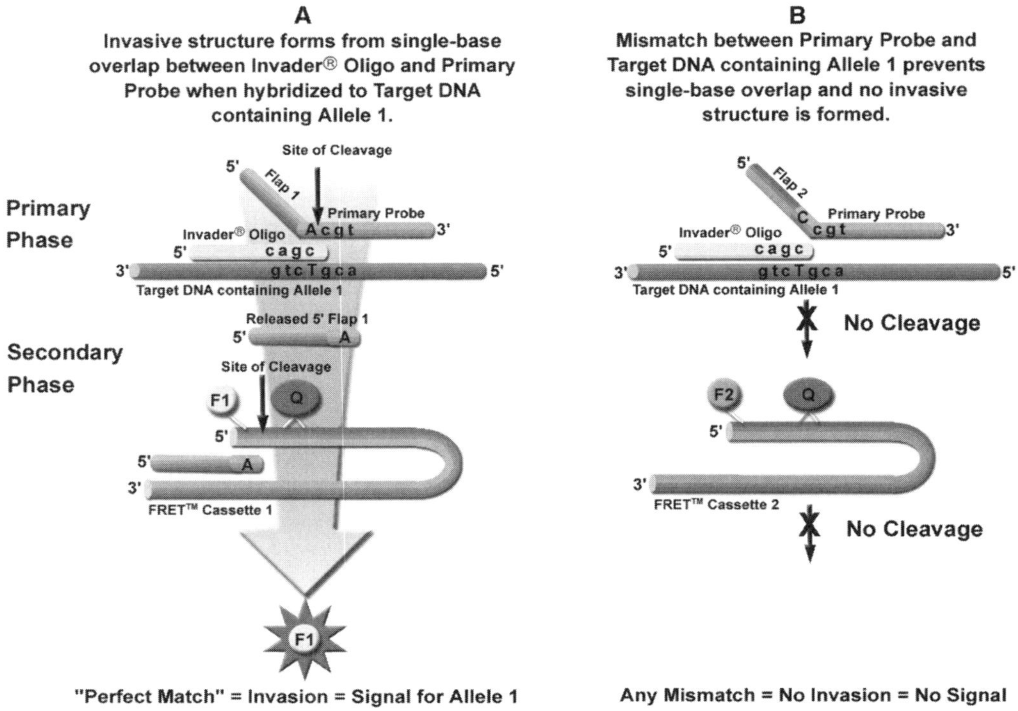

Fig. 4. The principle of the Invader technology. When transgenic DNA is present, fluorescence is produced as a result of linear multiplication of the signal in two phases of an Invader assay. (This diagram was obtained through the courtesy of Dr. Glen Donald of Third Wave Technology, Madison, WI.)

complex plant genomes without PCR-based amplification *(31)*. The principle of the technology relies on a flap endonuclease enzyme from archaea that recognizes a specific DNA structure formed by the hybridization of two overlapping oligonucleotides, invasive (or invader) oligo and primary probe, to the target DNA (Fig. 4). The enzyme then cleaves the resulting "flap" molecule, and no fluorescence is released in the first phase of the reaction. No flap molecules are cleaved in the absence of specific flap structure.

In the second phase of the reaction, the flap molecule anneals to another universal DNA oligonucleotide cassette. The 5' end of this cassette is fluorogenic and synthesized based on the fluorescence resonance energy transfer method of signal amplification *(32)*. On formation of the specific structure, cleavage of the flap DNA (Fig. 4) results in the release of fluorescence that is proportional to the number of original target molecules present in the genomic DNA. This quantitative fluorescent signal is read in a fluorometer. Both primary and secondary phase reactions occur simultaneously.

Invader assays require about 25–100 ng of genomic DNA, depending on the genome complexity. The reactions are isothermic and incubated at 64°C for 2–4 h in an oven. Unlike PCR, no temperature cycling is required. Furthermore, the assays are insensitive to DNA contamination at the level frequently observed in PCR-based assays. All of these attributes, combined with its quantitative nature, amenability to high-throughput assays, and minimum need for laboratory equipment make Invader a highly desirable detection method for event-sorting applications.

Like TaqMan assays, Invader assays are also carried out in a biplex format. The utility of Invader assays has been proven for zygosity determinations, copy number estimation, and transgene presence *(31,* Nirunsuksin et al., unpublished data*)*.

3.3. Transgene Expression Analysis

During event sorting, it is critical that gene products (ribonucleic acid [RNA] and protein) conferring a trait are accurately measured over several plant generations and across various genotypes, growth and development stages, and different zygosity states to find the best-performing events. Adequate and stable expression is the key to finding the candidate events. This necessitates the application of the most practical and efficient tools for characterizing gene products that signify expression.

3.3.1. RNA Analysis

Early in the event sorting process, messenger RNA (mRNA) encoded by the transgene is characterized for size and quantity with the Northern blotting technique *(33,34)*. This technique involves running either total RNA or poly A RNA fractions on a denaturing agarose gel, blotting of the RNA onto a nylon membrane, and subsequent hybridization of the blot to a radioactively labeled probe corresponding to the transgene sequence.

The blots are rehybridized with another probe corresponding to an endogenous gene constitutively expressed in plants for estimating the relative quantity of the transgenic mRNA. Although the technique is highly sensitive, it is labor intensive and not generally used to follow transgene expression during event sorting. Quantitative measurement of mRNA has been successful by the reverse transcriptase PCR (RT-PCR) technique *(35,36)*. It involves conversion of mRNA into complementary DNA (cDNA), the quantity of which can be accurately measured using the real-time RT-PCR approach of the TaqMan platform *(37)* or with any other detection methods. Alternatively, cleavage-based Invader technology has been successfully utilized for quantitation of mRNA *(38)*.

Generally, mRNA quantitation is not a preferred method for sorting events because RNA synthesis precedes by one or more steps the final molecules that confer traits. Furthermore, mRNA analysis may not be a practical approach, especially when plants are grown in the field and resources required for freezing the samples are not readily available.

3.3.2. Protein Analysis

When protein molecules confer a transgenic trait, enzyme-linked immunosorbent assay (ELISA) is the most practical and economical way of measuring qualitative and quantitative expression *(39)*. Inherent sensitivity of the assay is reflective of the antigen:antibody recognition mechanism. Complementing ELISA with the Western blotting techniques *(40,41)* in the early stages of product development allows detailed characterization of transgenic proteins. Western blots permit size fractionation of total proteins on acrylamide gels for accurate measurement of protein sizes. Any alarming protein degradation or multimer formation because of protein interactions can also be examined on the Western blots.

3.3.3. Gene-Silencing Analysis

ELISA analysis over two to three hemizygous and homozygous plant generations generally identifies events showing significant gene expression variability; such events are not advanced. However, minor and infrequent variability in expression may not be as obvious in the early stages of event sorting because of the confounding effects of variabilities inherent in the detection methods, plant sampling, and growth conditions. Such events warrant gene-silencing analysis at the nucleic acid level.

Posttranscriptional gene silencing has been highly correlated with the presence of 21- to 26-bp small interfering RNA (siRNA; *42*); both transcriptional gene silencing and posttranscriptional gene silencing have been associated with the methylation of transgene and other relevant sequences *(23,29)*. Therefore, either the detection of siRNA or methylation of the transgenic DNA has been the basis of examining gene silencing. Methods for siRNA detection are based on isolation of the total RNA fraction, enrichment for small RNA, size fractionations of small RNA on polyacrylamide gels, and subsequent hybridization to appropriate nucleic acid probes *(43–45)*. Because this method is labor intensive, its application is limited to a selected set of events.

DNA methylation in plants occurs in the symmetrical (CpG and CpXpG) and asymmetrical cytosine bases *(29,45)*. Southern blot analysis is a frequently used method for DNA methylation analysis in transgenic plants *(46,47)*. The principle of the method relies on the use of restriction enzymes that are sensitive to DNA methylation. These enzymes do not cleave when cytosine residue in the recognition site is methylated. As a result, higher molecular weight DNA fragments are identified on Southern blots (Fig. 5). This method is conducive to identification of DNA methylation in different components of transgenes. For example, methylation in the coding and promoter sequences can be identified utilizing restriction enzymes that produce characteristic size bands encompassing those regions on hybridization with the corresponding DNA probes. DNA methylation analysis is also favorable for studies involving field-grown plants because of DNA stability in sample shipment.

Bisulfite treatment of genomic DNA provides an alternative method for the detection of methylated cytosines. This treatment causes conversion of unmethylated cytosines to uracil; methylated cytosines remain unchanged. Uracil is read as thymine during PCR amplification, and this change is easily read on sequencing gels. Based on this principle, several methods have been developed for identification of methylated bases. These include direct sequencing of the PCR products *(48–50)*, single-strand conformation polymorphism analysis followed by sequencing *(51)*, and size fractionation of PCR-amplified fragments with or without restriction enzyme digestion *(52–54)*. The success of any of these methods for a specific application can depend on several factors, including heterogeneity in the cytosine base methylation, complexity of transgenic events, and the presence of homologous genes or components native to the host plant.

4. Introgression of Transgenic Traits

Transgenic events showing desirable trait expression must be introgressed into several different elite lines for trait commercialization. Conventional breeding programs typically require six generations of backcrosses (BCs) for trait introgression. This takes 2 years in maize with the use of continuous nurseries. However, it may take longer with

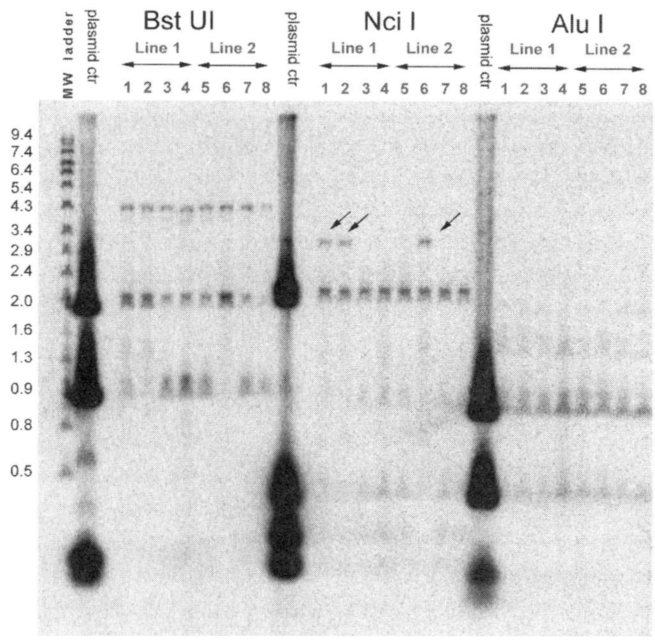

Fig. 5. Example of DNA methylation analysis in transgenic plants. Genomic DNA samples from eight homozygous plants of two genotypes, representing T2 generation, were digested with restriction enzymes *Bst*U I, *Nci* I, and *Alu* I. *Bst*U I and *Nci* I recognition sites were sensitive to CpG methylation; *Alu* I site was not. Samples 1, 2, and 6 showed very low transgenic protein expression as measured by ELISA. These samples also contained a band of higher molecular weight at the 3.4 kb position in the *Nci* I digests (arrows) that was not present in samples 3, 5, 7, and 8 expressing the expected amount of transgenic protein. *Bst*U I digest did not show any additional band correlated with low expression. However, it did show a band of increased intensity at the 2 kb position and a decreased intensity band at the 0.9 kb position in lanes 1, 2, and 6. It appears that the 0.9 kb band has shifted up to the 2 kb position as a result of DNA methylation. No altered banding pattern was seen with *Alu* I restriction enzyme. The very first lane on the gel represents a DNA molecular weight standard; the lanes labeled as plasmid control represent nontransgenic plant DNA spiked with the plasmid used in transformation. These plasmid DNA samples were also digested with the three enzymes and are present accordingly in different panels. MW ladder: DNA molecular weight size standard; fragment sizes are in kilobases.

other crops. Marker-assisted selection (MAS) of plants can reduce this time by half *(55)*. The concept here is to identify and select fewer plants that contain a higher percentage of the desirable recurrent parent genome than expected. For example, the expected averages of the recurrent parent genome at the BC_1 and BC_2 generations are 75 and 87.5%, respectively. With the application of MAS at the BC_1 and BC_2 generations, plants converted for the genome content very close to that of recurrent parent could be obtained *(56)*. When BC_2 is the last breeding generation, MAS must be applied at BC_1 for the most impact.

A successful MAS program requires an efficient marker system (*see* Sections 4.1–4.4) coupled with a marker application strategy. The marker application strategy involves choosing markers at equally spaced intervals throughout the genome; these markers are polymorphic between two parents of the cross. An interval of 20 cM (or map units) has been proposed as ideal to tag any gene of interest along the chromosomes with selection fidelity of 99% *(58)*, with the exception of the regions flanking the gene of interest. Accordingly, 60 markers will be required to cover a 1200-cM genome. Based on computer simulation and MAS based on restriction fragment length polymorphism (RFLP) in tomato, analysis of only 30 plants in each of the three generations has been reported as sufficient for the whole recurrent parent genome recovery *(55)*.

During MAS, the chromosomal region flanking the transgene is of special focus because of its tight linkage to the trait. Any undesirable set of genes located in this region can adversely influence the trait or plant phenotype. Thus, the effects associated with this undesirable flanking region are generally referred to as *linkage drag (58)*. This region apparently is not the focus of as much attention if the original transformation was carried out in an elite background. Sizable linkage drags of 10 cM can persist even after 20 rounds of BC breeding *(59)*. MAS can effectively remove the linkage drag *(55)*.

Transgenic traits to date are generally comprised of single-locus traits, and their introgression can be straightforward with MAS. However, the inherent value of MAS is further realized when multiple single-locus or multiloci traits are stacked during trait pyramiding strategies. In these instances, markers flanking the particular genes can be closely monitored for precise genome recovery along with the ones employed in the whole genome recovery.

The most effective marker systems of today are comprised of DNA markers. Such systems offer unlimited opportunities for creating new markers by exploiting the genome structure and sequence information contained in it. Development of a DNA marker system requires significant effort and multiple years of development. The success of a marker system is determined by the informativeness of markers, availability of a dense genetic linkage map, amenability to high-throughput screening, and low cost per assay. Several types of DNA marker systems have been developed *(60)*; those already applied or have the potential to be applied for MAS are described in this section.

DNA markers are heritable and allow unbiased scoring because of their insensitivity to the environment. Selections based on plant phenotypes can be adversely affected by stress conditions, such as disease and water stress, and may hinder plant selection *(55)*. In addition, desirable phenotypes may not be as obvious in the early BC generations because of transgenes residing in inferior nonelite genetic backgrounds. Under all of these situations, MAS provides an unequivocal means of selecting and advancing the most converted plants.

4.1. RFLP Marker System

RFLP markers differentiate individuals by recognizing differences in the restriction enzyme fragment lengths identified with the Southern blotting technology *(61,62)*. Such differences are characteristics of mutations that occur within the restriction enzyme recognition sites. In addition, large insertions and deletions contributing to the length differences are also detected. However, both types of polymorphism are not distinguishable from one another in the final results. RFLP markers are created by precisely fragmenting the genomic DNA, typically with 6 bp recognizing restriction enzymes. This DNA is size fractionated on an agarose gel, blotted onto a nylon membrane, and sequentially hybridized to cloned genomic or cDNA probes. Such probes generally occur as one to two copies within the same species genomic DNA *(63)*. The blots are exposed to X-ray films or Phosphor Imager, and it can take 1–3 days to visualize the results.

RFLP was the first DNA-based marker technology capable of representing the whole genome in marker-assisted breeding or introgression programs. Because of this virtue, RFLP genetic linkage maps of many crops were developed *(64)*. Another highly desirable attribute of the RFLP markers is their codominant nature, meaning alleles from both parents are simultaneously scored in the final DNA patterns. Despite the high information content and construction of successful linkage maps in many of the crops, the technology is not currently a preferred method for MAS because of its low throughput and high cost per assay.

Nonetheless, RFLP linkage maps continue to be a valuable resource in understanding and developing other marker systems. RFLPs provide a framework of markers for constructing simple sequence repeat (SSR) and amplified fragment length polymorphism (AFLP) (described next) genetic maps *(65,66)*. RFLP markers can be converted into single-nucleotide polymorphisms (SNPs) or sequence tagged site markers *(67)* for higher throughput screening.

4.2. AFLP Marker System

The AFLP technology employs selective amplification of DNA fragments from genomic DNA samples that have been digested with combinations of restriction enzymes *(68)*. This selectivity comes from the presence of additional 1–3 bases at 3' ends of the primers, which recognize and amplify only a subset of DNA fragments. On size fractionation, the amplified DNA is resolved into 25–100 bands per sample, depending on the genome complexity.

Like RFLP markers, AFLP markers distinguish individuals based on the presence of point mutations in the restriction enzyme recognition sites and smaller insertions or deletions in the target DNA fragments. Most AFLPs are dominant markers, meaning alleles from only one of the parents can be identified and scored. Identification of heterozygous and homozygous alleles has been possible through quantitative scoring of the markers *(69)*. AFLP linkage maps are constructed *de novo* each time a new population is analyzed. This is because not every laboratory has the instrumentation to precisely measure AFLP band sizes represented in relatively complex DNA fingerprints. Secondly, identical size AFLP bands in two populations, generated with the same enzyme and primer combinations, may not map to the same location *(69)*. The latter uncertainty stems from the fact that AFLP markers are identified by the limited sequence homology during amplification. The homology of 1–3 bp located at each terminus of amplified fragments may not be sufficient to prove unequivocally the same sequence in two identical size fragments.

Given sufficient levels of DNA polymorphism within a population, a total of 8–10 primer combinations, applied to two or more enzyme combinations, may prove practical in MAS projects. Linkage maps are generally constructed in the first BC generation and then utilized in all of the generations for plant selections. In our experience, if plants were infested with bacterial or viral diseases or with aphids, then the DNA of those organisms will also be extracted along with the plant DNA and will result in artifactual bands in the final DNA patterns. Often, these bands can be of very high intensity and can mask other markers or not allow other markers to amplify with expected band intensity.

Because of the high multiplex ratio of the AFLP markers and no *a priori* sequence requirement *(65)*, AFLP markers are still an attractive system for many species for which no other high-throughput marker system has been developed.

4.3. SSR Marker System

SSRs are ubiquitous in all eukaryotic genomes. They are comprised of repeat motifs 1–6 nucleotides long and are found tandemly repeated in various genomes *(70)*. One SSR is found every 6–7 kb in plants, which is equivalent to the frequency found in mammals *(71)* There are several attributes of the SSR sequences that make them highly desirable markers:

1. SSRs are highly abundant and found interspersed with single-copy as well as repetitive sequences in genomes *(72–74)*.
2. They are associated with high levels of polymorphism *(75)*, which have been attributed to DNA polymerase slippage during chromosomal DNA replication *(76)*. As a result, the repeat numbers can either increase or decrease within a given sequence.
3. SSRs are multiallelic loci with more than five alleles typically observed in plants *(77)*.
4. SSR markers are amenable to high-throughput PCR assays *(78)*. Multiple (10–20) markers can be fractionated with high resolution within hours and semiautomatically scored.
5. SSRs are codominant markers that provide high information content per assay in genetic analysis.

The development of SSR markers requires flanking sequence information for the amplification of the intervening variable repeat. To accomplish this, SSR-enriched libraries for various repeat motifs are constructed, and SSR-containing clones are identified and sequenced for developing PCR assays *(79)*. This labor-intensive and expensive process has been a disadvantage of the marker system and its rapid application in many species. SSR-based linkage maps have been developed for many plant species *(66)*. In an effort funded by the National Science Foundation, a high-resolution SSR map of maize has been developed. This map is comprised of a total of 1734 SSR markers, of which 978 markers are new *(66)*. SSR maps can be used across populations and genotypes within a species. High-resolution maps are highly desirable not only for successful MAS, but also for directional genome walking, individual genotyping, and genome evolution studies *(80)*.

Often, SSR linkage maps may either lack or contain a limited number of markers in certain regions of the chromosomes. In addition, crops with narrow germplasm base may overall lack polymorphism with SSR. In such cases, a higher resolution marker technology is needed that can reveal polymorphism at the DNA sequence level.

4.4. SNP Marker System

An SNP marker assays point mutations in the genomic DNA and offers unlimited opportunities for developing high-resolution maps. Existing data on corn suggest that, on average, one SNP is found every 48 to 131 bp. Maize is predicted to have over 20 million informative SNPs that may be available for analysis *(81,82)*. Cereon Corporation a Monsanto company, (http://www.arabidopsis.org/cereon/) has identified over 37,344 SNPs in *Arabidopsis* between Columbia and Landsberg ecotypes. No comparable data exist in any other plant species. As efforts are invested by public and private institutions toward genome, Expressed Sequenced Tag, and cDNA sequencing from diverse genotypes within a species, a wealth of genomic sequence information will become available.

The current strategy for SNP identification is PCR amplification of the target DNA from many diverse genotypes, followed by cloning and sequencing of those fragments. The sequences are then aligned using software to find SNPs, which are then validated. Because of the extensive effort involved, which can be cost prohibitive, a full SNP marker system for any of the crops is not available yet for MAS applications.

An alternative approach for obtaining SNP information is the use of public databases, in which sequences of the gene homologs have been reported from multiple genotypes. All of these SNPs still have to be validated prior to use.

Although it is a daunting task to develop a large number of SNP markers from scratch, a more practical and faster approach is to convert the existing RFLP markers into SNP markers that represent defined map locations. A framework of such SNP markers, when complemented with another marker system such as SSR, will provide an effective tool for MAS. A similar approach can be applied when SSR markers are scanty in certain regions of the chromosomes. These "gaps" can be filled with SNP markers, rendering maps effective for MAS.

Several detection technologies are currently used for scoring SNP polymorphisms *(83)*. SNP markers are amenable to higher throughput because no gel-based separation of DNA fragments is required. Target DNA is generally amplified and then subjected to SNP analysis using any of the following approaches: oligonucleotide hybridization, nuclease cleavage of mismatches, oligonucleotide ligation, primer extension, or direct sequencing. No clearly preferred technologies considering the cost and throughput of SNP markers have emerged, although technologies such as pyrosequencing *(82,84)* and matrix-assisted laser desorption/ionization time-of-flight mass spectrometry (MALDI-TOF MS) seem to be commonly practiced in laboratories *(85,86)*.

Codominant SNPs are less polymorphic than SSRs because only two alleles of SNPs can be found in any given population. This disadvantage can be offset by analyzing haplotypes (or several SNPs present on a stretch of DNA) to give polymorphism equivalent to RFLP *(82)*. In maize, such stretches are reported to be 100–200 bp long before linkages are broken *(81)*. Because maize is considered a very polymorphic species, SNPs need to be understood in species with low polymorphism, such as soybean and melon, for its full utilization in MAS projects *(87)*.

Because genomes are organized such that clusters of genes are separated by the repeated sequences *(88)*, no one marker strategy may be able to provide the best genome coverage desired. Therefore, a combination of marker systems may prove most practical. A plethora of other strategies for creating new markers with random amplification of polymorphism (RAPD) and single-primer amplification reaction (SPAR) techniques *(89,90)* and for modifying the existing markers with sequence-characterized amplified regions (SCAR) and sequence-tagged site (STS) strategies *(60,67)* are available to provide higher efficiency screening. Such markers can be added, as necessary, to the marker systems described here for desired marker density or for filling up the "gaps" in the chromosomal regions. The overall goal is to obtain maximum polymorphism that is evenly spaced throughout the chromosomes for the most efficient marker-assisted introgression.

Following transgene introgression, events undergo several cycles of seed increases prior to final seed production. The product registration process begins many years prior to commercialization and is discussed in Chapter 11 in this book.

5. Product Quality

Quality of seed products has been an important aspect of the seed business in several countries. Variety identity and seed quality standards are obligatory for the seed producers. The introduction of GM products into genetic backgrounds essentially similar to their nontransgenic counterparts has elevated the purity standards to a new level *(91)*. Segregation of GM products in the marketplace has also led to revisiting the breeding and production practices across the industry, particularly in cross-pollinating crops such as maize *(92)*.

Driven by consumer concern and legislation regulating the levels of GM in food products in European countries *(93)*, the technology for detection of GM in non-GM backgrounds has advanced considerably *(10)*. The methods are generally based on PCR technology, and commercial kits are available to detect approved GM products *(94)*. Whereas the adventitious presence of GM in non-GM products has been the driver behind improved field production practices; the adventitious presence of GM is also important to maintain the integrity of the GM products. The threshold levels for the presence of unintentional GM are variable, depending on the country importing the grain.

6. Conclusion

Currently, 5–10 years are invested in developing and registering a GM product, which does not include the gene discovery time. Advances in several technological areas that will be necessary to minimize this time include improvement in gene delivery techniques leading to direct transformation of elite varieties bypassing the tissue culture steps, development of techniques for directing transgene insertions into desirable genomic locations *(95)* for stable and targeted transgene expression *(96)*, a chip-based marker system for rapid screening of genomes during trait introgression, and evolution of trait-specific identity preservation and detection methods to launch the products into the marketplace successfully.

The benefits of GM products are evidenced by the vast increase in the products and the planting acreage around the world. Introduction of GM products with new traits is expected to grow in the near future with advances in molecular tool development and gene discovery technologies *(97)*. In this dynamic situation, regulations and environmental risk studies will play a significant role *(98)*. Advances in analytical technologies are improving the ability to detect the adventitious presence of GM in non-GM grain. Without global acceptance of threshold limits for the presence of GM products, growers and food producers have to extend their resources to segregate products in the distribution channel.

Acknowledgments

We thank Don Blackburn, Jane Stautz, Tom Patterson, and David Stimpson for incisive discussions. We are also grateful to Steve Thompson, Sam Reddy, Wilas Nirunsuksiri, and Siva Kumpatla for their critical comments and suggestions on this chapter. The efforts of Greg Schulenberg and Jill Rader in creating figures are highly appreciated.

References

1. Vasil, I. K. (1988) Progress in the regeneration and genetic manipulation of cereal crops. *Biotechnology (NY)* **6**, 397–402.
2. Narasimhulu, S. B., Deng, X., Sarria, R., and Gelvin, S. B. (1996) Early transcription of Agrobacterium T-DNA genes in tobacco and maize. *Plant Cell* **8**, 873–876.
3. Potrykus, I. (1991) Gene transfer to plants: assessment of published approaches and results. *Annu. Rev. Plant Physiol. Plant Mol. Biol.* **42**, 205–225.
4. Hinchee, M. A. W., Corbin, D. R., Armstrong, C. L., et al. (1994) Plant transformation. In *Plant Cell and Tissue Culture*. (Vasil, I. K. and Thorpe, T. A., eds.). Kluwer, Dordrecht, The Netherlands, pp. 231–270.
5. Birch, R. G. (1997) Plant transformation: problems and strategies for practical application. *Annu. Rev. Plant Physiol. Plant Mol. Biol.* **48**, 297–326.
6. National Agricultural Statistics Service (NASS). (2003) Prospective Plantings. Available at: http://usda.mannlib.cornell.edu/reports/nass/field/pcp-bp/pspl0302.pdf. Accessed March 31, 2003.
7. Petolino, J. F., Roberts, J. L., and Jayakumar, P. (2003) Plant cell culture — a critical tool for agricultural biotechnology. In *Hand Book of Industrial Cell Culture: Mammalian, Microbial, and Plant Cells*. (Vinci, V. A. and Parekh, S. R., eds.). Humana, Totowa, NJ, pp. 243–258.
8. Merlo, D. (2003) Molecular tools for engineering plant cells. In *Hand Book of Industrial Cell Culture: Mammalian, Microbial, and Plant Cells*. (Vinci, V. A. and Parekh, S. R., eds.). Humana, Totowa, NJ, pp. 217–241.
9. Mumm, R. H. and Walters, D. S. (2001) Quality control in the development of transgenic crop seed products. *Crop Sci.* **41**, 1381–1389.
10. Gryson, N., Rooms, T., Messens, K., and Dewettinck, K. (2002) Labeling, detection and traceability of GMOs. *Cervisia* **27**, 38–40.
11. Anklam, E., Gadani, F., Heinze, P., Pijnenburg, H., and Van Den Eede, G. (2002) Analytical methods for detection and determination of genetically modified organisms in agricultural crops and plant-derived food products. *Eur. Food Res. Technol.* **214**, 3–26.
12. Han, K.-H., Ma, C., and Strauss, S. H. (1997) Matrix attachment regions (MARs) enhance transformation frequency and transgene expression in poplar. *Transgenic Res.* **6**, 415–420.
13. Kado, C. I. (1991) Molecular mechanisms of crown gall tumorigenesis. *Crit. Rev. Plant Sci.* **10**, 1–32.
14. Zymbryski, P. C. (1992) Chronicles from the *Agrobacterium*-plant cell DNA transfer story. *Annu. Rev. Plant Physiol. Plant Mol. Biol.* **43**, 465–490.
15. Frame, B. R., Shou, H., Chikwamba, R.K., et al. (2002). *Agrobacterium tumefaciens*-mediated transformation of maize embryos using a standard binary vector system. *Plant Physiol.* **129**, 13–22.
16. Hei, Y., Ohita, S., Komari, T., and Kumashiro, T. (1994) Efficient transformation of rice (*Oryza sativa* L.) mediated by *Agrobacterium* and sequence analysis of the boundaries of the T-DNA. *Plant J.* **6**, 271–282.
17. Petolino, J. F. (2001) Direct DNA delivery into intact cells and tissues. In *Transgenic Plants and Crops*. (Hui, Y. H., Khachatourians, G. G., McHughen, A., Nip, W. K., and Scorza, R., eds.). Marcel Dekker, New York, pp. 137–143.
18. Song, P., Cai, C. Q., Skokut, M., Kosegi, B. D., and Petolino, J. F. (2002) Quantitative real-time PCR as a screening tool for estimating transgene copy number in WHISKERS™ derived transgenic maize. *Plant Cell Rep.* **20**, 948–954.
19. Gelfand, D. H., Holland, P. M., Saiki, R. K., and Watosno, R. M. (1993) Homogenous Assay System Using the Nuclease Activity of a Nucleic Acid polymerase. US patent 5,210,015, May 11.
20. German, M. A., Kandel-Kfir, M., Swarzberg, D., Matsevitz, T., and Granot, D. (2003) A rapid method for the analysis of zygosity in transgenic plants. *Plant Sci.* **164**, 183–187.

21. Hobbs, S. L. A., Warkentin, T. D., and DeLong, C. M. O. (1993) Transgene copy number can be positively or negatively associated with transgene expression. *Plant Mol. Biol.* **21**, 17–26.
22. Kumpatla, S. P., Chandrasekharan, M. B., Iyer, L. M., Li, G. and Hall, T. C. (1998) Genome intruder scanning and modulation systems and transgene silencing. *Trends Plant Sci.* **3**, 97–104.
23. Matzke, M. A., Aufsatz, W., Kanno, T., Mette, M. F., and Matzke, J. M. (2002) Homology-dependent gene silencing and host defense in plants. *Adv. Genet.* **46**, 235–275.
24. Ingham, D. J., Beer, S., Money, S., and Hansen, G. (2001) Quantitative real-time PCR assay for determining transgene copy number in transformed plants. *BioTechniques* **3**, 132–140, 2001
25. Lee, L. G., Connell, C. R., and Bloch, W. (1993) Allelic discrimination by nick-translation PCR with fluorogenic probes. *Nucleic Acids Res.* **21**, 3761–3766.
26. Livak, K. J., Flood, S. J., Marmaro, J., Giusti, W., and Deetz, K. (1995) Oligonucleotides with fluorescent dyes at opposite ends provide a quenched probe system useful for detecting PCR product and nucleic acid hybridization. *PCR Methods Appl.* **4**, 357–362.
27. Kononov, M. E., Bassuner, B., and Gelvin, S. B. (1997) Integration of T-DNA binary vector "backbone" sequences into the tobacco genome: evidence for multiple complex patterns of integration. *Plant J.* **11**, 945–957.
28. Register, J. C. (1997) Approaches to evaluating the transgenic status of transformed plants. *Trends Biotechnol.* **15**, 141–146.
29. Wassenegger, M. (2002) Gene Silencing. *Int. Rev. Cytol.* **219**, 61–113.
30. James, V. A., Avart, C., Worland, B., Snape, J. W., and Vain, P. (2002) The relationship between homozygous and hemizygous transgene expression levels over generations in populations of transgenic plants. *Theor. Appl. Genet.* **104**, 553–561.
31. Lyamichev, V., Mast, A. L., Hall, J. G., et al. (1999) Polymorphism identification and quantitative detection of genomic DNA by invasive cleavage of oligonucleotide probes. *Nat. Biotechnol.* **17**, 292–296.
32. Ghosh, S. S., Eis, P. S., Blumeyer, K., Fearon, K., and Millar, D. P. (1994) Real time kinetics of restriction endonuclease cleavage monitored by fluorescence resonance energy transfer. *Nucleic Acids Res.* **22**, 3155–3159.
33. Thomas, P. S. (1980) Hybridization of denatured RNA and small DNA fragments transferred to nitrocellulose. *Proc. Natl. Acad. Sci. USA* **77**, 5201–5205.
34. Nagy, F., Kay, S. A., and Chua, N.-H. (1988) Analysis of gene expression in transgenic plants. In *Plant Molecular Biology Manual.* (Gelvin, S. B. and Schilperoort, R. A., eds.). Kluwer, Dordrecht, The Netherlands, B4, pp. 1–29.
35. Celi, F. S., Mentuccia, L., Proietti-Pannunzi, L., di Gioia, C. R. T., and Andreoli, M. (2000) Preparing polyA-containing RNA internal standard for multiplex competitive RT-PCR. *BioTechniques* **29**, 454–458.
36. Hoegh, A. M. and Hviid, T. V. F. (2001) Design of an internal DNA standard for competitive RT-PCR using partial intron sequence and an artificial linker sequence. *BioTechniques* **31**, 730–734.
37. Asselbergs, F. A. M. and Widmer, R. (2003) Rapid detection of apoptosis through real-time reverse transcriptase polymerase chain reaction measurement of the small cytoplasmic RNA Y1. *Anal. Biochem.* **318**, 221–229.
38. Eis, P. S., Olson, M. C., Takoa, T., et al. (2001) An invasive cleavage assay for direct quantitation of specific RNAs. *Nat. Biotechnol.* **19**, 673–676.
39. Winston, S. E., Fuller, S. A., and Hurrell, J. G. R. (1988) Enzyme-linked immunosorbent assays (ELISA) for detection of antigens. In *Current Protocols in Molecular Biology*, Vol. 2. (Ausubel, F. M., Brent, R., Kingston, R. E., et al., eds.). Greene and Wiley-Interscience, New York, pp. 11.2.1– 11.2.9.
40. Burnette, W. H. (1981) Western blotting: electrophoretic transfer of proteins from SDS-polyacrylamide gels to unmodified nitrocellulose and radiographic detection with antibody and radioiodinated protein A. *Anal. Biochem.* **112**, 195–203.

41. Towbin, H., Staehelin, T., and Gordon, J. (1979) Electrophoretic transfer of proteins from polyacrylamide gels to nitrocellolose sheets: procedure and some applications. *Proc. Natl. Acad. Sci. USA* **76**, 4350–4354.
42. Hamilton, A. J. and Baulcombe, D. C. (1999) A species of small antisense RNA in posttranscriptional gene silencing in plants. *Science* **286**, 950–952.
43. Sijen, T., Vijn, I., Rebocho, A., et al. (2001) Transcriptional and posttranscriptional gene silencing are mechanistically related. *Curr. Biol.* **11**, 436–440.
44. Goto, K., Kanazawa, A., Kusaba, M., and Masuta, C. (2003) A simple and rapid method to detect plant siRNAs using nonradioactive probes. *Plant Mol. Biol. Rep.* **21**, 51–58.
45. Gruenbaum, Y., Naveh-Many, T., Cedar, H., and Razin, A. (1981) Sequence specificity of methylation in higher plant DNA. *Nature* **292**, 860–862.
46. Ingelbrecht, I., Van Houdt, H., Van Montagu, M., and Depicker, A. (1994) Posttranscriptional silencing of reporter transgenes in tobacco correlates with DNA methylation. *Proc. Natl. Acad. Sci. USA* **91**, 10,502–10,506.
47. Jones, L., Hamilton, A. J., Voinnet, O., Thomas, C. L., Maule, A. J., and Baulcombe, D. C. (1999) RNA–DNA interactions and DNA methylation in post-transcriptional gene silencing. *Plant Cell* **11**, 2291–2301.
48. Clark, S. J., Harrison, J., Paul, C. L., and Frommer, M. (1994) High sensitivity mapping of methylated cytosines. *Nucleic Acids Res.* **22**, 2990–2997.
49. Myohanen, S., Wahlfors, J., and Janne, J. (1994) Automated fluorescent genomic sequencing as applied to the methylation analysis of the human ornithine decarboxylase gene. *DNA Seq.* **5**, 1–8.
50. Oakeley, E. J. (1999) DNA methylation analysis: a review of current methodologies. *Pharmacol. Ther.* **84**, 389–400.
51. Burri, N. and Chaubert, P. (1999) Complex methylation patterns analyzed by single-strand conformation polymorphism. *BioTechniques* **26**, 232–234.
52. Herman, J. G., Graff, J. R., Myohanen, S., Nelkin, B. D., and Baylin, S. B. (1996) Methylation-specific PCR: a novel PCR assay for methylation status of CpG islands. 1996. *Proc. Natl. Acad. Sci.* **93**, 9821–9826.
53. Velinov, M., Gu, H., Genovese, M., Duncan, C., Brown, W. T., and Jenkins, E. (2000) The feasibility of PCR-based diagnosis of Prader–Willi and Angelman syndromes using restriction analysis after bisulfite modification of genomic DNA. *Mol. Genet. Metab.* **69**, 81–83.
54. Xiong, Z. and Laird, P. W. (1997) COBRA: a sensitive and quantitative DNA methylation assay. *Nucleic Acids Res.* **25**, 2532–2534.
55. Tanksley, S. D., Young, N. D., Paterson, A. H., and Bonierbale, M. W. (1989) RFLP mapping in plant breeding: new tools for an old science. *Biotechnology* **7**, 257–264.
56. Hillel, J., Schaap, T., Haberfeld, A., et al. (1990) DNA fingerprints applied to gene introgression in breeding programs. *Genetics* **124**, 783–789.
57. Tanksley, S. D. (1983) Molecular markers in plant breeding. *Plant Mol. Biol. Rep.* **1**, 3–8.
58. Zeven, A. C., Knott., D. R., and Johnson, R. (1983) Investigation of linkage drag in near isogenic lines of wheat by testing for seedling reaction to races of stem rust, leaf rust and yellow rust. *Euphytica* **32**, 319–327.
59. Stam, P. and Zeven, A. C. (1981) The theoretical proportion of the donor genome in near-isogenic lines of self-fertilizers bred by backcrossing. *Euphytica* **30**, 227–238.
60. Staub, J. E., Serquen, F. C., and Gupta, M. (1996) Genetic markers, map construction, and their application in plant breeding. *Hort. Sci.* **31**, 729–741.
61. Botstein, D., White, R. L., Skolnick, M., and Davis, R. W. (1980) Construction of a genetic linkage map in man using restriction length polymorphisms. *Am. J. Hum. Genet.* **32**, 314–331.
62. Southern, E. M. (1975) Detection of specific sequences among DNA fragments separated by gel electrophoresis. *J. Mol. Biol.* **98**, 503–517.

63. Murray, M. G., Chyi, Y. S., Cramer, J. H., et al. (1991) Application of restriction fragment length polymorphism to maize breeding. *Plant Mol. Biol.* **2**, 249–261.
64. Burow, M. D. and Blake, T. K. (1998) Molecular tools for the study of complex traits. In *Molecular Dissection of Complex Traits*. (Paterson, A. H., ed.). CRC, New York, pp. 13–29.
65. Haanstra, J. P. W., Wye, C., Verbakel, H., et al. (1999) An integrated high-density RFLP-AFLP map for tomato based on two *Lycopersicon esculentum L. pennellii* F2 populations. *Theor. Appl. Genet.* **99**, 254 – 271.
66. Sharopova, N., McMullen, M. D., Schultz, L., et al. (2002) Development and mapping of SSR markers for maize. *Plant Mol. Biol.* **48**, 463–481.
67. Olson, M., Hood, L., Cantor, C., and Botstein, D. (1989) A common language for physical mapping of the human genome. *Science* **245**, 1434–1435.
68. Vos, P., Hogers, R., Bleeker, M., et al. (1995) AFLP: a new technique for DNA fingerprinting. *Nucleic Acids Res.* **23**, 4407–4414.
69. Mank, M. V. R., Antonise, R., Bastiaans, E., et al. (1999) Two high-density AFLP linkage maps of *Zea Mays* L.: analysis of distribution of AFLP markers. *Theor. Appl. Genet.* **99**, 921–935.
70. Tautz D. and Renz, M. (1984) Simple sequence repeats are ubiquitous components of eukaryotic genomes. *Nucleic Acids Res.* **12**, 4127–4138.
71. Cardle, L., Macaulay, M., Marshall, D. F., Milbourne, D., Ramsay, L., and Waugh, R. (2000) *SSR Frequency and Occurrence in Plant Genomes.* Annual report, Scottish Crop Research Institute, No. 1999/2000, pp. 108–110.
72. Condit, R. and Hubbell, S. P. (1991) Abundance and DNA sequence of two-base repeat regions in tropical tree genomes. *Genome* **34**, 66–71.
73. Röder, M. S., Plashke, J., König, S. U., et al. (1995) Abundance, variability and chromosomal location of microsatellites in wheat. *Mol. Gen. Genet.* **246**, 327–333.
74. Tamarino, G. and Tingey, S. (1996) Simple sequence repeats for germplasm analysis and mapping in maize. *Genome* **39**, 277–287.
75. Schug, M. D., Hutter, C. M., Wetterstrand, K. A., Gaudette, M. S., Mackay, T. F. C., and Aquadro, C. F. (1998) The mutation rates of di-, tri- and tetra-nucleotide repeats in *Drosophila melanogaster*. *Mol. Biol. Evol.* **15**, 1751–1760.
76. Levinson, G. and Gutman, G. A. (1987) Slipped-strand mispairing: a major mechanism for DNA sequence evolution. *Mol. Biol. Evol.* **4**, 203–221
77. Senior, M. L., Murphy, J. P., Goodman, M. M., and Stuber, C. W. (1998) Utility of SSR for determining genetic similarities and relationship in maize using an agarose gel system. *Crop Sci.* **38**, 1088–1098.
78. Mitchell, S. E., Kresovich, S., Jester, C. A., Javir Hernandez, C., and Szewc-McFadden, A. K. (1997) Application of multiplex PCR and fluorescence-based, semi-automated allele sizing technology for genotyping plant genetics resources. *Crop Sci.* **37**, 617–624.
79. Connell, J. P., Pammi, S., Iqbal, M. J., Huizinga, T., and Reddy, A. S. (1998) A high throughput procedure for capturing microsatellites from complex plant genomes. *Plant Mol. Biol. Rep.* **16**, 341–349.
80. Matsuoka, Y., Mitchell, S. E., Kresovich, S., Goodman, M., and Doebley, J. (2002) Microsatellites in Zea—variability, patterns of mutations and use for evolutionary studies. *Theor. Appl. Genet.* **104**, 436–450.
81. Tanaillon, M. T., Sawkins, M. C., Long, A. D., Gaut, R. L., Doebley, J. F. and Gaut, B. S. (2001) Patterns of DNA sequence polymorphism along chromosome 1 of maize (*Zea mays* spp *mays* L.). *Proc. Natl. Acad. Sci. USA* **98**, 9161–9166.
82. Rafalski, J. A. (2002) Novel genetic mapping tools in plants: SNPs and LD-based approaches. *Plant Sci.* **162**, 329–333.
83. Gut, I. G. (2001) Automation in genotyping single nucleotide polymorphisms. *Hum. Mutat.* **17**, 475–492.

84. Ronaghi, M., Uhlen, M., and Nyren, P. (1998) A sequencing method based on real-time pyrophosphate. *Science* **281**, 363–365.
85. Ross, P., Hall, L., and Haff, L. A. (2000) Quantitative approach to single-nucleotide polymorphism analysis using MALDI-TOF mass spectrometry. *BioTechniques* **29**, 620–629.
86. Pusch, W., Kraeuter, K.-O., Froehlich, T., Stalgies, Y., and Kostrzewa, M. (2001) Genotools SNP manager: a new software for automated high-throughput MALDI-TOF mass spectrometry SNP genotyping. *BioTechniques* **30**, 210–215.
87. Shattuck-Eidens, D. M., Bell, R. N., Neuhausen, S. L., and Helentjaris, T. (1990) DNA sequence variation within maize and melon: observation from polymerase chain amplification and direct sequencing. *Genetics* **126**, 207–217.
88. Walbot, V. and Petrov, D. A. (2001) Gene galaxies in the maize genome. *Proc. Natl. Acad. Sci.* **98**, 8163–8164.
89. William, J. G. K., Kubelik, A. R., Livak, K. J., Rafalski, J. A., and Tingey, S. V. (1990) DNA polymorphisms amplified by arbitrary primers are useful genetic markers. *Nucleic Acids Res.* **18**, 6531–6535.
90. Gupta, M., Chyi, Y.-S., Romero-Severson, J., and Owen, J. L. (1994) Amplification of DNA markers from evolutionarily diverse genomes using single primers of simple-sequence repeats. *Theor. Appl. Genet.* **89**, 998–1006.
91. Lang, L. (2002) Seed production aspects of genetically modified crop varieties. *Acta Agron. Hung.* **50**, 313–319.
92. Jemison, J. M., Jr. and Vayda, M. E. (2001) Cross-pollination from genetically engineered corn: wind transport and seed source. *AgBioForum* **4**, 87–92.
93. Nap, J., Metz, P. L. J., Escaler, M., and Conner, A. J. (2003) The release of genetically modified crops into the environment. Part 1. Overview of current status and regulations. *Plant J.* **33**, 1–18.
94. Kuribara, H., Shindo, Y., Matsuoka, T., et al. (2002) Novel reference molecules for quantitation of genetically modified maize and soybean. *J. AOAC Int.* **85**, 1077–1089.
95. Ow, D. W. (1996) Recombinase directed chromosome engineering in plants. *Curr. Opin. Biotechnol.* **7**, 181–186.
96. Srivastava, V., Anderson, O. D., and Ow, D. W. (1999) Single-copy transgenic wheat generated through the resolution of complex integration patterns. *Proc. Natl. Acad. Sci. USA* **96**, 11,117–11,121.
97. Reddy, S., Larrinua, I. M., Ruegger, M. O., Shukla, V. K., and Sun, Y. (2003) Functional genomics for plant trait discovery. In *Handbook of Industrial Cell Culture* (Vinci, V. A. and Parekh, S. R., Eds.). Humana, Totowa, NJ, pp. 197–216.
98. Conner, A. J., Glare, T. R., and Nap, J-P. (2003) The release of genetically modified crops into the environment. Part II. Overview of ecological risk assessment. *Plant J.* **33**, 19–46.

9

Gene Silencing in Plants

Nature's Defense

W. Michael Ainley and Siva P. Kumpatla

1. Introduction

Plants have evolved mechanisms to limit viral infections and genomic damage that can occur by the invasion, proliferation, and expression of viruses and mobile genetic elements such as retroelements and transposons *(1)*. Up to 95% of a plant's genome is comprised of repetitive elements. The mechanisms involved with limiting expression of this "junk" deoxyribonucleic acid (DNA) have significantly hindered progress in agricultural biotechnology because DNA carrying genes of interest is often subjected to the same protective surveillance mechanisms and their expression shut down. This phenomenon, known as gene silencing, can occur immediately following integration of transgenes or over several generations. Gene silencing can affect some or all plants derived from a transgenic event, and expression can be partially or fully turned off.

Gene silencing in plants has been divided into transcriptional gene silencing (TGS) and posttranscriptional gene silencing (PTGS) (*see* Table 1 and refs. *2–7*). The primary distinction between TGS and PTGS relates to the different mechanisms responsible for reduced messenger ribonucleic acid (mRNA) levels. In TGS, transcription is completely or partially turned off and can be correlated with methylation of the promoter driving the silenced gene. Genes affected by PTGS have a normal rate of transcription (as measured by nuclear run-ons), but a reduced steady-state mRNA level. Hence, PTGS acts posttranscriptionally and is characterized by mRNA degradation. The protein-coding regions of genes silenced by PTGS are usually methylated, at least in the 3' end of the gene. PTGS is dependent on the gene that is silenced being transcriptionally active initially. TGS is mitotically and meiotically heritable whereas PTGS can be reversed between generations (i.e., expression of silenced genes is switched on at meiosis).

Transgenic events prone to silencing are characterized by recombined, complex DNA integration, particularly if an inverted repeat is present *(8)*. When an inverted repeat of sufficient length (greater than 21 bp) is transcribed, the homologous DNA can become methylated and silenced. This is true for both TGS (promoter region transcribed) and PTGS (protein-coding region transcribed). Importantly, the resultant silencing can act in *trans*, that is, silence other genes containing the homologous sequence (*see* Section 7).

From: *The GMO Handbook: Genetically Modified Animals, Microbes, and Plants in Biotechnology*
Edited by: S. R. Parekh © Humana Press Inc., Totowa, NJ

Table 1
Comparison of Transcriptional and Posttranscriptional Gene Silencing

	TGS	PTGS
Cause of reduced mRNA	Transcription is reduced	Transcription rate is not affected; RNA is degraded
Methylation pattern	Symmetrical C sites (CG and CNG); can have all Cs methylated	All Cs
Inheritance	Silencing is inherited by progeny	Expression resets at meiosis
Triggers	Insertion in heterochromatic region; inverted repeat structure of promoter, particularly if promoter region is transcribed	High expression; aberrant transcript production; inverted repeat of the coding sequence
Maintenance of silencing	Methylation and silencing maintained by maintenance DNA methyltransferase	Trigger must be available for re-establishing silencing in the next generation
Transport of silencing signal	Signal is not transported	A signal can be transported to other parts of the plant
Mechanism of *trans*-silencing	Current view is that transcription of inverted repeat of the promoter is necessary for *trans*-silencing	Should always occur in *trans*-

2. Transcriptional Gene Silencing

Although somewhat dependent on the method of DNA delivery, all transformation methods generate a significant proportion of transgenic events that do not express the gene of interest, even though they contain an intact copy of the gene (*see* refs. *4, 9,* and *10*). Nonexpression of some of these inserts is related to their chromosomal environment (position effect). Inactivation by methylation of a transgene can result from insertion into an inactive chromatin site, such as found in highly condensed heterochromatin. Methylation can also spread from flanking genomic DNA into transgenic sequences. Heterochromatic spreading, and hence methylation, of genes was found for genes bordering a DNA element in petunia that has an inverted repeat *(3)*. Spreading of a repressed chromatin state has been particularly well documented in *Drosophila (11)*. If the region of an insert that is methylated includes a promoter expressing the gene of interest, promoter activity will be reduced.

The degree of methylation can vary per individual plant, depending on a stochastically determined probability of occurrence, the environment, or zygosity (hemizygosity or homozygosity). Other than chromosomal environment, the other principle determinant of TGS appears to be insert structure. Inverted repeats of promoter regions that are transcribed are silenced by mechanisms described for PTGS in the next section *(8)*. Inverted repeats that are not transcribed are thought to be methylated by a yet to be elucidated mechanism by which plants appear to sense inverted repeat structures.

3. Posttranscriptional Gene Silencing

Posttranscriptional gene silencing is induced by RNA virus infection *(12)*, very high levels of expression of transgenes *(13,14)*, or expression of inverted repeats of transgene coding regions (*see* refs. *2, 4, 5, 7,* and *8*). Proposed models suggest that transgene and viral-induced PTGS are related mechanistically, but each has unique features. Aberrant RNAs *(15)* and nonpolyadenylated and double-strand RNAs (dsRNAs) *(16)* can also induce PTGS. Aberrant RNAs include RNAs that are transported to the cytoplasm, but are improperly transcribed or processed. In addition, it has been shown that RNAs that lack an intact open reading frame may trigger PTGS. Endogenous genes have also been observed to be silenced when released from normal expression controls because of high expression *(17)*.

3.1. PTGS During Plant Development and Growth

PTGS has been described as a stochastic event; that is, its occurrence within a population of individual plants is more or less random. Its level within a population can be altered by environmental changes. One of the best examples of the dependence of PTGS on environmental changes is from one of the pioneering studies of gene silencing. In this study, a petunia line transformed with a dihydroflavonol reductase gene, a gene conferring flower pigmentation, showed a moderate level of silencing, as indicated by production of white flowers, when maintained in a normal greenhouse environment *(18)*. However, when plants representing the same event were grown in high light, silencing was much more pronounced. Temperature has also been shown to influence the level of gene silencing *(19)*. How environmental inputs interact with the various steps of the PTGS pathway has not been thoroughly investigated. Nonetheless, these observations underscore the importance of wide geographic testing of transgenic crops.

Early work in tobacco suggested that, unlike TGS, PTGS is reversed during meiosis *(20–22)*. More recent data support a different explanation for lack of PTGS in young seedlings of an event prone to silencing. Mitsuhara et al. *(23)* used a luciferase reporter gene to show that PTGS does not occur in proliferating tissue.

3.2. Silencing Can Spread Throughout the Plant

Several lines of evidence suggest that a mobile signal of PTGS is responsible for the spread of silencing via phloem transport and cells connected by plasmadesmata *(24)*. The first indication that PTGS could spread from its site of onset came from studies demonstrating a visible phenotype of genes cosuppressed (silencing of endogenous genes) in cells that were not clonally related *(25,26)*. Further data supporting a mobile signal were obtained from experiments in which different grafting combinations of transgenic and nontransgenic scions and root stocks were tested for spreading of the silencing signal *(27,28)*.

These results clearly showed that silencing could be transmitted to a nonsilenced scion from a silenced rootstock. In addition, the silencing was sequence specific: only those genes silenced in the source of the silencing signal became silenced in the scion. Rootstocks containing silenced transgenes could silence nonsilenced scions expressing the same transgene, but at a different genomic locus *(29)*. Interestingly, in this study, the silenced state of the scions was maintained when grafted onto a nonsilenced rootstock. This result contrasts with the work of Palauqui and Vacheret *(17)*, in which the silencing was not maintained in all silenced scions after regrafting onto wild-type rootstocks. Only those scions that could trigger silencing themselves maintained a silenced state after regrafting.

Although systemic silencing can spread extensively in a plant, this phenomenon appears to be excluded from meristematic regions *(30)*. It is not clear if this exclusion is because the signal is not transmitted to these regions or if the cells contained in the meristem lack the cellular machinery necessary to respond to the signal. The silencing signal also appears excluded from seed, possibly because of its inability to cross the parent placenta. Although several models have been proposed, the nature of the silencing signal is still unknown *(24,31)*.

3.3. Mechanism of PTGS

Although many of the details are unclear, considerable progress in understanding the mechanism of PTGS has been made. For transgenes, it is currently believed that aberrant RNAs are converted to dsRNAs by cytoplasmic RNA-dependent RNA polymerases (RdRP) *(32,33)*. RNA viruses carry their own RdRP for producing dsRNAs, which is a normal part of their mode of propagation *(12)*. The dsRNAs, or their fragments, are mobile and transported through plasmadesmata and phloem, possibly resulting in the spread of silencing through the plant (*see* Section 3.2).

dsRNAs are degraded into 21–23 nucleotide small interfering RNAs (siRNAs) by an RNase thought to be similar to DICER in *Drosophila (34)*. This enzyme has RNase III homology, which is consistent with its ability to degrade dsRNAs. The siRNAs become attached to a protein contained in an RNA-induced silencing complex (RISC)

(35). A helicase is thought to convert the double-strand siRNA to a single-strand RNA molecule, which remains attached to the RISC. The attached siRNA provides the specificity of the RNase activity of the RISC to those molecules homologous to the siRNAs. Thus, once initiated, PTGS will cause the degradation of both dsRNAs and single-strand mRNAs. Although still not completely resolved, the current data suggest that the RISC cuts mRNA at a limited number of sites; the resultant free ends then serve as substrates for rapid degradation.

3.4. The Relationship of PTGS and TGS

An emerging concept mechanistically links PTGS and TGS by virtue of the requirement of dsRNAs in initiating silencing *(3,36–38)* (*see* Fig. 1). The presence of an inverted repeat in the protein-coding sequence of a gene results in a transcript containing a double-strand region. This dsRNA triggers PTGS, and in this case, the need for an RdRP to synthesize a dsRNA is circumvented.

In a similar way, an inverted copy of a promoter can be transcribed if it inserts next to a functional promoter (*see* Section 7). The dsRNA produced from the promoter is also "diced" by a DICER-like activity. As a consequence, the presence of either the dsRNA or its derivative siRNAs results in the silencing of any promoter that shares homology.

The silenced promoter becomes methylated by *de novo* methylation, with its expression partially or completely turned off. The methylation pattern of the promoter regions can be passed on to subsequent generations by maintenance DNA methyltransferases (*see* Section 4). The coding sequence also becomes methylated in transgenic events that express transcripts containing inverted repeats of this region; however, it is not likely the methylation is involved in the initiation of silencing, but could be important to its maintenance.

4. Gene Silencing and DNA Methylation

DNA methylation has been associated with both PTGS and TGS. The importance of methylation to the mechanism of gene silencing is still under investigation. Although a vast majority of the evidence shows clear correlation of methylation and silencing, it is not known if it is causally related or a secondary consequence.

Several experiments using mutations of DNA methyltransferases have shed light on the role of DNA methylation in gene silencing *(39,40)*. Plant DNA methylation can occur at sites that are symmetric between the two strands of DNA, specifically the cytosine in cysteine guanine (CG) and CNG, or at nonsymmetric sites, typically any cytosine and adenine residue. Methyltransferases that methylate CG or CNG include maintenance methyl-transferases that duplicate the methylation pattern of the template DNA strands into the newly synthesized strands of DNA during replication. These methyltransferases are important for heterochromatin formation after replication. Recent reviews provide a more extensive compilation of mutants affecting DNA methylation that relate to silencing *(39,40)*. The focus here is on the most thoroughly characterized mutant families.

Inverted repeats in transgenes can be methylated by *de novo* methyltransferases; however, it may be that only inverted repeats that are transcribed are methylated (*see*

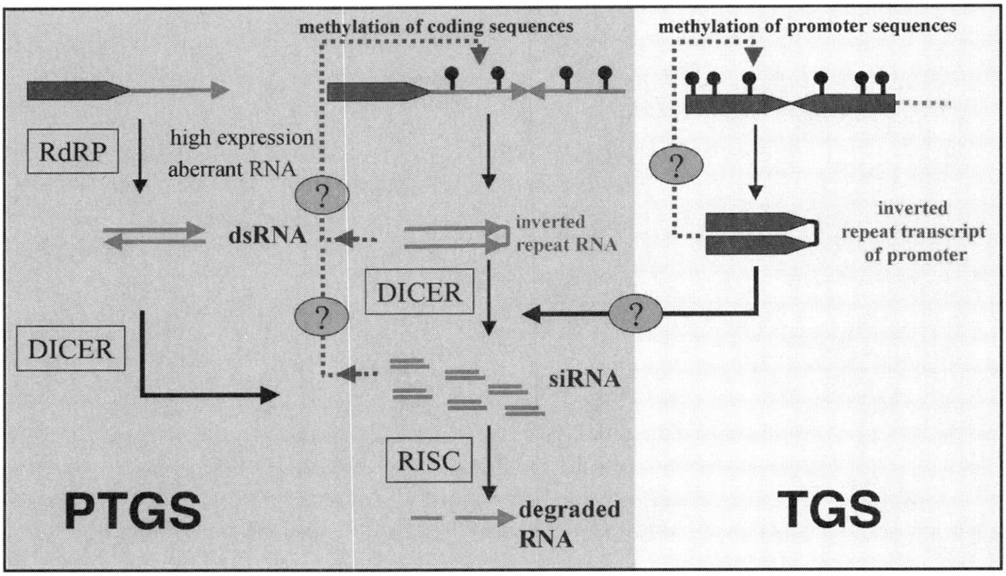

Fig. 1. RNA-dependent gene silencing: common mechanism for TGS and PTGS. Transcription of inverted repeats of the transcript coding region (*left panel: thin arrows*) and promoter region (*right panel: thick arrows*) generate double-strand RNA (dsRNA) that is digested by an RNase III-type of activity (DICER). The digestion of the promoter transcript may occur in the nucleus; the digestion of dsRNA of the coding region takes place in the cytoplasm or nucleus. Either the dsRNA or the small RNA products are thought to target methylation of the homologous genomic DNA. High levels of expression of a transgene or production of an aberrant RNA also trigger the synthesis of a dsRNA by an RNA-dependent RNA polymerase (RdRP) (*left panel*). The dsRNAs produced in the cytoplasm are digested by DICER to small interfering RNAs (siRNAs). siRNAs then bind to an RNA-induced silencing complex (RISC) and provide the sequence specificity of the RNase activity present in the RISC toward mRNAs homologous to the dsRNAs.

Section 7). Such loci can silence homologous genes at difference loci. Once the silencing locus is removed by segregation in progeny, silencing can remain *(41)*. In contrast to methylation at both asymmetric and symmetric cytosines in the inverted copy of the transgene, the copies segregated from the silencing locus are only methylated at symmetric cytosines, consistent with methylation by maintenance methyltransferases. This is in agreement with a study by Dieguez et al. *(42)*, in which a 35S promoter (modified to remove all CGs and CNGs) could be silenced in *trans*, but the silenced state was not maintained in the progeny lacking the *trans*-silencing copy of the promoter.

The high sequence and structural similarities between prokaryotic and eukaryotic DNA methyltransferases allowed the isolation of the first plant DNA methyltransferase from *Arabidopsis*, MET1 *(43)*. Inactivation of MET1 by antisense constructs resulted in pleotrophic effects *(44,45)*. One of the most notable effects was alteration of homeotic gene expression, leading to an altered floral development pattern. After several generations of inbreeding, cytosine methylation was shown to decrease by up to 90% in plants with reduced MET1 activity. Methylation of both repetitive and single-copy genes was affected. Evidence suggests that MET1 methylates cytosines in the CpG sequence *(44)*, although some data indicate that it may have a wider sequence specificity *(40)*. When a mutated MET1 is segregated away, methylation is restored, but slowly.

The role of MET1 in PTGS and TGS was recently investigated. Jones et al. *(46)* demonstrated that RNA-directed DNA methylation (RdDM) using a viral system (viral-induced gene silencing) that carries DNA fragments of a region homologous to either the promoter or the coding sequence induced methylation in those respective areas of homology. Unlike PTGS, the TGS induced in those plants carrying the promoter fragment was inherited. If the MET1 gene was suppressed, methylation of the promoter was lost in subsequent generations. Although initiation of RdDM was independent of MET1, maintenance of TGS was dependent on the MET1 gene product.

Plants are unique in having a DNA methyltransferase that methylates CpNpG, CHROMOMETHYLASE 3 (CMT3). CMT3 belongs to a family of DNA-binding proteins characterized as having a chromodomain and often associated with heterochromatin. Mutants of this gene were isolated in *Arabidopsis (47,48)* and showed reduced CpNpG methylation and reactivated expression of endogenous retrotransposon sequences. CMT3 mutants are morphologically and developmentally normal.

4.1. Genes Other Than Methyltranferases Involved With DNA Methylation

Mutants that have reduced methylation of genomic DNA in *Arabidopsis* were identified by direct screening of plants with altered methylation of centromeric regions *(49)*. These mutants, *ddm1* (DNA demethylation 1), have retained as little as 30% of cytosine methylation found in wild-type plants. Interestingly, phenotypic changes are minimal in these mutants. Further characterization demonstrated that DNA methyltransferase activity was not changed *(50)*. Once isolated, the gene was shown to have homology to the SWI2/SNF2 family of proteins *(51)*, which have an ATPase-dependent chromatin remodeling activity *(52)*.

Both *DDM1* and *MET1* are necessary for maintenance of TGS *(53)*. In *met1* mutants, most, but not all, previously transcriptionally silenced transgenes are expressed. PTGS-silenced loci are differentially released from silencing in *ddm1* and *met1* mutants *(53)*. In a *met1* background, PTGS is inhibited in a stochastic manner and continues to be released during plant development, whereas in a *ddm1* background, some plants show restoration of gene expression in F3 plants, but the level of PTGS is constant throughout development. This suggests that DDM1 could be acting early in development or could be affecting establishment, but not maintenance, of PTGS.

A gene was identified by Amedeo at el. *(54)* that releases transcriptional silencing of methylated genes, but does so without modifying the methylation sites or levels in the genes. This gene, known as the Morpheus molecule or *mom1* gene, has sequence similarities to the SWI2/SNF2 family of proteins *(52)*. Interestingly, *mom1* mutant plants displayed no obvious abnormal phenotype, even after several generations of inbreeding. Silencing of loci that had been reactivated by *mom1* mutants was restored immediately after introduction of a wild-type MOM gene.

In experiments using plants containing mutations in both the *mom1* and *ddm1* genes were tested for their effects on gene silencing, there was an additive or synergistic effect of the combined mutants on expression of transgenes *(55)*. Interestingly, the centromeric regions of the double mutants were characterized by apparent disintegration of their organization and even a tendency to aggregate into a superheterochromatic structure. One explanation considered by the authors for the extreme phenotypes of the plants was that the functions of the two genes are complementary for normal chromatin structure and the maintenance of epigenetic states in plants. Alternatively, independent pathways for gene silencing with overlapping targets or a limited set of chromatin regions affected by the *mom* gene product, such as multicopy, inverted repeat regions may explain these results *(56)*.

4.2. De Novo *Methyltransferases*

In addition to maintenance methylation activities, plants and other organisms have a second class of DNA methyltransferases that methylate DNA at each round of replication. This activity, known as *de novo* methylation, can methylate DNA at both symmetrical and nonsymmetrical sites and does not require hemi-methylated DNA to direct where the methylation occurs. Based on homology with proteins serving similar functions in mice, plant genes have been isolated and shown to have *de novo* methylation activity.

In an elegant set of experiments, Cao and Jacobsen *(57,58)* characterized what are likely the genes responsible for the most important *de novo* methylation activities in *Arabidopsis*. These two genes, DOMAINS REARRANGED METHYLASE 1 and 2 (DRM1 and DRM2) had previously been identified as having homology to the mammalian Dnmt3 methyltransferases. Cao and Jacobsen developed *Arabidopsis* lines containing *drm1* and *drm2* mutants, singly or in combination, and used a *cmt3* mutant together with *drm1 drm2* double mutant lines. The double-homozygous *drm1 drm2* lines were indistinguishable from wild-type *Arabidopsis* even after five generations of inbreeding.

Cao and Jacobsen studied the *de novo* methylation of two *Arabidopsis* genes, FWA and Superman (SUP). The FWA gene contains a direct repeat that is normally methylated and silenced, but epigenetic variants have been isolated that are not methylated. The hypomethylation of these variants is stable. When transformed with a transgenic copy *(57)* of the gene, the *fwa* epigenetic variant becomes methylated by *de novo* methylation. In *drm1 drm2* mutant lines, the *fwa* gene did not become methylated when using the FWA transformation protocol. Interestingly, when the nonmethylated epigenetic variants were outcrossed to lines wild type for the DRM genes, the FWA gene remained hypomethylated even in the presence of the transgenic copy. They found similar results with SUP, a homeotic gene, using a transgenic copy of the gene that contained an inverted repeat to initiate methylation and thus silencing of the native gene. The authors suggested that these results indicate that the *de novo* methylation of the DNA occurs prior to insertion of the gene into the genome or within the first generation; hence, in the absence of the initial methylation event, methylation will not occur. Support that these genes are not involved with maintenance of methylation is that there is no change in the methylation pattern in *drm1 drm2* plants after several generations of inbreeding.

Cao and Jacobsen extended their initial study to include the *cmt3* mutant and defined what context of cytosine sites (symmetric or asymmetric) are affected in the various mutant combinations *(58)*. They found that, in some genes analyzed, *drm1 drm2* mutants lacked methylation at all asymmetric and CpNpG sites. The *cmt3* eliminated some, but not all, methylation of CpNpG sites. In *drm1 drm2 cmt3* plants, all methylation was absent except at CpG sites. Although none of the *drm1*, *drm2*, or *cmt3* mutants had any phenotypic changes from wild-type plants, the triple mutants showed developmental retardation, reduced plant size, and partial sterility. Together, these results suggest that the DRM and CMT3 genes act redundantly during development. Based on sequence specificity and types of phenotypic changes seen in mutant plants, these genes are likely to affect a different set of genes or DNA than the DDM1 and MET1 genes.

4.3. Chromatin-Related Changes Associated With Gene Silencing

Data from the yeast *Schizosaccharomyces pombe* suggest a relationship between RNA-dependent silencing and chromatin structure *(59)*. The formation of silent chromatin requires histone H3 to be deacetylated and then methylated on lysine 9. Heterochromatin protein 1 binds to the methylated lysine 9 residue. Methylation of DNA then converts the chromatin into a silenced state. It was found that disruption of RNA-dependent gene silencing reduces methylation of lysine 9. One possible explanation is that dsRNAs may facilitate Clr4 histone H3 methylase via its chromodomain to bind to homologous DNA and methylate histone H3 lysine 9.

4.4. Use of Mutations Affecting DNA Methylation

The pattern emerging in studies of DNA methylation in plants is that the proteins that influence methylation have overlapping and redundant specificities. To understand gene silencing fully, mutants or methods of down-regulation of key genes involved with methylation need to be used singly and in combination with others. The use of some gene mutants, such as *mom1*, may restore gene expression of a silenced gene and may be phenotypically neutral. Further work will be necessary to show if this would be a viable approach to limit TGS or PTGS. Importantly, this gene or other genes alone would not necessarily work with all transgenes.

One approach for using multiple mutants having serious long-term detrimental effects when expressed in whole plants would be to express the mutant phenotypes in a tissue or developmentally specific manner. In the case of TGS, it would most likely be simpler to produce transgenic events that lack DNA structures prone to activate methylation, such as inverted repeats.

5. Suppressors of Silencing

PTGS is postulated to have arisen as a defense mechanism against viruses and transposons. Many viruses have evolved repression mechanisms to prevent or reduce PTGS; in many cases, these mechanisms allow infection by a broad range of viruses when PTGS has been inactivated *(60–64)*. The best studied viral-encoded repressors are helper component-proteinase (HC-Pro) from potyviruses *(65–67)*, PVX p25 from potato virus X *(68)*, CMV 2b from cucumber mosaic virus *(66)*, and p19 from tombusviruses *(69,70)*. Another repressor activity is a plant-derived calmodulin protein rgs-CaM *(71)*, which was discovered because of its interaction with HC-Pro.

HC-Pro prevents or reverses both transgene-induced and virus-induced silencing *(65)* and thereby acts at steps of the PTGS process in common between these two pathways of silencing. Studies have demonstrated that HC-Pro prevents accumulation of siRNAs, but has no effect on mobilization of the silencing signal or DNA methylation *(72,73)*. Possible mechanisms of how HC-Pro reduces silencing include (1) preventing DICER from binding to dsRNA; (2) preventing cleavage of the dsRNA by DICER; or (3) interfering with incorporation of the siRNAs into RISC, making the siRNAs unstable *(72)*. Interestingly, HC-Pro has no effect on TGS *(74)*, even if the TGS results from transcription of inverted copies of promoter sequences *(75)*. This apparent contradiction is explained if the dsRNA produced by the transcript of a promoter region remains in the nucleus and is degraded to small RNAs (smRNAS) *(75)*, possibly by a gene similar to DICER, such as the CARPEL FACTORY (CAF) gene from *Arabidopsis (35)*. In this way, the cytoplasmically located HC-Pro would never interact with the components responsible for TGS.

The repressors CMV 2b, PVX p25, and p19 target different parts of the PTGS process than HC-Pro. Although CMV 2b was suggested to interfere with PTGS initiation in new growth in infected plants *(66,76)*, studies showed that it inhibits spread of the silencing signal *(77)*. Although other alternatives are possible, the most likely mechanism for CMV 2b inhibition of PTGS is that it interacts with the silencing signal directly. Interestingly, the protein has a nuclear targeting signal that may be important for its inhibition of reinitiation of genomic DNA methylation at each generation.

PVX p25, although not as successful as HC-Pro in reducing silencing, prevents the spread of the systemic silencing signal *(68)*. The effectiveness of a viral infection depends on the relative speed of viral spread and transport of the silencing signal *(60)*. If the silencing signal precedes the virus into uninfected cells, the viral infection is attenuated. Hence, the effectiveness of silencing suppressors that limit the systemic silencing signal depends on the rate of spread of the signal relative to the availability of the suppressor.

p19 also was characterized as involved with suppression of the movement of the systemic silencing signal *(69)*, but plants expressing the protein showed altered morphology *(70)*. This protein was shown to bind to two nucleotide overhangs of 21–23 nucleotide siRNAs. Interestingly, when used in *Agrobacterium* coinfection experiments, p19 enhanced expression of a reporter gene more efficiently than p25, 2b, or HC-Pro *(78)*.

5.1. Strategies for Using Silencing Repressors in Transgenic Plants

The primary value of silencing repressors is to allow the production of very high levels of protein. This includes transgenic events containing genes of interest that are expressed from a strong promoter or, alternatively, viral-based expression systems in which very high levels of mRNA or virus RNA are made. Based on currently available information, HC-Pro appears to be the best candidate for preventing silencing because it is independent of the upstream initiator of silencing (transgene or virus). However, overexpression of HC-Pro or rgs-CaM results in abnormal growth of plants *(65)*. A mutated form of HC-Pro has been identified that has significantly reduced phenotypic effects *(79)*. If used in whole plants, the expression of native HC-Pro should be limited temporally or spatially to prevent detrimental effects.

Overexpression of rgs-CaM is an option with similar benefits and limitations of Hc-Pro, but the gene may act in a host plant-specific manner. CMV 2b, p19, and PVX p25 could also be of some benefit, depending on the specific gene, its expression pattern, and host plant. CMV 2b reduced, but did not eliminate, silencing in plants in which it was coexpressed with a silencing transgene *(77)*.

6. Mutations of Genes Involved in Silencing

Screens for mutants of proteins involved with PTGS were undertaken in the laboratories of Vaucheret *(33,80)* and Baulcombe *(32)*. This work identified key genes involved with PTGS of transgenes and one gene involved with PTGS of both transgenes and virus-induced silencing. The screens did not lead to the isolation of genes responsible for several activities known to be important for PTGS. The Vaucheret laboratory used ethyl methylsulfonate (EMS) mutagenesis of plants containing a direct repeat of a reporter gene that was silenced at a high frequency. Three genetic loci were identified with various numbers of alleles for each. In the presence of these mutant backgrounds, coined *sgs* (suppressor of gene silencing), a reporter gene was not silenced and contained a lower level of methylation. A fourth class of genes, *ago*, is related to PAZ/PIWI proteins, and most alleles displayed significant developmental abnormalities and were sterile *(81)*. A few alleles, referred to as hypomorphic *ago1* mutants, were obtained that reduced PTGS, but lacked the same level of structural abnormalities.

The screen carried out in the Baulcombe laboratory used a line carrying two transgenic inserts, which were partially silenced when present individually, but when together, were severely silenced *(32)*. One insert expressed a replicating virus that coded for a reporter gene. The second insert contained a transgenic cassette that expressed the same reporter gene using a strong constitutive promoter. Four complementation groups of silencing defective (*sde*) mutants were found, none of which displayed any observable differences from wild-type plants. The first two classes, *sde1* and *sde2*, reversed PTGS completely; the third (*sde3*) was effective in restoring expression in true leaves and flowers, but not tissues developed earlier in plant growth (cotyledons) and hypocotyls. Plants from all classes of *sde* mutants lack DNA methylation associated with PTGS and have a reduced level of siRNAs.

Progress has been made recently in elucidating the function of some of the genes identified in the mutant screens. The genes corresponding to the SDE1 and SGS2 loci were isolated and were the same gene *(32,33)*. The coding sequence was highly homologous to genes encoding RdRP from several organisms. Interestingly, mutants of this gene completely inhibit PTGS of transgenes, but not silencing caused by viral infections. The lack of a requirement of this activity by viruses is explained by the fact that their genomes contain their own RdRP. RdRP activity for transgene PTGS converts single-strand mRNA (presumably aberrant in some way) into the dsRNA that initiates PTGS. It has been postulated that the siRNAs that are the end product of PTGS can serve as primers for dsRNA synthesis of all homologous mRNAs *(32)*. Thus, once initiated, a self-propagating cycle can quickly reduce cytoplasmic RNA levels of the affected mRNA.

A second gene that may also be important for dsRNA synthesis is the product of the SDE3 gene *(82)*. This gene has homology to RNA helicases. Similar to the SDE1 locus, mutants of this gene preferentially affect PTGS of transgenes. How an RNA helicase activity is used in the PTGS process is not known at present. One proposal is that the RNA helicase is important for converting aberrant single-strand RNA into dsRNA *(82)*.

It is not clear if any plants bearing mutations equivalent to those described here could be used to reduce transgene-induced PTGS in whole plants. Although they are critical to transgene-induced PTGS, the *sde1*, *sde3*, and *sgs3* mutants are more susceptible to some, but not all, viruses, suggesting different mechanisms of virus defense *(32,33,82)*. The use of these mutations will be dependent on the plant host and the particular viruses that infect the host.

7. *Trans*-Silencing: Phenomenon and Features That Contribute to the *Trans*-Silencing Capability of a Locus

Trans-silencing is a phenomenon in which a silenced locus inactivates an unlinked target locus with which it shares sequence homology *(10,83)*. Homology of promoter sequences leads to TGS, whereas homology in coding regions leads to PTGS. A growing body of evidence regarding TGS indicates that *trans*-silencing and methylation of a target locus in the presence of a stably methylated silencing locus relies on a promoter sequence-specific signal that originates at the silencing locus and directs *de novo* methylation of homologous promoter(s) at the target locus.

Such a signal could be the result of a DNA–DNA pairing or RNA–DNA interaction. Although DNA–DNA pairing as a mechanism to initiate inactivation or to impose a silencing state from one sequence to the other has been well documented in the filamentous fungus *Ascobolus immersus (84)*, currently very few studies exist that suggest such a possibility in higher plants *(85)*. To date, no direct evidence has been obtained demonstrating the physical existence of DNA–DNA pairing because of a dearth of suitable techniques *(10)*.

Matzke et al. *(85)* designed an experimental system in transgenic tobacco to mimic transvection-like phenomenon known in *Drosophila* for DNA–DNA interaction. It was observed that a transgene locus capable of pairing, as revealed by *trans*-activation of a hetero allele, could also be inactivated. Although the transgene locus tested in this study was complex and contained inverted repeat, no siRNAs were detected. It is not known whether the siRNA concentration was too low to be detected or not present. If, indeed, it can be conclusively demonstrated that no siRNAs are made in this system of silencing, this would suggest the existence of DNA–DNA interactions.

The possibility that DNA–RNA interaction could induce *de novo* methylation of plant nuclear genes was first suggested by Wassenegger et al. *(86)*, by which nuclear cyclic DNA copies of a viroid (a plant pathogen that contains an untranslated, highly base-paired RNA molecule as the sole genetic material) became methylated only during replication. This work suggested that, because the replicating viroid was restricted to the nucleus, it somehow interacted with the corresponding "homologous" chromosomal DNA copies and triggered methylation.

During this time, parallel research efforts in posttranscriptional inactivation mechanisms have unequivocally demonstrated the existence and the crucial role of siRNAs corresponding to coding regions in establishing PTGS *(87)*. Because promoter homology is required for TGS and coding region homology is essential for PTGS, in light of Wassenegger's results, it was hypothesized that, for an RNA molecule to be involved in promoter homology-dependent gene silencing, promoter sequences at a silencing locus need to be transcribed.

The hypothesis that promoter-specific RNA is crucial for setting up transcriptional silencing was further strengthened by results from Park et al. *(88)*, who demonstrated that a locus that became methylated and silenced following interaction with a silencing locus failed, in turn, to act as a silencing locus. Dieguez et al. *(89)* also demonstrated that cytosine methylation at symmetrical CG and CNG sites is required for the maintenance, but not for the establishment, of TGS in plants.

Based on these studies it became apparent that methylation itself is not sufficient for inducing *trans*-silencing, and that another molecule, perhaps an RNA, is needed for its establishment. As discussed next, convincing evidence for the involvement of RNA in TGS was provided by Mette et al. *(36,37)* and Sijen et al. *(38)*, who intentionally transcribed promoter sequences in transgenic plants and showed that the targets containing homologous promoter sequences are silenced in the presence of transcribed RNAs.

Mette et al. *(37)* demonstrated the involvement of RdDM in TGS by using nopaline synthase promoter (NOSpro) sequences intentionally transcribed by the cauliflower mosaic virus 35S promoter to produce NOSpro RNAs. In one transgenic tobacco line, a nonpolyadenylated NOSpro RNA that deviated from the expected size was able to induce methylation and transcriptional inactivation of homologous NOSpro copies in *trans*.

To confirm that this aberrant NOSpro RNA was required for the TGS, Mette et al. *(37)* introduced the 271 locus *(90)*, a general suppressor of 35S promoters, to suppress the 35S promoter, thereby hindering the synthesis of NOS RNAs. This resulted in the alleviation of silencing and reduced methylation of target NOSpro, providing further evidence that NOS RNA initiated the silencing process.

Follow-up studies by Mette et al. *(36)* demonstrated the presence of dsRNA corresponding to NOSpro and their approximately 23 nt cleavage products. They have also shown that *de novo* methylation of the target promoter affected by TGS can be triggered by a dsRNA containing the promoter sequences. These and other studies *(91)* concluded that RNA hairpins transcribed from inverted DNA repeats were most effective as *trans*-acting signals in establishing TGS.

Sijen et al. *(38)* conducted similar studies investigating the role of both promoter and coding region repeats in instigating TGS and PTGS, respectively. By targeting flower pigmentation genes in petunia, they demonstrated that transgenes expressing dsRNA corresponding to coding regions can induce PTGS, whereas expression of dsRNA corresponding to promoter sequences can cause TGS. Silencing was accompanied by the methylation of sequences homologous to dsRNA in both TGS and PTGS, suggesting that TGS and PTGS are mechanistically related.

De Buck et al. *(92)* and De Buck and Depicker *(93,94)* analyzed the correlation between transgene silencing and the presence of inverted repeats in *Arabidopsis* using an inverted or single copy of β glucuronidase (GUS) reporter gene. In transformants in which GUS genes were present as inverted repeats separated by a 732-bp palindromic sequence spacer, GUS expression was significantly reduced, and the locus was heavily methylated. This locus could also induce silencing of both allelic and nonallelic copies of the GUS gene. In contrast, GUS expression in transformants containing two inverted repeats of GUS gene separated by an 826-bp nonrepetitive spacer remained high, suggesting the importance of spacer regions in facilitating the formation of dsRNAs necessary for initiating and maintaining silencing. Removal of one of the copies from the inverted GUS repeat locus using the Cre recombinase system restored consistently high expression in both hemi- and homozygous states and a decrease in GUS gene methylation.

8. Susceptibility of Loci to *Trans*-Silencing

A comparison of a silencing-susceptible locus and a resistant locus provided some clues as to the propensity for silencing *(95)*. Not all loci are equally susceptible to *trans*-silencing. Although it is very difficult to understand all of the critical chromosome interactions involved with silencing, one locus that was resistant to *trans*-silencing contained a simple structure and was embedded in a stretch of flanking plant DNA that lacked repetitive sequences and transposable elements (TEs).

9. Transgene Expression and Insertion Site Characteristics

There is a growing body of evidence indicating that not all transgene loci containing repeats become silenced, and repetition *per se* is not always sufficient to induce methylation. These data and the demonstration that dsRNA and short siRNAs are responsible for silencing are contributing to the emerging theme that repeats, especially inverted repeats, that could potentially be transcribed are most efficient in establishing silencing. An analysis of 12 distinct transgene loci uncovered several "extrinsic" flanking sequence motifs, such as tandem repeats, retroelement remnants, microsatellites, matrix attachment regions, and the like that could have various effects on transgene expression (reviewed in ref. *95*). Similarly, "intrinsic" features of transgene loci that appear to promote instability and methylation included bacterial vector sequences that link transgene DNA with plant DNA. These observations suggest that the more closely transgene loci resemble transposons, repetitive elements, and the like of internal or external origin, the higher their susceptibility for silencing.

An analysis of PTGS *(91)* revealed that the three most common situations in which PTGS is triggered are (1) single copy of a "sense" transgene transcribed at high level; (2) loci consisting of an inverted repeat of a sense transgene transcribed at low level or of a promoterless transgene; and (3) loci consisting of a single copy of transgene carrying internal inverted repeats. Thus, inverted repeats capable of transcription are the most common anomalous features directly associated with a majority of TGS and PTGS cases. The possibility that complex loci that are not capable of forming dsRNA but still are silenced (perhaps because of the insertion in a heterochromatic region or because of the *de novo* heterochromatinization of the locus) still exists, but needs further study. Chromatin configuration is one of the two mechanisms of epigenetic inheritance at the chromosomal level; the other is methylation *(1)*. Conventional wisdom would suggest, however, that transgene loci that are less complex and are devoid of inverted repeats have less risk of silencing, provided they integrate into a "good" chromosomal location.

Suppression of an unwanted gene or element could be achieved at the transcriptional level through TGS without the need for posttranscriptional inactivation mechanisms. However, both TGS and PTGS exist in the systems studied so far. It has been proposed that TGS has evolved to control the copy number of TEs and retroviruses indirectly, whereas PTGS has been in place to protect against viruses and overproduction of individual RNAs *(4)*. Thus, it is possible that the more complex a locus is, the more it resembles TEs and other repeat structures perceived as foreign or deleterious by the genome and are perhaps targeted for silencing by genome intruder scanning and modulation systems *(6)*.

10. Conclusions

Plants have evolved sophisticated mechanisms to fend off foreign genes. Although not initially appreciated, the impact of gene silencing has made product development considerably more time consuming and expensive. From the discussion in this chapter, the most important lesson learned is the need to produce low copy, minimally rearranged transgenic insertions. Beyond that, depending on the specific use of the transgenic plant materials, other approaches may be valuable. In particular, the use of mutants of genes involved with the gene-silencing process or suppressors of PTGS has the most potential. None of these appear to be a panacea for all potential needs or uses. Many uses will depend on the characteristics of the particular gene in the genetic background of interest and the growing conditions of the crop.

Acknowledgments

We would like to thank Joe Petolino, Mike Murray, and Anne Gregg for their excellent input and for critically reading the manuscript.

References

1. Matzke, M. A., Mette, M. F., Aufsatz, W., Jakowitsch, J., and Matzke, A. J. M. (2000) Host defenses to parasitic sequences and the evolution of epigenetic control mechanisms. *Genetica* **107**, 271–287.
2. Matzke, M. A., Matzke, A. J. M., Pruss, G. J., and Vance, V. B. (2001) RNA-based silencing strategies in plants. *Curr. Opin. Genet. Dev.* **11**, 221–227.
3. Fagard, M. and Vaucheret, H. (2000) (Trans)gene silencing in plants: how many mechanisms? *Annu. Rev. Plant Physiol. Plant Mol. Biol.* **51**, 167–194.
4. Wassenegger, M. (2002) Gene silencing. *Int. Rev. Cytol.* **219**, 61–113.
5. Waterhouse, P. M., Wang, M. B., and Finnegan, E. J. (2001) Role of short RNAs in gene silencing. *Trends Plant Sci.* **6**, 1360–1385.
6. Kumpatla, S. P., Chandrasekharan, M. B., Iyer, L., Li, G., and Hall, T. C. (1998) Genome intruder scanning and modulation systems and transgene silencing. *Trends Plant Sci.* **3**, 97–104.
7. Meins, F., Jr. (2000) RNA degradation and models for post-transcriptional gene silencing. *Plant Mol. Biol.* **43**, 261–273.
8. Muskens, M. W. M., Vissers, A. P. A., Mol, J. N. M., Kooter, J. M., Matzke, M. A., and Matzke, A. J. M. (2000) Role of inverted DNA repeats in transcriptional and post- transcriptional gene silencing. *Plant Mol. Biol.* **43**, 243–260.
9. Vaucheret, H. and Fagard, M. (2001) Transcriptional gene silencing in plants: targets, inducers and regulators. *Trends Genet.* **17**, 29–35.
10. Matzke, M. A., Aufsatz, W., Kanno, T., Mette, M. F., and Matzke, A. J. M. (2002) Homology-dependent gene silencing and host defense in plants. *Adv. Genet.* **46**, 235–275.
11. Grewal, S. I. S. and Elgin, S. C. R. (2002) Heterochromatin: new possibilities for the inheritance of structure. *Curr. Opin. Genet. Dev.* **12**, 178–187.
12. Baulcombe, D. C. (1999). Fast forward genetics based on virus-induced gene silencing. *Curr. Opin. Plant Biol.* **2**, 109–113.
13. Lindbo, J. A., Silva-Rosales, L., Proebsting, W. M., and Dougherty, W. G. (1993) Induction of a highly specific antiviral state in transgenic plants: implications for regulation of gene expression and virus resistance. *Plant Cell* **5**, 1749–1759.

14. Smith, H. A., Swaney, S. L., Parks, S. L., Wernsman, E. A., and Dougherty, W. G. (1994) Transgenic plant virus resistance mediated by untranslatable sense RNAs: expression, regulation, and fate of nonessential RNAs. *Plant Cell* **6**, 1441–1453.
15. English, J. J., Mueller, E., and Baulcombe, D. C. (1996) Suppression of virus accumulation in trangenic plants exhibiting silencing of nuclear genes. *Plant Cell* **8**, 179–188.
16. Mezlaff, M., O'Dell, M., Cluster, P. D., and Flavell, R. B. (1997) RNA-mediated RNA degradation and chalcone synthase A silencing in *petunia*. *Cell* **88**, 845–854.
17. Palauqui, J. C. and Vaucheret, H. (1998) Transgenes are dispensable for the RNA degradation step of cosuppression. *Proc. Natl. Acad. Sci. USA* **95**, 9675–9680.
18. Meyer, P., Linn, F., Heidmann, I., Meyer, H., Niedenhof, I., and Saedler, H. (1992) Endogenous and environmental factors influence 35S promoter methylation of a maize A1 gene construct in transgenic petunia and its colour phenotype. *Mol. Gen. Genet.* **231**, 345–352.
19. Meza, T. J., Kamfjord, D., Hakelien, A.-M., et al. (2001) The frequency of silencing in *Arabidopsis thaliana* varies highly between progeny of siblings and can be influenced by environmental factors. *Transgen. Res.* **10**, 53–67.
20. Kunz, C., Schöb, H., Stam, M., Kooter, J. M., and Meins, F., Jr. (1996) Developmentally regulated silencing and reactivation of tobacco chitinase transgene expression. *Plant J.* **10**, 437–450.
21. Balandin, T. and Castresana, C. (1997) Silencing of a β-1,3-glucanase transgene is overcome during seed formation. *Plant Mol. Biol.* **34**, 125–137.
22. Dehio, C. and Schell, J. (1994) Identification of plant genetic loci involved in a posttranscriptional mechanism for meiotically reversible transgene silencing. *Proc. Natl. Acad. Sci. USA* **91**, 5538–5542.
23. Mitsuhara, I., Shirasawa-Seo, N., Iwai, T., Nakamura, S., Honkura, R., and Ohashi, Y. (2002) Release from post-transcriptional gene silencing by cell proliferation in transgenic tobacco plants: possible mechanism for noninheritance of the silencing. *Genetics* **160**, 343–352.
24. Mlotshwa, S., Voinnet, O., Mette, M. F., et al. (2002) RNA silencing and the mobile silencing signal. *Plant Cell* **14**, S289–S301.
25. Boerjan, W., Bauw, G., van Montagu, M., and Inze, D. (1994) Distinct phenotypes generated by overexpression and suppression of *S*-adenosyl-L-methionine synthetase reveal developmental patterns of gene silencing in tobacco. *Plant Cell* **6**, 1401–1414.
26. Palauqui, J. C., Elmayan, T., Dorlhac de Borne, F., Crete, P., Charles, C., and Vaucheret, H. (1996) Frequencies, timing, and spatial patterns of cosuppression of nitrate reductase and nitrite reductase in transgenic tobacco plants. *Plant Physiol.* **112**, 1447–1456.
27. Palauqui, J. C., Elmayan, T., Pollien, J. M., and Vaucheret, H. (1997) Systemic acquired silencing: transgene-specific post-transcriptional silencing is transmitted by grafting from silenced stocks to non-silenced scions. *EMBO J.* **16**, 4738–4745.
28. Voinnet, O. and Baulcombe, D. C. (1997) Systemic signaling in gene silencing. *Nature* **389**, 553.
29. Sonoda, S. and Nishiguchi, M. (2000) Graft transmission of post-transcriptional gene silencing: target specificity for RNA degradation is transmissible between silenced and non-silenced plants, but not between silenced plants. *Plant J.* **21**, 1–8.
30. Mitsuhara, I., Shirasawa-Seo, N., Iwai, T., Nakamura, S., Honkura, R., and Ohashi, Y. (2002) Release from post-transcriptional gene silencing by cell proliferation in transgenic tobacco plants: possible mechanism for noninheritance of the silencing. *Genetics* **160**, 343–352.
31. Ueki, S. and Citovsky, V. (2001) RNA commutes to work: regulation of plant gene expression by systemically transported RNA molecules. *BioEssays* **23**, 1087–1090.
32. Dalmay, T., Hamilton, A., Rudd, S., Angell, S., and Baulcombe, D. C. (2000) An RNA-dependent RNA polymerase gene in *Arabidopsis* is required for posttranscriptional gene silencing mediated by a transgene but not by a virus. *Cell* **101**, 543–553.

33. Mourrain, P., Beclin, C., Elmayan, T., et al. (2000) Arabidopsis SGS2 and SGS3 genes are required for posttranscriptional gene silencing and natural virus resistance. *Cell* **101**, 533–542.
34. Bernstein, E., Caudy, A. A., Hammond, S. M., and Hannon, G. J. (2001) Role for a bidentate ribonuclease in the initiation step of RNA interference. *Nature* **409**, 363–366.
35. Jacobsen, S. E., Running, M. P., and Meyerowitz, E. M. (1999) Disruption of an RNA helicase/RNase III gene in *Arabidopsis* causes unregulated cell division in floral meristems. *Development* **126**, 5231–5243.
36. Mette, M. F., Aufsatz, W., Winden, J. v. d., Matzke, M. A., Matzke, A. J. M., and van der Winden, J. (2000) Transcriptional silencing and promoter methylation triggered by double-stranded RNA. *EMBO J.* **19**, 5194–5201.
37. Mette, M. F., van der Winden, J., Matzke, M. A., and Matzke, A. J. M. (1999) Production of aberrant promoter transcripts contributes to methylation and silencing of unlinked homologous promoters in *trans*. *EMBO J.* **18**, 241–248.
38. Sijen, T., Vijn, I., Rebocho, A., et al. (2001) Transcriptional and posttranscriptional gene silencing are mechanistically related. *Curr. Biol.* **11**, 436–440.
39. Paszkowski, J. and Whitham, S. A. (2001) Gene silencing and DNA methylation processes. *Curr. Opin. Plant Biol.* **4**, 123–129.
40. Finnegan, E. J. and Kovac, K. A. (2000) Plant DNA methyltransferases. *Plant Mol. Biol.* **43**, 189–201.
41. Luff, B., Pawlowski, L., and Bender, J. (1999) An inverted repeat triggers cytosine methylation of identical sequences in *Arabidopsis*. *Mol. Cell* **3**, 505–511.
42. Diéguez, M. J., Vaucheret, H., Paszkowski, J., and Mittelsten Scheid, O. (1998) Cytosine methylation at CG and CNG sites in not a prerequisite for the initiation of transcriptional gene silencing in plants, but it is required for its maintenance. *Mol. Gen. Genet.* **259**, 207–215.
43. Finnegan, E. J. and Dennis, E. S. (1993) Isolation and identification by sequence homology of a putative cytosine methyltransferase from *Arabidopsis thaliana*. *Nucleic Acids Res.* **21**, 2383–2388.
44. Finnegan, E. J., Peacock, W. J., and Dennis, E. S. (1996) Reduced DNA methylation in *Arabidopsis thaliana* results in abnormal plant development. *Proc. Natl. Acad. Sci. USA* **93**, 8449–8454.
45. Ronemus, M. J., Galbiati, M., Ticknor, C., Chen, J., and Dellaporta, S. L. (1996) Demethylation-induced developmental pleotropy in *Arabidopsis*. *Science* **273**, 654–657.
46. Jones, L., Ratcliff, F., and Baulcombe, D. C. (2001) RNA-directed transcriptional gene silencing in plants can be inherited independently of the RNA trigger and requires Met1 for maintenance. *Curr. Biol.* **11**, 747–757.
47. McCullum, C. M., Comai, L., Greene, E. A., and Henikoff, S. (2000) Targeted screening for induced mutations. *Nat. Biotechnol.* **18**, 457.
48. Lindroth, A. M., Cao, X., Jackson, J. P., et al. (2001) Requirement of CHROMOMETHYLASE3 for maintenance of CpXpG methylation. *Science* **292**, 2077–2080.
49. Vongs, A., Kakutani, T., Martienssen, R. A., and Richards, E. J. (1993) *Arabidopsis thaliana* DNA methylation mutants. *Science* **260**, 1926–1928.
50. Kakutani, T., Jeddeloh, J. A., and Richards, E. J. (1995) Characterization of an *Arabidopsis thaliana* DNA hypomethylation mutant. *Nucleic Acids Res.* **23**, 130–137.
51. Jeddeloh, J. A., Stokes, T. L., and Richards, E. J. (1999) Maintenance of genomic methylation requires a SWI2/SNF2-like protein. *Nat. Genet.* **22**, 94–97.
52. Kingston, R. E. and Narlikar, G. J. (1999) ATP-dependent remodeling and acetylation as regulators of chromatin fluidity. *Genes Devel.* **13**, 2339–2352.
53. Morel, J. B., Mourrain, P., Beclin, C., and Vaucheret, H. (2000) DNA methylation and chromatin structure affect transcriptional and post-transcriptional transgene silencing in *Arabidopsis*. *Curr. Biol.* **10**, 1591–1594.

54. Amedeo, P., Habu, Y., Afsar, K., Scheld, O. M., and Paszkowski, J. (2000) Disruption of the plant gene MOM releases transcriptional silencing of methylated genes. *Nature* **405**, 203–206.
55. Mittelsten Scheid, O., Probst, A. V., Afsar, K., and Paszkowski, J. (2002) Two regulatory levels of transcriptional gene silencing in *Arabidopsis*. *Proc. Natl. Acad. Sci. USA* **99**, 13,659–13,662.
56. Stokes, T. L. and Richards, E. J. (2000) Mum's the word: MOM and modifiers of transcriptional gene silencing. *Plant Cell* **12**, 1003–1006.
57. Cao, X. and Jacobsen, S. E. (2002) Role of the Arabidopsis DRM methyltransferases in *de novo* DNA methylation and gene silencing. *Curr. Biol.* **12**, 1138–1144.
58. Cao, X. and Jacobsen, S. E. (2002) Locus-specific control of asymmetric and CpNpG methylation by the *DRM* and *CMT3* methyltransferase genes. *Proc. Natl. Acad. Sci. USA* **99**, 16,491–16,498.
59. Allshire, R. (2002) RNAi and heterochromatin: a hushed-up affair. *Science* **297**, 1818–1819.
60. Vance, V. and Vaucheret, H. (2001) RNA silencing in plants defense and counterdefense. *Science* **292**, 2277–2280.
61. Mallory, A., Smith, T. H., Braden, R., Pruss, G., Bowman, L., and Vance, V. (2002) Suppression of RNA silencing in plants. *Mol. Plant Microbe Interact.* **3**, 141–145.
62. Li, H. W. and Ding, S. W. (1999) Viral suppressors of RNA silencing. *Curr. Opin. Biotechnol.* **12**, 150–154.
63. Carrington, J. C., Kasschau, K. D., and Johansen, L. K. (2001) Activation and suppression of RNA silencing by plant viruses. *Virology* **281**, 1–5.
64. Voinnet, O. (2001) RNA silencing as a plant immune system against viruses. *Trends Genet.* **17**, 449–459.
65. Anandalakshmi, R., Pruss, G., Ge, X., Marathe, M., Smith, T. H., and Vance, V. (1998) A viral suppressor of gene silencing in plants. *Proc. Natl. Acad. Sci. USA* **95**, 13,079–13,084.
66. Brigneti, G., Voinnet, O., Li, W. X., Ji, L. H., Ding, S. W., and Baulcombe, D. C. (1998) Viral pathogenicity determinants are suppressors of transgenic silencing in *Nicotiana benthamiana*. *EMBO J.* **17**, 6739–6746.
67. Kasschau, K. D. and Carrington, J. C. (1998) A counterdefensive strategy of plant viruses: suppression of post-transcriptional gene silencing. *Cell* **95**, 461–470.
68. Voinnet, O., Lederer, C., and Baulcombe, D. C. (2000) A viral movement protein prevents spread of the gene silencing signal in *Nicotiana benthamiana*. *Cell* **103**, 157–167.
69. Voinnet, O., Pinto, Y. M., and Baulcombe, D. C. (1999) Suppression of gene silencing: a general strategy used by diverse DNA and RNA viruses of plants. *Proc. Natl. Acad. Sci. USA* **96**, 14,147–14,152.
70. Silhavy, D., Molnar, A., Lucioli, A., et al. (2002) A viral protein suppresses RNA silencing and binds silencing-generated, 21- to 25-nucleotide double-stranded RNAs. *EMBO J.* **21**, 3070–3080.
71. Anandalakshmi, R., Marathe, M., Ge, X., et al. (2000) A calmodulin-related protein that suppresses posttranscriptional gene silencing in plants. *Science* **290**, 142–144.
72. Mallory, A. C., Ely, L., Smith, T. H., et al. (2001) HC-Pro suppression of transgene silencing eliminates the small RNAs but not transgene methylation or the mobile signal. *Plant Cell* **13**, 571–583.
73. Llave, C., Kasschau, K. D., and Carrington, J. C. (2000) Virus-encoded suppressor of post-transcriptional gene silencing targets a maintenance step in the silencing pathway. *Proc. Natl. Acad. Sci. USA* **97**, 13,401–13,406.
74. Marathe, M., Smith, T. H., Anandalakshmi, R., et al. (2000) Plant viral suppressors of post-transcriptional silencing do not suppress transcriptional silencing. *Plant J.* **22**, 51–59.
75. Mette, M. F., Matzke, A. J. M., and Matzke, M. A. (2001) Resistance of RNA-mediated TGS to HC-Pro, a viral suppressor of PTGS, suggests alternative pathways for dsRNA processing. *Curr. Biol.* **11**, 1119–1123.

76. Beclin, C., Berthome, R., Palauqui, J. C., Tepfer, M., and Vaucheret, H. (1998) Infection of tobacco or *Arabidopsis* plants by CMV counteracts systemic post-transcriptional silencing of nonviral (trans) genes. *Virology* **252**, 313–317.
77. Guo, H. S. and Ding, S. W. (2002) A viral protein inhibits the long range signaling activity of the gene silencing signal. *EMBO J.* **21**, 398–407.
78. Voinnet, O., Rivas, S., Mestre, P., and Baulcombe, D. C. (2003) An enhanced transient expression system in plants based on suppression of gene silencing by the p19 protein of tomato bushy stunt virus. *Plant J.* **33**, 949–956.
79. Mallory, A. C., Parks, G., Endres, M. W., et al. (2002) The amplicon-plus system for high-level expression of transgenes in plants. *Nature Biotechnol.* **20**, 622–625.
80. Elmayan, T., Balzergue, S., Béon, F., et al. (1998) Arabidopsis mutants impaired in cosuppression. *Plant Cell* **10**, 1747–1757.
81. Morel, J. B., Godon, C., Mourrain, P., et al. (2002) Fertile hypomorphic ARGONAUTE (ago1) mutants impaired in post- transcriptional gene silencing and virus resistance. *Plant Cell* **14**, 629–639.
82. Dalmay, T., Horsefield, R., Braunstein, T. H., and Baulcombe, D. C. (2001) SDE3 encodes an RNA helicase required for post-transcriptional gene silencing in Arabidopsis. *EMBO J.* **20**, 2069–2077.
83. Matzke, M. A., Matzke, A. J. M., and Mittelsten Scheid, O. (1994) Inactivation of repeated genes–DNA–DNA interactions? In *Homologous Recombination and Gene Silencing in Plants.* (Paszkowski, J., ed.). Kluwer, Dordrecht, The Netherlands, pp. 271–307.
84. Colot, V., Maloisel, L., and Rossignol, J. (1996) Interchromosomal transfer of epigenetic states in *Ascobolus*: transfer of DNA methylation is mechanistically related to homologous recombination. *Cell* **86**, 855–864.
85. Matzke, M., Mette, M. F., Jakowitsch, J., et al. (2001) A test for transvection in plants: DNA pairing may lead to trans-activation or silencing of complex heteroalleles in tobacco. *Genetics* **158**, 451–461.
86. Wassenegger, M., Heimes, S., Riedel, L., and Sänger, H. L. (1994) RNA-directed *de novo* methylation of genomic sequences in plants. *Cell* **76**, 567–576.
87. Hamilton, A. and Baulcombe, D. C. (1999) A species of small antisense RNA in posttranscriptional gene silencing in plants. *Science* **286**, 950–952.
88. Park, Y.-D., Papp, I., Moscone, E. A., et al. (1996) Gene silencing mediated by promoter homology occurs at the level of transcription and results in meiotically heritable alterations in methylation and gene activity. *Plant J.* **9**, 183–194.
89. Dieguez, M. J., Vaucheret, H., Paszkowski, J., and Mittelsten Scheid, O. (1998) Cytosine methylation at CG and CNG sites is not a prerequisite for the initiation of transcriptional gene silencing in plants, but it is required for its maintenance. *Mol. Gen. Genet.* **259**, 207–215.
90. Vaucheret, H. (1993) Identification of a general silencer for 19S and 35S promoters in a transgenic tobacco plant: 90 bp of homology in the promoter sequence are sufficient for trans-activation. *C. R. Acad. Sci. Paris* **316**, 1471–1483.
91. Mourrain, P., Beclin, C., and Vaucheret, H. (2000) Are gene silencing mutants good tools for reliable transgene expression or reliable silencing of endogenous genes in plants? *Genet. Eng.* **22**, 155–170.
92. de Buck, S., Montagu, M. v., Depicker, A., de Buck, S., and van Montagu, M. (2001) Transgene silencing of invertedly repeated transgenes is released upon deletion of one of the transgenes involved. *Plant Mol. Biol.* **46**, 433–445.
93. de Buck, S., Depicker, A., and de Buck, S. (2001) Disruption of their palindromic arrangement leads to selective loss of DNA methylation in inversely repeated gus transgenes in Arabidopsis. *Mol. Genet. Genom.* **265**, 1060–1068.

94. de Buck, S., Depicker, A., and de Buck, S. (2001) Silencing of invertedly repeated transgenes in *Arabidopsis thaliana. Meded. Faculteit Landbouwkundige en Toegepaste Biologische Wetenschappen* **66**, 393–399.
95. Matzke, M. A., Mette, M. F., Kunz, C., Jakowitsch, J., and Matzke, A. J. M. (2000) Homology-dependent gene silencing in transgenic plants: links to cellular defense responses and genome evolution. *Stadler Genet. Symp. Ser.* **22**, 141–162.

10

Value Creation and Capture With Transgenic Plants

William F. Goure

1. Introduction

The aim of this chapter is not to review the many different transgenic crop products in development in academic and industrial laboratories throughout the world. These are adequately described in published reviews and literature articles and on Internet Web sites *(1)*. The following three Internet Web sites are good sources of reviews on agricultural biotechnology products in the market and in development: Biotechnology Industry Organization, www.bio.org/foodag/; Agbios, www.agbios.com; Ag Biotech Infonet, www.biotech-info.net/. The Web site of the Animal Health and Inspection Service of the US Department of Agriculture (USDA) (http://www.aphis.usda.gov/bbep/bp/status.html) provides a listing of many agricultural biotechnology products currently in development. Transgenic crops for the production of pharmaceutical products have also been reported in published reviews *(2)* and are not discussed here. This chapter also does not review the commercialization history of those transgenic crop products commercialized to date.

Instead, the aim of this chapter is to discuss the prospective challenges and opportunities regarding commercialization of the next generation of transgenic crops for the production of food, feed, and fiber. This discussion is organized to address the three fundamental issues associated with the commercialization of any new product: how much value is created, how much value is captured, and how the value is delivered to the marketplace. Although these issues seem simple, in reality they are very complex, and many products in development fail to reach the market, or fail in the market, because one or more of these fundamental issues was not adequately addressed or understood prior to commercialization. This is particularly the case for products such as transgenic crop products, which often result because of a "technology push" vs a "market pull."

A number of rigorous evaluations of future trends in agricultural demand and the impact of such demand on production have been completed (*3–5*; *see also* http://www.fertilizer.org/ifa/statistics/indicators/pocket_requirements.asp). The main conclusions of these studies can be summarized as follows:

From: *The GMO Handbook: Genetically Modified Animals, Microbes, and Plants in Biotechnology*
Edited by: S. R. Parekh © Humana Press Inc., Totowa, NJ

1. Demand for cereals will double by 2020. Demand for corn, driven primarily by the increased demand for meat, will increase much faster than the demand for other cereals.
2. To meet demand, a 40% increase in grain production by 2020 is needed. Of the increased production, 80% will come from increased crop yields, whereas only 20% will come from an increase in cultivated land.
3. Most of the increased demand for food will occur in developing countries. The primary factors driving the increased food demand will be population growth, rising incomes, and urbanization.
4. Net cereal imports by developing countries will double by 2020.
5. About 60% of the developing world's cereal imports will come from the United States, which is expected to increase net cereal exports to developing countries by over 33% to approximately 114 million metric tons by 2020.
6. Food prices will remain steady or decrease slightly between 1995 and 2020.

Because of these trends, agriculture is faced with some very interesting opportunities and very difficult challenges. On one hand, farmers will find expanding markets for their products. However, to satisfy this growth in demand, they will need to increase production significantly without increasing arable land. Moreover, because commodity grain prices are not expected to rise significantly, to remain economically viable, farmers must increase productivity without increasing overall costs. For farmers in developed countries, increasing productivity without increasing costs will also be necessitated by increased agricultural production in developing countries, where land and labor costs will be lower.

Regardless of the crop grown, major production expenditures fall into four main categories:

1. Input costs for seeds, fertilizers, and pesticides.
2. Labor costs.
3. Equipment ownership, operation, and maintenance costs.
4. The cost of land.

Labor and equipment costs are very difficult to reduce because a reduction in one is often caused by, or leads to, an increase in the other *(6)*. In addition, increased farm sizes *(7)*, decreased availability of farm labor *(2)*, and increased use of chemicals *(2)* will prevent significant reductions in equipment costs. The cost of land is likely to increase because of continued urbanization and an increase in the productive value of high-quality farmland. Thus, for farmers to improve the economic viability of farming, they must increase productivity while reducing input costs for seeds, fertilizers, or pesticides. Genetically enhanced transgenic crops can be powerful tools to help farmers meet these challenges.

It has been less than a decade since Flavr-Savr™ tomatoes, the first transgenic crop in the United States, was registered for commercialization *(8)*. Since 1996, when transgenic soybeans, cotton, corn, and canola were first commercialized, global crop area planted with transgenic crops has grown at a sustained rate of greater than 10% per year to almost 59 million hectares in 16 countries in 2002 *(9)*. This rate of growth has been touted as one of the highest for any technology in agriculture, using hybrid corn for the comparison *(10)*. The compounded annual growth rate for the global area planted to transgenic crops for the period 1996 to 2002 was approximately 7%. For the United

States, which has the largest acreage of transgenic crops, the compounded annual growth rate for planted acres for the period 1996 to 2002 was also almost 7% *(11)*.

Although the market growth of transgenic crops has been impressive, it has been small in comparison to the adoption of farm tractors, which fundamentally changed farming practices. Between 1928 and 1960, the compounded annual growth rate for the number of farm tractors in the United States was approximately 18% *(12)*.

It is also important to note that two countries, the United States and Argentina, account for 89% of global transgenic crop acreage; most of this acreage is because of transgenic soybeans *(11)*. Thus, the market growth of transgenic crops, although significant, has not been as great as hoped by proponents of the technology.

2. Creating Value

The cost and time to discover, develop, and commercialize a transgenic crop is reported to be $50–300 million and take 6–12 years *(13)*. Furthermore, only approximately 1 gene or trait becomes a commercial product for every 250 that are investigated in the discovery process. These cost, time, and frequency of success numbers have profound implications on value creation considerations. Table 1 shows the results of a net present value determination for a hypothetical new transgenic corn product that has the market penetration and value comparable to *Bacillus thuringiensis* (*Bt*) corn. Tables 2 and 3 show the affect of the cost and time to develop and commercialize this hypothetical transgenic corn product and the trait premium on the net present value of the product. Tables 4–6 display the same type of information for a hypothetical new transgenic soybean product that is comparable to Roundup Ready® (Monsanto Company) soybeans with respect to value and market penetration.

These analyses reveal a number of key points regarding the value amount that must be created by a new transgenic crop product to justify the costs and time needed for its development and commercialization.

1. The total value created by the product will typically have to exceed $500 million per year.
2. The minimum gross annual revenues at peak sales for the trait provider should be in the range $175–200 million.
3. Products with total discovery, development, and commercialization costs exceeding $125 million that take over 12 years to bring to market in most cases will fail to provide an acceptable return on investment and in many cases will have a negative net present value.
4. A trait with high acreage penetration and moderate value creation will generally provide a better return on investment than a product with small acreage penetration but high value creation per acre. For example, the trait premium for the hypothetical corn product in Table 1 is $8.50 per acre, or 30% greater than the trait premium of $6.50 for the hypothetical soybean product in Table 4. However, because the soybean product is planted on 58.6 million acres at maturity compared to 30.2 million acres for the corn product, the soy product has an estimated net present value that is over 50% greater than that of the corn product.

Relatively few transgenic crop product concepts can achieve these high hurdle rates for value creation. Table 7 shows the estimated value created by a range of product concepts for corn, soybeans, cotton, and wheat. Four of the possible product concepts fail to achieve the value creation hurdle of $500 million per year, even though several

Table 1
Discounted Cash Flow Calculation for a Hypothetical New Transgenic Corn Product

Key Assumptions											
GM trait premium ($ per acre)	$8.50										
SG&A costs[a]	10%										
Value share to distribution chain[a]	10%										
Farm price per bushel[b]	$2.25										
Total value created for grower	66%										
Expected gross value created for grower ($ per acre)	$25.00										
Total cost to bring trait to market ($ million)	$50										
Years to bring trait to market	9										
Discount rate	20%										
Yr	0	2	4	6	8	10	12	14	16	18	20
Harvested US corn acreage (million acres)[b]	69.3	71.5	72.0	72.8	73.3	73.7	74.1	74.5	74.9	75.3	75.7
Estimated actual yield (bushels per acre)[c]	130	142	145.4	148.7	152	155.2	158.4	161.6	164.8	168	171.2
Equivalent gross yield increase of trait premium	2.9%	2.7%	2.6%	2.5%	2.5%	2.4%	2.4%	2.3%	2.3%	2.2%	2.2%
Equivalent yield increase of total value created for the grower	8.5%	7.8%	7.6%	7.5%	7.3%	7.2%	7.0%	6.9%	6.7%	6.6%	6.5%
Market penetration[d]	0%	0%	0%	0%	0%	9%	25%	34%	40%	40%	40%
Total GM trait acres (million acres)	0.0	0.0	0.0	0.0	0.0	6.6	18.5	25.3	29.9	30.1	30.2
Gross trait fees to trait provider ($ million)	0.0	0.0	0.0	0.0	0.0	56.3	157.1	214.8	254.1	255.5	256.8
Annual costs to develop and commercialize the trait ($ million)	$5.6	$5.6	$5.6	$5.6	$5.6	$0.0	$0.0	$0.0	$0.0	$0.0	$0.0
SG&A costs ($ million)	$0.0	$0.0	$0.0	$0.0	$0.0	$5.6	$15.7	$21.5	$25.4	$25.5	$25.7
Value share with distributors ($ million)	$0.0	$0.0	$0.0	$0.0	$0.0	$5.6	$15.7	$21.5	$25.4	$25.5	$25.7
Net trait provider revenues ($ million)	($5.6)	($5.6)	($5.6)	($5.6)	($5.6)	$45.0	$125.7	$171.9	$203.3	$204.4	$205.5
NPV of net trait provider revenues ($ million)	$82.3										

Abbr: SG & A, Sales, general and administrative; FAPRI, food and agricultural policy research institute.
[a]Percentage of gross trait premium.
[b]Average of prospective farmgate prices of US soybeans based on FAPRI 2003 data *(13)*.
[c]Based on FAPRI 2003 data *(13)*.
[d]Based on the market penetration of GM corn in the United States for the period 1996 to 2003 *(13)*.

Table 2
Effect of Time and Cost to Develop a New Transgenic Corn Product on the Net Present Value of the Discounted Cash Flow of the Net Trait Provider's Revenues (in $ million)[a]

Total Cost to Develop and Commercialize the Trait ($ million)	Years to Develop and Commercialize						
	6	7	8	9	10	11	12
$25	$182	$147	$118	$94	$74	$57	$43
$50	$167	$134	$106	$82	$63	$47	$33
$75	$152	$120	$93	$71	$52	$37	$24
$100	$137	$106	$80	$59	$41	$26	$14
$125	$122	$92	$68	$47	$30	$16	$5
$150	$107	$79	$55	$36	$20	$6	($5)
$175	$92	$65	$43	$24	$9	($4)	($14)
$200	$77	$51	$30	$12	($2)	($14)	($23)
$225	$62	$38	$17	$1	($13)	($24)	($33)
$250	$47	$24	$5	($11)	($24)	($34)	($42)
$275	$32	$10	($8)	($23)	($35)	($44)	($52)
$300	$17	($3)	($20)	($34)	($45)	($54)	($61)

[a]For these values, all assumptions in Table 1 are fixed except the years to develop and commercialize the trait and the total cost to develop and commercialize the trait. Values in parenthesis are negative net present values.

Table 3

Effect the Cost to Develop a New Biotechnology Transgenic Corn Product and the Amount of the Trait Fee or Seed Premium for the Trait on the Net Present Value of the Discounted Cash Flow of the Net Trait Provider's Revenues ($ million)[a]

Total Cost to Develop and Commercialize the Trait ($ million)	Amount of Trait Premium						
	$5.50	$6.50	$7.50	$8.50	$9.50	$10.50	$11.50
$25	$57	$69	$81	$94	$106	$119	$131
$50	$45	$57	$70	$82	$95	$107	$120
$75	$33	$46	$58	$71	$83	$95	$108
$100	$22	$34	$47	$59	$71	$84	$96
$125	$10	$22	$35	$47	$60	$72	$85
$150	($2)	$11	$23	$36	$48	$61	$73
$175	($13)	($1)	$12	$24	$36	$49	$61
$200	($25)	($12)	($0)	$12	$25	$37	$50
$225	($37)	($24)	($12)	$1	$13	$26	$38
$250	($48)	($36)	($23)	($11)	$2	$14	$26
$275	($60)	($47)	($35)	($23)	($10)	$2	$15
$300	($71)	($59)	($47)	($34)	($22)	($9)	$3

[a]For these values, all assumptions in Table 1 are fixed except the amount of the trait fee and the total cost to develop and commercialize the trait. Values in parenthesis are negative net present values.

Table 4
Discounted Cash Flow Calculation for a Hypothetical New Transgenic Soybean Product

Key Assumptions											
GM trait premium ($ per acre)	$6.50										
SG&A costs[a]	10%										
Value share to distribution chain[a]	10%										
Farm price per bushel[b]	$5.21										
Total value created for grower	66%										
Expected gross value created for grower ($ per acre)	$19.12										
Total cost to bring trait to market ($ million)	$50										
Years to bring trait to market	9										
Discount rate	20%										
Yr	0	2	4	6	8	10	12	14	16	18	20
Harvested US soybean acreage (million acres)[b]	72.2	72.3	72.3	72.4	72.3	72.4	72.3	72.3	72.3	72.3	72.3
Estimated actual yield (bushels per acre)[c]	37.8	40.3	41.2	42	42.9	43.7	44.5	45.3	46.1	46.9	47.7
Equivalent gross yield increase of trait premium	3.3%	3.1%	3.0%	3.0%	2.9%	2.9%	2.8%	2.8%	2.7%	2.7%	2.6%
Equivalent yield increase of total value created for the grower	9.7%	9.1%	8.9%	8.7%	8.6%	8.4%	8.2%	8.1%	8.0%	7.8%	7.7%
Market penetration[d]	0%	0%	0%	0%	0%	13%	54%	75%	81%	81%	81%
Total GM trait acres (million acres)	0.0	0.0	0.0	0.0	0.0	9.4	39.0	54.2	58.6	58.6	58.6
Gross trait fees to trait provider ($ million)	$0.0	$0.0	$0.0	$0.0	$0.0	$61.1	$253.8	$352.5	$380.7	$380.7	$380.7
Annual costs to develop and commercialize the trait ($ million)	$5.6	$5.6	$5.6	$5.6	$5.6	$0.0	$0.0	$0.0	$0.0	$0.0	$0.0
SG&A costs ($ million)	$0.0	$0.0	$0.0	$0.0	$0.0	$6.1	$25.4	$35.2	$38.1	$38.1	$38.1
Value share with distributors ($ million)	$0.0	$0.0	$0.0	$0.0	$0.0	$6.1	$25.4	$35.2	$38.1	$38.1	$38.1
Net trait provider revenues ($ million)	($5.6)	($5.6)	($5.6)	($5.6)	($5.6)	$48.9	$203.0	$282.0	$304.5	$304.5	$304.5
NPV of net trait provider revenues ($ million)	$127										

[a]Percentage of gross trait premium.
[b]Average of prospective farmgate prices of US soybeans based on FAPRI 2003 data *(13)*.
[c]Based on FAPRI 2003 data *(13)*.
[d]Based on the market penetration of GM soybeans in US for the period 1996 to 2003 *(13)*.

Table 5
Effect of Time and Cost to Develop a New Transgenic Soybean Product on the Net Present Value of the Discounted Cash Flow of the Net Trait Provider's Revenues ($ million)[a]

Total Cost to Develop and Commercialize the Trait ($ million)	Years to Develop and Commercialize						
	6	7	8	9	10	11	12
$25	$269	$218	$175	$139	$109	$84	$64
$50	$254	$204	$162	$127	$98	$74	$54
$75	$239	$190	$149	$116	$87	$64	$45
$100	$224	$177	$137	$104	$77	$54	$35
$125	$209	$163	$124	$92	$66	$44	$26
$150	$194	$149	$112	$81	$55	$34	$17
$175	$179	$135	$99	$69	$44	$24	$7
$200	$164	$122	$87	$57	$33	$14	($2)
$225	$149	$108	$74	$46	$23	$4	($12)
$250	$134	$94	$61	$34	$12	($6)	($21)
$275	$119	$81	$49	$22	$1	($17)	($31)
$300	$104	$67	$36	$11	($10)	($27)	($40)

[a]For these values, all assumptions in Table 4 are fixed except the years to develop and commercialize the trait and the total cost to develop and commercialize the trait. Values in parenthesis are negative net present values.

Table 6

Effect the Cost to Develop a New Transgenic Soybean Product and the Amount of the Trait Fee or Seed Premium for the Trait on the Net Present Value of the Discounted Cash Flow of the Net Trait Provider's Revenues ($ million)[a]

Total Cost to Develop and Commercialize the Trait ($ million)	Amount of Trait Premium						
	$2.50	$3.50	$4.50	$5.50	$6.50	$7.50	$8.50
$25	$46	$69	$93	$116	$139	$162	$185
$50	$35	$58	$81	$104	$127	$150	$174
$75	$23	$46	$69	$92	$116	$139	$162
$100	$11	$34	$58	$81	$104	$127	$150
$125	($0)	$23	$46	$69	$92	$115	$139
$150	($12)	$11	$34	$58	$81	$104	$127
$175	($24)	($0)	$23	$46	$69	$92	$115
$200	($35)	($12)	$11	$34	$57	$81	$104
$225	($47)	($24)	($1)	$23	$46	$69	$92
$250	($59)	($35)	($12)	$11	$34	$57	$80
$275	($70)	($47)	($24)	($1)	$22	$46	$69
$300	($82)	($59)	($36)	($12)	$11	$34	$57

[a]For these values, all assumptions in Table 4 are fixed except the amount of the trait fee and the total cost to develop and commercialize the trait. Values in parenthesis are negative net present values.

Table 7
Estimated Value Creation for Various Transgenic Corn, Soybean, and Cotton Products

Product Concept or Trait	Estimated Gross Value per Year ($ million)	Reference
6.4% Yield improvement for US corn from resistance to pathogen stresses[a]	$1350	*14, 15*
11% Yield improvement for North American wheat from resistance to pathogen stresses	$1200	*15*
7.4% Yield improvement for US soybeans from resistance to pathogen stresses[a]	$1100	*14, 15*
8.4% Yield improvement for European sugar beets from resistance to pathogen stresses	$1070	*15*
5% Increase in US corn yield per year[a]	$1060	*14*
Nitrogen use efficiency in US corn	$890	*16*
High energy (oil) content corn	$820	*16*
5% Increase in US soybean yield per year[a]	$740	*14*
High phosphorous availability corn	$300	*16*
5% Increase in US cotton yield per year[a]	$160	*14*
4.6% Yield improvement for US upland cotton from resistance to pathogen stresses[a]	$150	*14, 15*
High protein/amino acid content corn	$40	*16*

[a]In general, values were estimated for US acres only because of opposition to genetically modified crops in Europe and Brazil and the difficulties of capturing value from genetically modified crops in Asia and Latin America.
[b]From original ref. *18*.

have been the subjects of intense research efforts. For example, transgenic products that would completely protect US cotton against pathogen losses or increase cotton yields by 5% fail to achieve the value creation hurdle rate that would justify an expenditure of $50 million and at least 6 years of research and development to bring the technologies to market.

It is important to note that the estimated value created by the transgenic crop products listed in Table 7 is the total value that could be created and not necessarily the actual value creation that would be achieved commercially. For example, high-oil corn was estimated in 1999 to have a total value creation potential of $820 million per year *(16)*. However, total acres of high-oil corn have steadily declined, from a high of approximately 900,000 acres in 1998 to 500,000 acres in 2002, with an estimated total value of only approx $20 million *(17)*. This clearly illustrates the importance of distinguishing between the hypothetical value that can be created by a particular technology and the actual value that can be created in a competitive marketplace.

2.1. Input Traits

Input traits are the focus of most current research efforts for the development of new genetically modified (GM) crops. In the United States, which accounts for over 70% of all global experimental field releases of GM plants *(18)*, 68% of the trials have been focused on input traits for the control of pests (Fig. 1) *(19)*. Herbicide tolerance and insect resistance accounted for 26 and 25%, respectively, of the total field releases. Viral, fungal, bacterial, and nematode resistance accounted for 11% of the total field releases. In the European Union, approximately 60% of experimental releases were for herbicide tolerance, and about 16% and 7%, respectively, were for insect resistance and virus resistance.

The differences between the types of traits that were the focus of field trials in the United States and Europe undoubtedly reflect differences in commercial needs between the two regions. Nonetheless, herbicide tolerance, insect resistance, and protection against pathogens are the major focus of research and development efforts directed toward new GM crops. Although this focus is influenced to some extent by the available technology, such as glyphosate tolerance and *Bt*-derived insect tolerance, it is also strongly influenced by the significant unmet agricultural needs for better tools and technologies to control losses caused by pests.

2.1.1. Pest Protection Products

Figure 2 shows the economic losses caused by various pests to major agronomic crops *(15)*. It has been estimated that 42% of total attainable production of eight major crops (rice, wheat, corn, potatoes, cotton, soybeans, barley, and coffee) are lost to insects, diseases, and weeds, which cause 16, 13, and 13% production losses, respectively. The global economic value of the lost production for these eight crops caused by pests is estimated as $250 billion annually. Total global economic losses for all crops because of insects, diseases, and weeds are estimated to be as high as $500 billion annually. Because of the significant and clearly defined economic value of crop losses caused by insects, diseases, and weeds, development of biotechnology products to protect crops against these pests will continue to be a dominant focus of transgenic crop research and development efforts.

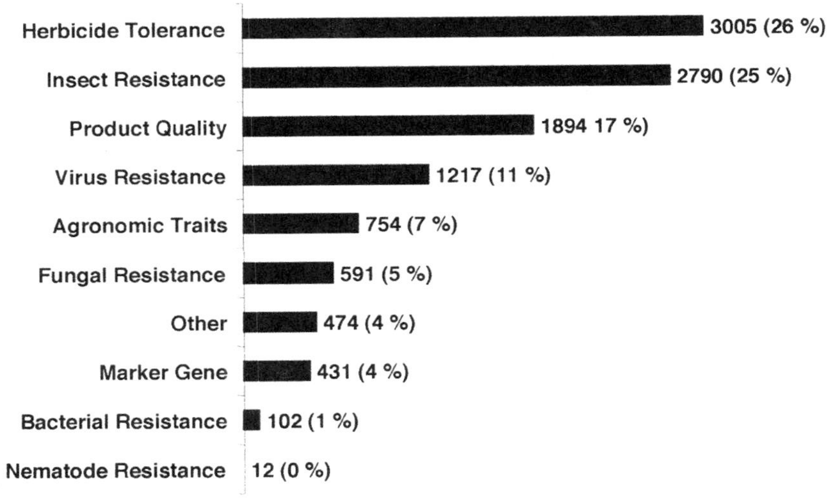

Fig. 1. Total number of approved US field releases since 1987 by phenotype category.

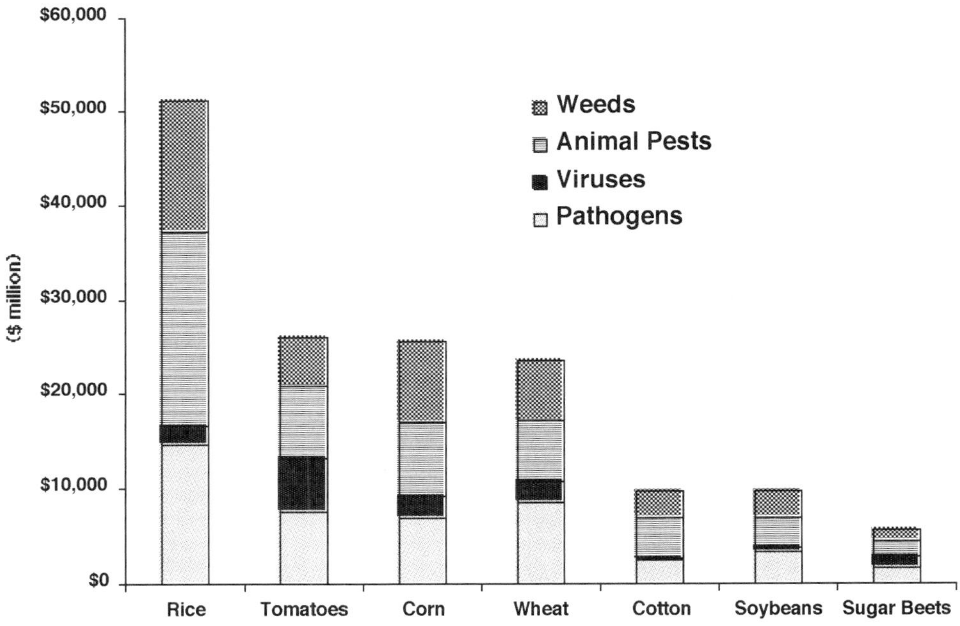

Fig. 2. Economic losses caused by various pests to major agronomic crops.

2.1.2. Drought Tolerance Products

Examination of the categories of phenotypes listed in the USDA database of field trial notifications and release permits reveals that general yield increases and drought, salt, and environmental stress tolerances are the other major input traits that are the focus of research and development efforts *(19)*.

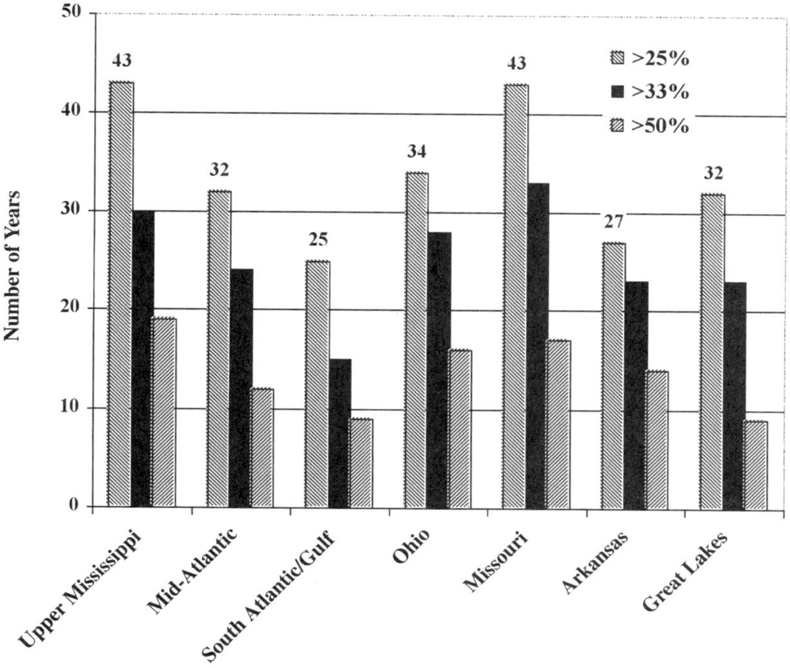

Fig. 3. Number of years between 1896 and 1995 that US corn-producing regions suffered from severe or extreme drought greater than 25, 33, or 50% of the year.

The value creation of general yield increases is readily determined by simply multiplying the percentage increase in yield on a per unit basis by the value of a unit of production. However, in contrast to the clear value creation potential for protection against pests and general yield increases, there is a paucity of quantitative data regarding value creation potential for tolerance to environmental stresses such as drought. Nonetheless, there is no doubt that drought losses have a significant economic impact on agriculture.

According to data provided by the USDA and the National Oceanographic and Atmospheric Administration, between 1896 and 1995, there were 25 to 33% of major US corn production areas that suffered from severe or extreme drought greater than 25% of the time (*see* Fig. 3) *(20,21)*. More detailed analysis of drought data compiled by the National Weather Service showed that, for 26 of the 44 years between 1960 and 2003, significant regions of US corn production suffered drought conditions *(20,22)*. These data provide qualitative evidence of crop yield losses because of drought in the United States. However, the data do not provide quantitative information regarding the economic magnitude of the losses, which is needed to assess the potential value of drought tolerance technology.

By correlating Palmer Drought Severity Index data compiled by the USDA *(23)* with US corn production areas *(24)* and comparing differences in USDA yield data *(25)* for years of severe or extreme drought to preceding or succeeding years without drought, a semiquantitative estimate of economic losses caused by severe or extreme

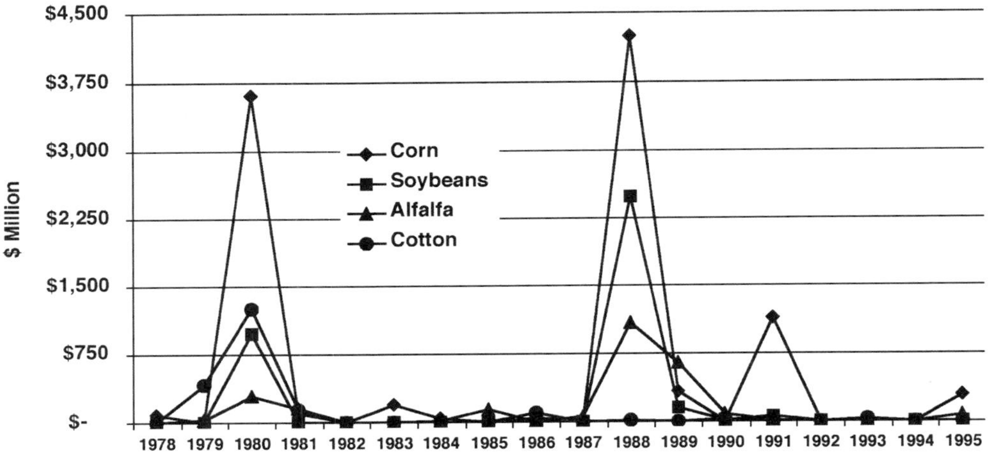

Fig. 4. Estimated economic losses in corn, soybeans, alfalfa, and cotton in the United States caused by severe or extreme drought.

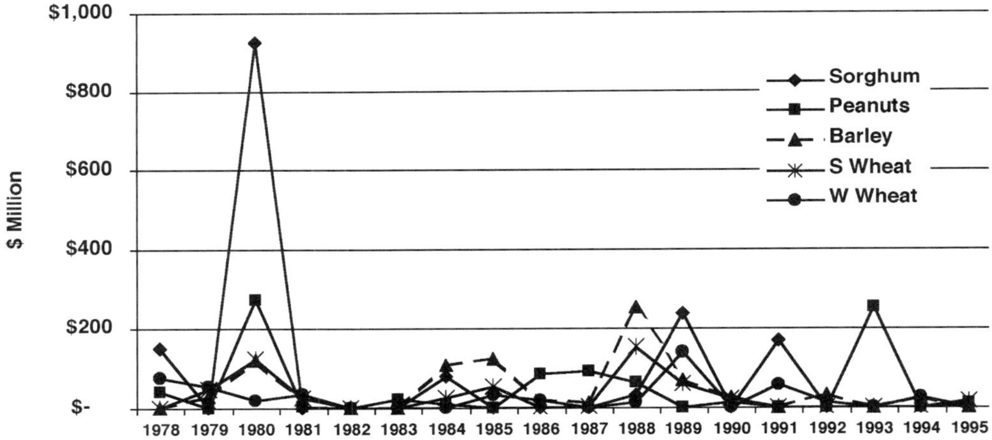

Fig. 5. Estimated economic losses in sorghum, peanuts, barley, spring wheat, and winter wheat in the United States caused by severe or extreme drought.

droughts can be determined (*see* Figs. 4 to 6 and Table 8). This analysis suggests crop losses averaging over $1 billion annually in the United States.

However, these estimates undoubtedly underestimate the actual losses caused by drought because they only consider losses caused by severe or extreme drought. For example, corn is particularly sensitive to drought damage, with yield losses of 5 to 50% caused by only four consecutive days of visible wilting *(26)*. Thus, even in years when the crop is not subject to severe or extreme droughts, yield losses likely occur because of moisture deprivation.

Based on available US government data, it is possible to gain more quantitative determinations of the economic losses in various crops because of drought. By correlating historical monthly Palmer Drought Severity Index data for different geographic

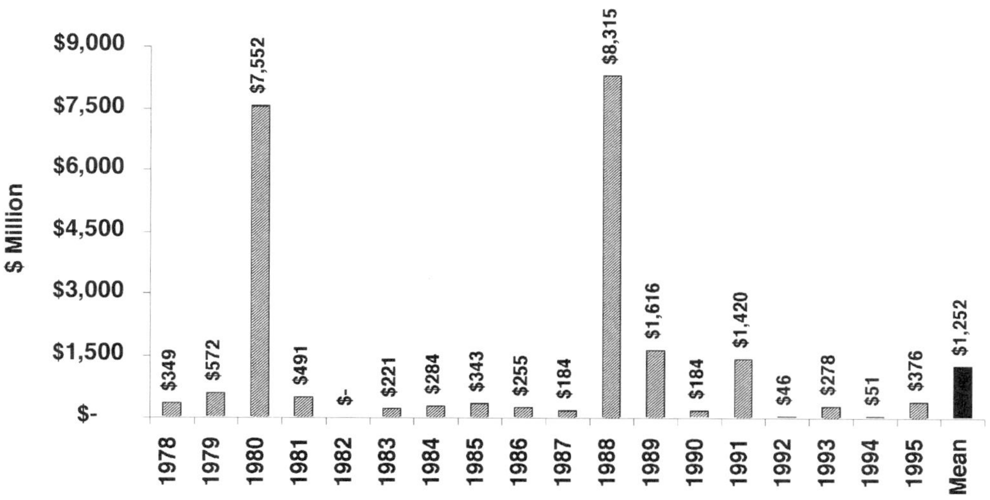

Fig. 6. Estimated cumulative economic losses in major US row crops caused by severe or extreme drought.

regions with crop production data for those regions, it is possible to gain a rather accurate estimate of the economic losses caused by mild droughts *(27)* and thereby estimate total value creation potential of drought tolerance technology.

Thus, for example, by comparing the difference in corn yield data for 1990 in Iowa climatic divisions 1, 7, and 9 (*see* Fig. 7), which were moisture stressed during part of the critical July-to-August silking period, with yield data for the other divisions, which were not moisture stressed, it is possible to gain accurate estimates of the economic losses for corn caused by moisture stress. Figure 8 shows similar data for 1995. By conducting such analysis for all major corn-producing states over a 5- or 10-year period, it is possible to gain an accurate estimate of the total annual economic losses for corn caused by moisture deprivation. Although such an analysis would be very time consuming to conduct, it is highly recommended for any company seeking to invest over $50 million and 6 years of research and development to develop and commercialize a transgenic drought tolerance product.

2.1.3. Reduced Fertilizer Use Products

About 50% of the world's cropland is devoted to growing cereals (e.g., wheat, rice, corn, barley, etc.), which account for approx 70% of all human caloric intake *(28)*.

For most crops, expenditures for fertilizers are the largest variable costs of production *(29–32)*. For cereals, and essentially all nonlegume crops, nitrogen is the major fertilizer nutrient based on both weight and costs. For example, in the United States in 1999, total fertilizer usage of nitrogen, phosphorus, and potash was, respectively, 12.4, 4.3, and 5 million nutrient tons *(33)*. The costs for these nutrients were $5.2, $1.1, and $0.7 billion dollars, respectively. (These costs were calculated based on costs per pound for nitrogen, phosphate, and potash of $0.21, $0.25, and $0.13, respectively; *see* ref. *28*.) The United States is second only to China in total usage of fertilizers and accounts for over 15% of total global fertilizer consumption *(34)*.

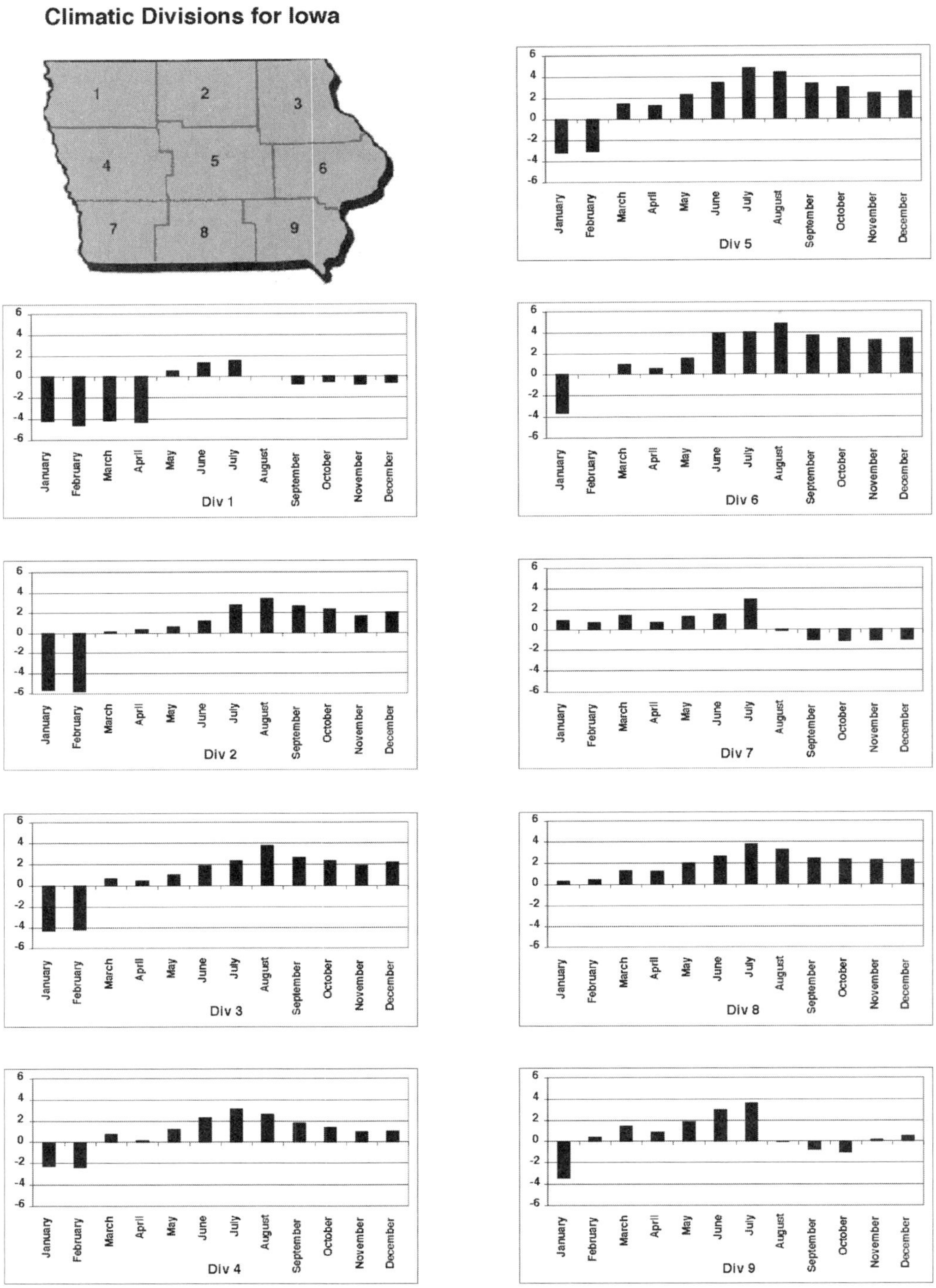

Fig. 7. Example of historical monthly Palmer Drought Severity Index data for Iowa for 1990.

Table 8
Calculated Economic Losses for Corn Producers Caused by Severe or Extreme Drought in the United States[a]

	1978	1980	1981	1983	1984	1986	1987	1988	1989	1990	1991	1995	
Iowa		578.6						1061.7	182.3			55.7	
Illinois		1213.2						1334.4	94.6		532.2	113	
Indiana					38.6			483			373.4	115.8	
Kansas	81.9	122.7											
Kentucky		180.3		198		36.9	5.4	101.3			56		
Michigan							19.5	129					
Minnesota		272.9						427.6					
Missouri		329.2						134.6	29.1				
Nebraska		741.7	66.6										
Ohio								242.7			168.8		
South Dakota		153.4	35.6					120.8	23.6				
Wisconsin								207.8					
Total losses	$82	$3592	$102	$198	$39	$37	$25	$4243	$330	–	$1130	$285	Mean $559

[a]All values are in millions of dollars. Economic losses were determined by comparing differences in yields between years of severe or extreme drought and preceding and succeeding years without severe or extreme drought and then multiplying yield losses by the price of corn grain as reported by the USDA for a particular year.

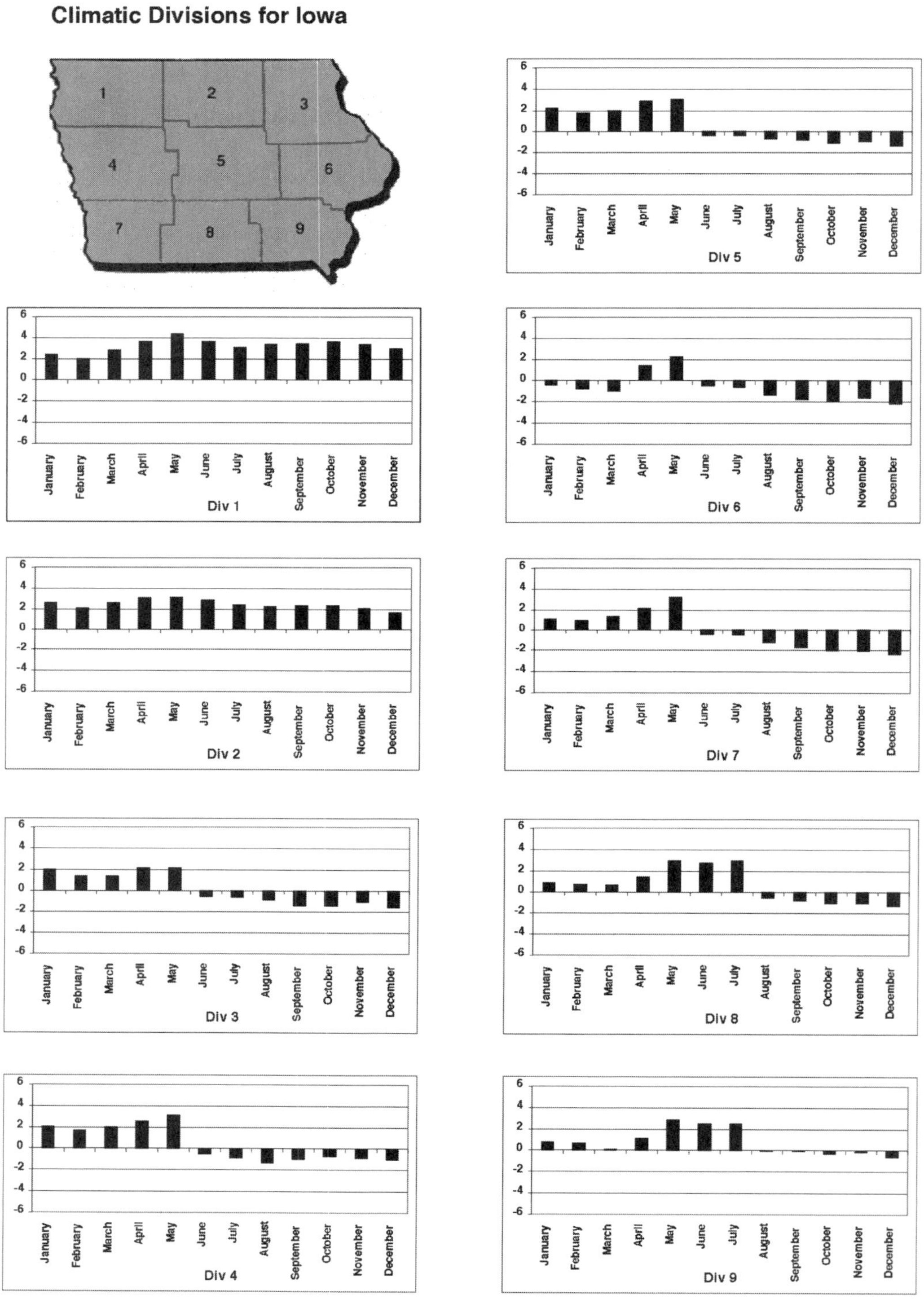

Fig. 8. Example of historical monthly Palmer Drought Severity Index data for Iowa for 1995.

Worldwide, cereal production consumes approximately 65% of the total fertilizers applied and 70% of the total nitrogen fertilizers applied *(35)*. Corn production is the largest user of fertilizers in the United States and consumes over 45% of total fertilizers used *(36)*. Of the corn grown in the United States, 98% receives fertilizer applications *(35)*. Excluding application costs, US corn producers spend approximately $2.3 billion each year for nitrogen, $528 million for phosphate, and $350 million for potash fertilizers.

New transgenic crop technology that reduces nitrogen fertilizer applications in corn by as little as 25%, although still allowing a grower to achieve normal production, would result in value creation estimated as $100–250 million annually in the United States alone based on reduced expenditures for nitrogen fertilizers.

Technology that reduces application rates of nitrogen fertilizers would also have significant environmental and societal benefits. Although there is debate regarding the exact magnitude of the impact of nitrogen fertilizer use on the environment, there is broad consensus that the impact is detrimental *(37–41)*. Detrimental impacts because of nitrogen fertilizer use include pollution of surface waters caused by runoff, pollution of groundwaters caused by leaching, green house gas emissions, and acid rain. Thus, in addition to direct cost saving by growers, reducing nitrogen fertilizer applications in corn and other crops will benefit the environment.

Given the potential economic and environmental benefits of reduced nitrogen fertilizer use, continued research and development efforts on this trait seems justified.

2.2. Output Traits

Some have suggested that the development and commercialization of second-generation transgenic crop products should be focused on providing direct benefits to consumers in the form of better nutrition, better taste, or longer shelf life to facilitate public acceptance of the technology. However, it is not clear that output traits have the potential to create the economic value needed to justify the costs and time necessary to bring them to market. Current value-enhanced grain crops on the market are niche products that in total are planted on less then 4 million acres in the United States *(42)*.

Table 9 provides a list of output traits named in the USDA database of field test notifications and release permits *(19)*. However, as shown in Table 10, only a relatively small number of the traits are the subject of active investigation by the major agricultural biotechnology and food companies *(19)*. Moreover, as shown in Table 10, of the 10 crops for which field trial notifications or release permits were requested for output traits, humans directly consume only rice, potato, and tomato in significant amounts. In contrast, corn and soybeans, which accounted for the majority of notifications and release permits for output traits, are primarily consumed as animal feeds or are only indirectly consumed by humans. Thus, the output traits that are the focus of most ongoing research and development are unlikely to benefit consumers directly.

It is also interesting to note that, with the exception of only two companies, none of the major food companies engaged in field trials of tomatoes and potatoes with enhanced output traits in the late 1990s are continuing trials today.

Output traits, in general, can be categorized into four groups: nutraceutical and nutritional enhancements, feed traits, improved processing characteristics, and "biofactory" traits that increase the production of plant-derived materials.

Table 9
Output Traits Listed in US Field Test Notifications and Release Permits

Improved animal feed quality	Altered fiber quality	Altered pigment metabolism or composition
Anthocyanin production in seeds	Altered fiber strength	Altered polyphenyl oxidase levels
Increased antioxidant enzyme	Enhance flavor	Altered processing characteristics
Increased β-carotene	Reduced flower and fruit abscission	Prolonged shelf life
Reduced fruit bruising	Altered flower color	Increased or altered praline levels
Reduced caffeine levels	Increased fruit firmness	Altered protein levels
Increased carbohydrate levels	Decreased fruit invertase	Increased rubber yield
Altered carbohydrate metabolism	Delayed or altered fruit ripening	Reduced salicylic acid
Altered carotenoid content	Increased fruit solids	Increased secondary metabolites
Altered carotenoid metabolism	Altered fruit sugar and sweetness	Increased seed composition
Reduced catalase levels	Improved grain processing	Increased seed quality
Altered cell walls	Improved breading making	Altered seed storage protein
Altered color	Improved fruit quality	Altered senescence
Delayed fruit softening	Increased phosphorous	Increased solids
Improved digestibility	Increased protein	Increased stanols
Increased dry matter content	Industrial enzyme production	Increased, decreased, or altered starch metabolism
Altered erucic acid	Increased iron levels	Increased or modified sterols
Altered ethylene metabolism, production, and synthesis	Increased or altered lysine levels	Increased sugar alcohol
Reduced ethylene	Melanin produced in cotton fibers	Altered sugar content
Extended flower life	Reduced levels of nicotine	Increased tryptophan levels
Altered fatty acid levels	Imporved or altered nutrition quality	Increased tuber solids
Altered fatty acid metabolism	Altered oil profile or quality	Increased tyrosine levels
Altered feed properties	Increased peroxidase levels	Increased vitamin C

2.2.1. Nutraceutical and Nutritional Enhancements

The global nutraceuticals industry has a market value exceeding $50 billion *(43)*. The primary markets are vitamins, minerals, and herbal supplements *(43)*, none of which are a major focus of current transgenic crop research and development efforts. Vitamins are derived primarily by synthetic means; minerals are not produced in plants.

Investigation of herbals and botanicals is a possible area for biotechnology applications. However, because of the great diversity of herbal plants, their often exotic nature, and the likely consumer rejection of GM herbals and botanicals, nutraceuticals are not a compelling area for the development of transgenic crops with enhanced output traits.

2.2.2. Feed Traits

As shown by the data in Table 10, corn and soybeans engineered to produce high levels of lysine, methionine, or tryptophan or low levels of phytate are areas of active research and development. The primary customers for such output traits are poultry, swine, beef, and dairy producers. If successfully brought to market, these products are likely to face a very challenging competitive response from current feed additive suppliers, such as Archer Daniels Midland, BASF, Roche, Novus International, Sumitomo, Degussa, and many others.

The competitive response by suppliers of oil and fat supplements for the feeds industry was an important contributor to the reduced commercial success of high-oil corn. Similar competitive pressures are expected for transgenic crops with high lysine, methionine, tryptophan or low phytate traits. Thus, any company seeking to develop value-enhanced transgenic crops with altered amino acid, vitamin, or phosphorous profiles for the animal feeds industry should conduct an extensive and detailed competitive analysis to ensure that the potential value creation of the products can actually be realized.

2.2.3. Processor Traits

Processor traits are directed toward increasing the production efficiency of commodity products such as vegetable oils, protein meal, starch, or flour by reducing costs or increasing output. Figure 9 shows the US sales for plant-derived materials (for all chemicals except tocopherols, *see* ref. *44*; for tocopherol, *see* ref. *45*), which makes it possible to estimate the potential value for many processor traits.

For example, sucrose, vegetable oils, and starch are the three largest value plant-derived food commodities. Assuming the need to exceed an annual gross revenues hurdle rate of $200 million per year for the trait developer and a 2:1 value share with growers, processors, and other participants in the commodity chain, a total annual value creation of at least $600 million is needed. Thus, a percentage value added of 10 to over 50% to the current value of the sucrose, vegetable oils, or starch commodity markets is needed to achieve an acceptable return on investment. Although such increases in value creation are possible for sucrose and vegetable oils, they will be extremely difficult to achieve for other plant-derived products. Thus, it can be concluded that few transgenic crops engineered with processor traits are likely to justify the high costs and time necessary for their development and commercialization.

2.3. Value Creation Conclusions

Because of the high costs, time, and risks associated with the development of the second-generation transgenic crops, only a few traits have the potential to achieve a value creation in the $200–500 million per year range, which is necessary to provide the trait developers with an acceptable return on investment. Achieving this high value capture hurdle rate is even more difficult because the major market for transgenic crops is likely to be primarily North America for the foreseeable future (*see* Section 3).

Table 10
Output Trait Field Test Notifications and Release Permits Sorted by Selected Companies

Company	Output Trait	Crops
Aventis (now part of BASF)	Altered carbohydrate metabolism	Corn and rice
BASF	Increased starch levels	Corn and rice
Bayer CropSciences	Altered carbohydrate metabolism	Corn and cotton
BHN Research	Increased solids	Tomato
	Altered fruit sugar profile	Tomato
Biogemma	Altered starch metabolism	Corn and wheat
	Altered seed quality	Corn
	Altered erucic acid	Rapeseed
Cambell (1997[a])	Altered or delayed fruit ripening	Tomato
	Altered fruit sugar	Tomato
	Improved fruit quality	Tomato
Cargill	Altered fatty acid metabolism	Rapeseed
	Altered storage proteins	Corn
	Altered amino acid composition	Rapeseed
Dekalb (now part of Monsanto)	Increased tryptophan levels	Corn
	Increased lysine levels	Soybean
	Increased methionine levels	Corn
	Altered carbohydrate metabolism	Corn

Dow AgroSciences	Reduced phytate levels	Corn
DuPont	Improved animal feed quality	Corn
	Altered carbohydrate metabolism	Corn and soybean
	Altered protein quality	Corn and soybean
	Increased phosphorous levels	Corn
	Increased lysine levels	Corn
	Altered oil quality	Corn and soybean
Forage Genetics	Decreased lignin levels	Alfalfa
Frito Lay (1997[a])	Altered carbohydrate metabolism	Potato
Heinz (1999[a])	Increased solids	Tomato
	Delayed fruit ripening	Tomato
Hunt-Wesson (1998[a])	Reduced pectin esterase levels	Tomato
	Reduced polygalacturonase levels	Tomato
	Increased fruit solids	Tomato
	Altered fruit ripening	Tomato
J. R. Simplot Company (2003[a])	Altered storage protein	Potato
Limagrain	Altered starch metabolism	Corn
	Lipase expressed in seed	Corn
	Altered nutritional quality	Rapeseed
Lipton (2000[a])	Increased anti-oxidant enzyme	Tomato

(continued)

Table 10 *(continued)*
Output Trait Field Test Notifications and Release Permits Sorted by Selected Companies

Monsanto	Altered amino acid content	Corn
	Increased tryptophan levels	Corn
	Altered oil profile	Corn, soybean, and rapeseed
	Altered seed composition	Corn, soybean, and rapeseed
	Altered protein	Corn
	Increased protein levels	Corn
	Increased lysine levels	Corn
	Altered nutritional quality	Corn
	Altered senescence	Corn
	Altered color	Corn
	Altered carbohydrate metabolism	Corn
	Altered storage protein	Corn
National Starch and Chemical (2003[a])	Altered starch metabolism	Corn
Nestle (1995[a])	Altered carbohydrate metabolism	Tomato
Pioneer (now part of DuPont)	Improved animal feed quality	Corn, soybean, sunflower
	Improved nutritional quality	Soybean
	Altered oil profile or quality	Corn and soybean
	Production of novel proteins	Corn and soybean
	Improved grain processing	Corn and soybean
	Increased secondary metabolites	Soybean

	Altered feed properties	Soybean
	Antiprotease producing	Soybean
	Mycotoxin degradation	Corn
	Altered cell wall	Corn
	Increased lysine levels	Corn
	Increased methionine levels	Corn and soybean
	Increased phosphorous	Corn
	Altered carbohydrate metabolism	Corn
	Altered protein quality	Corn
R. J. Reynolds	Altered carotenoid content	Tobacco
Syngenta	Altered seed composition	Corn
	Altered carbohydrate metabolism	Potato
Van den Bergh Foods (1994[a])	Altered fruit ripening	Tomato
Zeneca (now part of Syngenta)		
	Altered carotenoid content	Tomato
	Increased dry matter	Tomato
	Increased solids	Tomato
	Altered carbohydrate metabolism	Corn
	Altered ethylene metabolism	Tomato
	Delayed fruit ripening	Tomato
	Reduced pectin esterase levels	Tomato
	Altered processing characteristics	Tomato
	Altered protein quality	Tomato

[a]The most recent year notifications or release permits were recorded for this company by the Animal and Plant Health Inspection Service of the USDA.

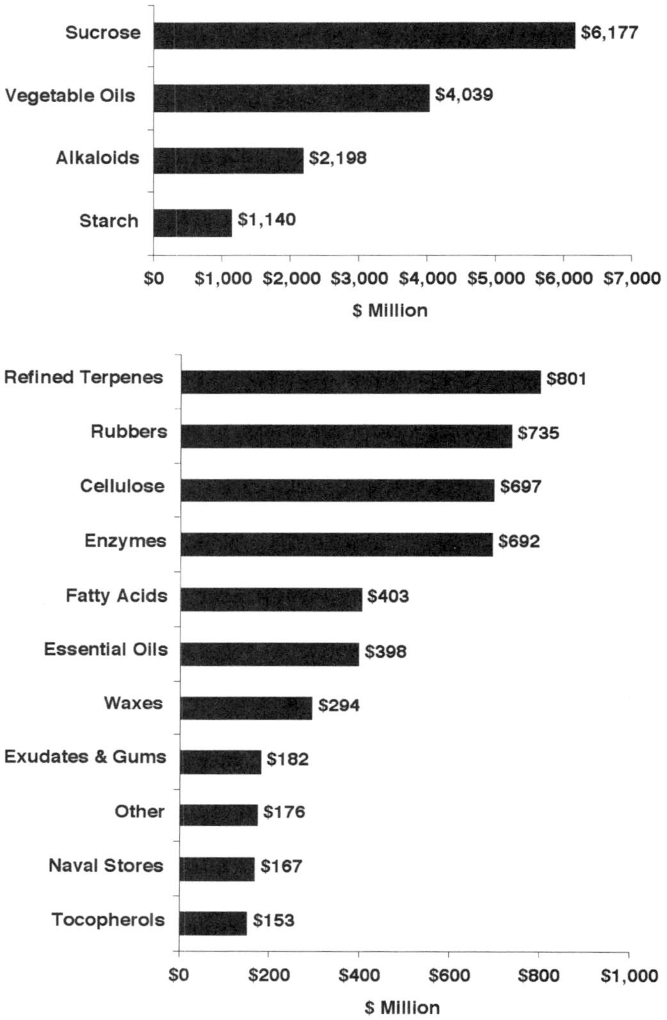

Fig. 9. US sales of plant-derived products in 1989.

Transgenic crops engineered with pathogen tolerance, increased yield, drought tolerance, or the ability to grow and develop normally with reduced inputs of fertilizers will likely create the most value for both trait developers and farmers. In contrast, development of transgenic crops with output traits that provide a direct benefit for consumers and also generate sufficient value to justify development costs will be very challenging.

3. Capturing the Value Created

Value capture has been recognized for many years as one of the most important issues facing agricultural biotechnology *(46)*. Unfortunately, rather than improving, some could argue that the situation with respect to value capture has deteriorated significantly based on the rapid spread of illegal, black market, or farmer saved "brown bag" seed for transgenic crops. Black market glyphosate-tolerant soybeans are widespread in Argentina and Brazil *(47–50)*, which account for approximately 40% of the total global soybean production.

The US General Accounting Office, in an investigation of the difference in prices for GM seeds in the United States and Argentina *(47)*, concluded that the lower price for Roundup Ready soybeans in Argentina was because of the stronger control over patented seed technology in the United States and the extensive sales of black market soybean seed in Argentina. It is estimated that as much as 60% of the soybean seed used in Argentina is black market seed *(51)*. In Brazil, sales of black market Roundup Ready soybean seed is as high as 60% in the important Rio Grande do Sul production area, and black market seed is moving north into the Parana production region *(49)*.

China, with approximately 10% of the global production, is the only major producer, other than the United States, still reported to be free of black market transgenic soybean plantings.

Black market trade in transgenic cotton seed is also widespread. The five largest cotton-producing countries are the United States, China, India, Pakistan, and Uzbekistan, which together produce approximately 70% of all cotton *(52)*. With the exception of the United States and Uzbekistan, black market *Bt* cotton is planted on significant areas in the other major cotton-producing countries *(53–56)*. Black market *Bt* cotton is also reported to be expanding to other cotton-producing countries, such as Thailand and Indonesia *(57)*.

Even transgenic hybrid corn seed has been stated to be black-marketed in Latin America *(58,59)*.

Black market or brown bag seeds significantly reduce revenues for developers of transgenic crops. Until intellectual property protection systems improve in many developing countries in Latin America and Asia that are important producers of agricultural products, value capture for transgenic technologies will be largely limited to developed countries. Because of the strong opposition to transgenic crops in Europe, developers of transgenic crops will primarily depend on North America for the majority of their revenues from transgenic crops. This greatly reduces the possibilities for the development and commercialization of new transgenic crop products. For example, a transgenic trait that conferred broad-spectrum pathogen tolerance in rice, wheat, tomato, and corn is estimated to create up to $14.7, $8.5, $7.6, and $7.0 billion, respectively, of potential value *(15)*.

However, consideration of the geographic distribution of pathogen caused losses for these four crops (*see* Fig. 10) shows that only about $3 billion of the potential value creation occurs in the North American market, in which trait developers would be able to capture an acceptable amount of the value created. Thus, for rice and tomato, the majority of the economic losses caused by pathogens occur in Asia, which is a very low value capture market. Even though an effective transgenic solution to pathogen-caused losses in rice and tomato could create billions of dollars of value, because of the low value capture potential in the Asian markets, the return on the investment of over $50 million and 6 years to develop and commercialize such technology will likely be very poor. Alternatively, pathogen protection technology for wheat and corn for the North American market has the potential to create several billion dollars of total value and provide the trait developer with sufficient revenues to provide an adequate return on investment.

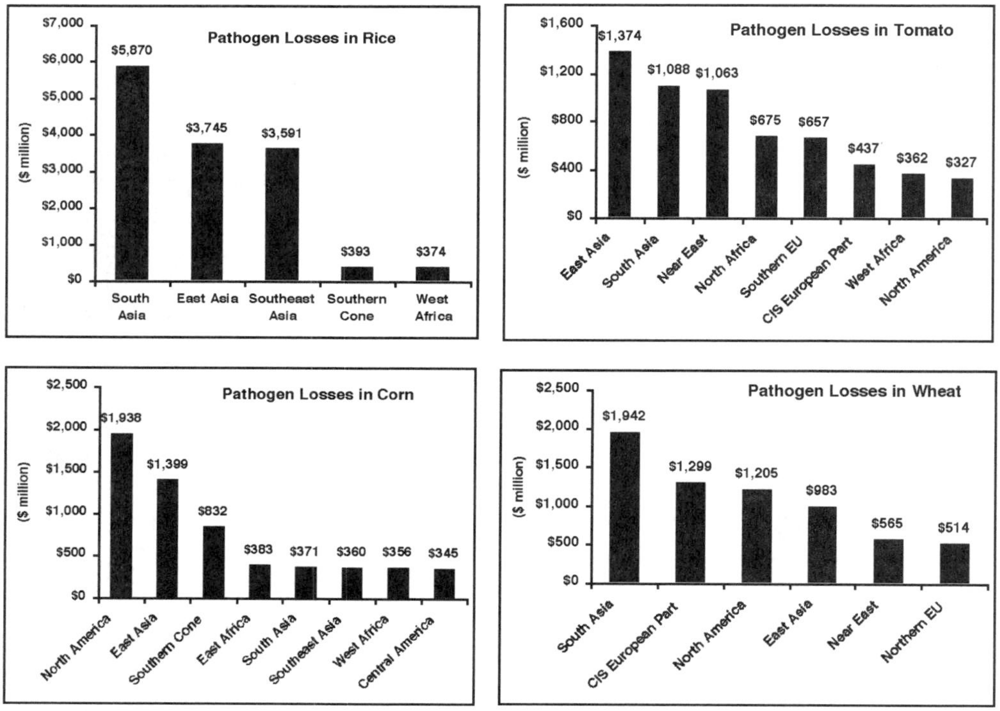

Fig. 10. Geographic distribution of pathogen losses in rice, tomatoes, corn, and wheat.

The agricultural biotechnology industry must redouble its efforts to develop effective value capture mechanisms to obtain adequate returns for the significant investment required to develop and commercialize the next generation of transgenic crops. Technology and science may contribute to improved value capture systems; however, it is more likely that marketing, distribution chain management, and customer service will play a larger role in improving value capture systems. The major producers of agricultural products will continue to be the United States, China, India, Brazil, Argentina, and western and central Europe. If the potential of agricultural biotechnology is to be realized, the industry must develop more effective ways of protecting intellectual property rights, minimizing black market seed, and capturing value for transgenic crops in developing countries.

4. Delivering the Value to the Market

Perhaps the biggest challenge facing the agricultural biotechnology industry is delivering the value of the technology to the market.

4.1. Enabling Assets

For any company to be able to develop and commercialize a transgenic crop, they must possess, or have access to, transformation technologies, germplasm, and traits. For example, licenses to 70 patents owned by approximately 30 different entities in addition to germplasm assets are needed to develop GoldenRice™ *(60)*. The high costs to gain access to the intellectual property and assets needed to develop and commercialize transgenic crops have been a significant factor in the consolidation in the agricultural biotechnology industry *(61,62)*. More significantly, only a very few large, multinational companies have the assets and intellectual property rights required to develop and commercialize transgenic crops and therefore deliver such products to the market.

4.2. Consumer Acceptance

At the end of the day, or rather the end of the commodity food chain, the objective of the industry should be to provide a product that is desired by consumers. It has been recognized for many years that the customer for transgenic crop products is not necessarily the farmer *(63)*. Nonetheless, the agricultural biotechnology industry has failed to make significant progress to increase consumer acceptance of foods derived from transgenic crops *(64–67)*. The industry has urged that the debate be "science based," and proponents of the technology have argued that the transgenic crop products are among the most thoroughly tested food products ever and have been shown to be safe *(68)*.

However, the debate has already largely shifted to a nonscientific debate about labeling and the right to choose *(69,70)*. Labeling of foods containing material from transgenic crops is mandatory in many countries outside North America and is under consideration by others *(71)*. It seems that the harder the industry has fought to keep the debate science based, to convince consumers the technology is safe, and to prevent labeling, the more the debate has shifted from science, the greater the concern by consumers, and the greater the demand for labeling.

5. Conclusions

The promise of agricultural biotechnology articulated when the first transgenic crops were commercialized was that the technology would create greater value for technology providers from input traits by shifting value from chemicals to seeds. The technology was projected to cause a reduction in research and development costs to commercialize new agricultural products, reduce the time needed to develop and commercialize such products, and provide growers and consumers with environmental and safety benefits. Agricultural biotechnology was also projected to create new market opportunities for output traits by integrating the food chain to enable capturing value from traits that benefited processors and consumers rather than just the farmer. Output traits were also projected to "decommoditize" the commodity food chain to create bigger margins for growers and technology providers.

Almost a decade after the first transgenic crops were commercialized, success for the technology has largely been limited to North America. Total development costs and the time required to develop and commercialize transgenic crop products are essentially the same as for traditional agricultural–chemical products *(72)*. Many consumers question the environmental and safety benefits of transgenic crops. Because of consumer acceptance concerns and the refusal by many food processors and retailers to utilize or sell products derived from transgenic crops *(73–75)*, it is necessary to segregate many transgenic crop products. This has caused "dis-integration" of the commodity food chain rather than integration. Finally, some non-GM products sell at a premium to the corresponding GM products.

Does this mean that agricultural biotechnology has failed? The answer is no. As a report sponsored by the United Nations noted, "Biotechnology has the potential to become a powerful tool in meeting the challenges posed by food insecurity, industrial underdevelopment, environment degradation and disease" *(76)*. The obstacles to the success of agricultural biotechnology are not technical. Rather, the principle obstacle is gaining public acceptance for the potential of the technology to provide safe and nutritious food products produced with less environmental impact than traditional agriculture.

To convince consumers of this will certainly take time. It will also likely require a significant increase in resources dedicated to public education, consultation, and listening. There is an old saying in business that "the customer is always right." To be successful, the industry must listen to the concerns and fears of consumers, who are the ultimate customers for the technology, and genuinely address and resolve those concerns. In the absence of such efforts, agricultural biotechnology will likely fail to achieve the great potential that is possible.

References

1. Rodermeyer, M. (2001) Harvest on the Horizon: Future Uses of Agricultural Biotechnology. Prepared by the Pew Initiative on Food and Biotechnology. Available on-line at: http://pewagbiotech.org/research/harvest/. Accessed September 2003.
2. Fernandez, M., Crawford, L., and Hefferan, C. (2003) Pharming the Field: A Look at the Benefits and Risks of Bioengineering Plants to Produce Pharmaceuticals. The Pew Initiative on Food and Biotechnology. Available on-line at: http://pewagbiotech.org/newsroom/releases/022803.php3. Accessed September 2003.
3. Pinstrup-Andersen, P., Pandya-Lorch, P., and Rosegrant, M. W. (1999) *World Food Prospects: Critical Issues for the Early 21st Century*. Vision 2020 Food Policy Report. International Food Policy Research Institute, Washington, DC. Available on-line at: http://www.ifpri.org/pubs/fpr/fpr29.pdf. Accessed September 2003.
4. Food and Agriculture Organization. (2002) World Agriculture: Towards 2015/2030, Summary Report. Food and Agriculture Organization of the United Nations. Available on-line at: http://www.fao.org/docrep/004/y3557e/y3557e00.htm. Accessed September 2003.
5. Food and Agriculture Organization. (2000) Fertilizer Requirements in 2015 and 2030. Food and Agriculture Organization of the United Nations, Rome. Available on-line at: ftp://ftp.fao.org/agl/agll/docs/barfinal.pdf. Accessed September 2003.
6. Ahearn, M., Yee, J., Ball, E., and Nehring, R. (1998) Agricultural Productivity in the United States. US Department of Agriculture, Agricultural Information Bulletin No. 740. Available on-line at: http://www.ers.usda.gov/publications/AIB740/. Accessed September 2003.
7. Huges, J. C. (2002) Change in the Average Size of US Farms, 1900–2001. Available on-line at: http://www1.sru.edu/gge/faculty/hughes/100/100-5/d-5-7.htm. Accessed September 2003.
8. James, C. and Krattiger, A. F. (1996) *Global Review of the Field Testing and Commercialization of Transgenic Plants, 1986 to 1995: The First Decade of Crop Biotechnology*. ISAAA Briefs No. 1. International Service for the Acquisition of Agri-biotech Applications, Ithaca, NY. Available on-line at: http://www.isaaa.org/Publications/briefs/briefs_1.htm. Accessed September 2003.
9. James, C. (2003) *Global Status of Commercialized Transgenic Crops: 2002*. ISAAA Briefs No. 27. International Service for the Acquisition of Agri-biotech Applications, Ithaca, NY. Available on-line at: http://www.isaaa.org/Publications/briefs/briefs_27.htm. Accessed September 2003.

10. Kalaitzaandonakes, N. (1999) A farm level perspective on agrobiotechology: how much value and for whom? *AgBioForum* **2**, 61–64.
11. Council for Biotechnology Information. (2003) US Farmers Increasingly Embrace Biotech Crops, Report Shows. Available on-line at: www.whybiotech.com/index.asp?id=3140. Accessed September 2003.
12. White, W. J., (2001) Economic history of tractors in the United States. In *EH.Net Encyclopedia*. (Whaples, R., ed.). Available on-line at: http://www.eh.net/encyclopedia/white.tractors.history.us.php. Accessed September 2003.
13. Graff, G. D. and Newcomb, J. (2003) *Agricultural Biotechnology at the Crossroads, Part 1: The Changing Structure of the Industry*. Bio Economic Research Associates, Cambridge, MA, p. 16. Available on-line at: http://www.bio-era.net/pdf/bioera_agbio_report.pdf. Accessed September 2003.
14. Food and Agricultural Policy Research Institute. (2003) FAPRI 2003 US Baseline Briefing Book, FAPRI-UMC Technical Data Report 04-03. Available on-line at: www.fapri.missouri.edu. Accessed September 2003.
15. CAB International. (2003) Crop Protection Compendium, 2003 Edition. Available on-line at: www.cabicompendium.org/cpc/home.asp. Accessed September 2003.
16. Fuller, P. A. (1998) Application of the Science of Genomics in Ag Biotech: The Importance of a Comprehensive Approach. Strategic Partnerships to Successfully Commercialize Agricultural Products. Global Business Research, July 23–24, Chicago, IL
17. Frerichs, R., Nafziger, E. D., Eckhoff, S., and Lattz, D. (2003) Illinois Specialty Farm Products, High Oil Corn–Updated for 2003. Available on-line at: http://web.aces.uiuc.edu/value/factsheets/corn/fact-oil-corn.htm. Accessed September 2003.
18. Brandt, P. (2003) Overview of the current status of genetically modified plants in Europe as compared to the USA. *J. Plant Physiol.* **160**, 735–742.
19. Information Systems for Biotechnology. (2003) Total Number of Approved Releases by Phenotype Category 1987–Present. Charts for Field Test Releases in the US. Available on-line at: www.nbiap.vt.edu/cfdocs/biocharts2.cfm. Accessed September 2003.
20. National Drought Mitigation Center. (2003) Historical Graphs of the Palmer Drought Index. Available on-line at: www.drought.unl.edu/whatis/palmer/riverbasin.htm. Accessed September 2003.
21. National Drought Mitigation Center. (2003) Understanding Your Risk, A Comparison of Droughts, Floods, and Hurricanes in the United States. Available on-line at: www.drought.unl.edu/risk/us/compare.htm. Accessed September 2003.
22. National Weather Service. (2003) Climate Prediction Center, Drought Monitoring, Past Palmer Drought Severity Index Maps by Week for 1998 Through 2003. Available on-line at: www.cpc.ncep.noaa.gov/products/monitoring_and_data/drought.html. Accessed September 2003.
23. Cook, E., Meko, D., Stahe, D., and Cleaveland, M. (1999) Reconstruction of Past Drought Across the Coterminous United States from a Network of Climatically Sensitive Tree-Ring Data. NOAA Paleoclimatology Program–PDSI Annual Plots Page. Available on-line at: www.ngdc.noaa.gov/paleo/pdsiyear.html. Accessed September 2003.
24. US Department of Agriculture. (2003) Major World Crop Areas and Climatic Profiles. Available on-line at: www.usda.gov/oec/waob/jawf/profiles/graphs/usa/usacrn.gif. Accessed September 2003.
25. US Department of Agriculture. (2003) USDA–National Agricultural Statistics Service Historical Data. Available on-line at: www.usda.gov/nass/pubs/histdata.htm. Accessed September 2003.
26. Hesterman, O. B. and Carter, P. R. (1988) *Utilizing Drought-Damaged Corn*. National Corn Handbook, NCH-58. Purdue University Cooperative Extension Service, West Lafayette, IN. Available on-line at: http://www.agcom.purdue.edu/AgCom/Pubs/NCH/NCH-58.html. Accessed September 2003.

27. National Climatic Data Center. (2003) Climate Division Drought Data–Graphing Options. Available on-line at: http://lwf.ncdc.noaa.gov/oa/climate/onlineprod/drought/main.html. Accessed September 2003.
28. Dyson, T. (1998) World Food Trends and Prospects to 2025. NAS Colloquium, Plants and Population: Is There Time? December 5–6. Available on-line at: http://www.lsc.psu.edu/nas/Speakers/Dyson%20manuscript.html. Accessed September 2003.
29. Iowa State University Extension. (2002) Estimated Costs of Crop Production in Iowa–2002. Report No. FM 1712. Available on-line at: www.econ.iastate.edu/faculty/duffy/2002CROPS_FM1712.pdf. Accessed September 2003.
30. University of Idaho. (2001) Northern Idaho Crop Production Costs. Available on-line at: links to Crop Costs and Returns Estimates (Enterprise Budgets) at http://www.ag.uidaho.edu/aers/publications. Accessed September 2003.
31. Reinsel, B. (1998) The Agricultural Resource Management Study: Serving the Information Needs of Agriculture. Agricultural Outlook, Economic Research Service of the USDA. Available on-line at: http://www.ers.usda.gov/publications/agoutlook/sep1998/ao254b.pdf. Accessed September 2003.
32. AgBrazil. (2002) Dryland Crop Production Costs in US$ per hectare, Barreiras, Bahia, Brazil: Crop Year 2001–2002. Available on-line at: http://www.agbrazil.com/. Accessed September 2003.
33. The Fertilizer Institute. (2003) US Fertilizer Use. The Fertilizer Institute: Statistics. Available on-line at: http://207.254.26.214/Statistics/Usfertuse.asp. Accessed September 2003.
34. Major Fertilizer Consuming Countries. (2002) The Fertilizer Institute: Statistics. Available on-line at: http://www.fti.org/Statistics, 2002. Accessed September 2003.
35. Harris, G. (1998) An Analysis of Global Fertilizer Application Rates for Major Crops. IFA Annual Conference, Toronto, Canada. Available on-line at: http://www.fertilizer.org/ifa/publicat/PDF/1998_biblio_18.pdf. Accessed September 2003.
36. Economic Research Service, United States Department of Agriculture. (2003) Fertilizer Use and Price Statistics. Available on-line at: http://usda.mannlib.cornell.edu/data-sets/inputs/86012/. Accessed September 2003.
37. Food and Agriculture Organization. (2003) Prospects for the Environment: Agriculture and the Environment. World Agriculture: Towards 2015/2030, Summary Report. Available on-line at: http://www.fao.org/docrep/004/y3557e/y3557e00.htm. Accessed September 2003.
38. Snyder, C. S., Bruulssema, T. W., Ludwick, A. E., et al. (2000) Hypoxia in the Gulf of Mexico and Fertilization Facts. IFA Technical Conference, New Orleans, Louisiana, October 2000.
39. World Resources Institute. (2000) Intensification of Agriculture: Chemical Inputs. World Resources 1998–99. Available on-line at: http://www.igc.org/wri/wr-98-99/agrichem.htm. Accessed September 2003.
40. International Fertilizer Industry Association. (1998) The Fertilizer Industry, World Food Supplies and the Environment. Available on-line at: http://www.uneptie.org/pc/agri-food/library/fi_food.htm. Accessed September 2003.
41. US Geological Survey. (1995) Chesapeake Bay: Measuring Pollution Reduction. Available on-line at: http://water.usgs.gov/wid/html/chesbay.html. Accessed September 2003.
42. US Grains Council. (2002) 2000–2001 Value-Enhanced Grains Quality Report. Available on-line at: www.vegrains.org/documents/2001veg_report/veg_report.htm. Accessed September 2003.
43. Gruenwald, J. and Herzberg, F. (2002) The Global Nutraceuticals Market. Available on-line at: www.bbriefings.com/pdf/foodingredients_2002/publication/gruenwald.pdf. Accessed September 2003.
44. Rotheim, P. (1990) *Plant-Derived Chemicals: Performance Requirements, Applications and Manufacturing*. Business Communications Company, Norwalk, CT.

45. Herbers, K. (2003) Vitamin production in transgenic plants. *J. Plant Physiol.* **160**, 821–829.
46. Burrill, G. S. (1999) Agbio Circa "99": Where Have We Been? Where Are We Going? North American AGBIO Expo, Chicago, IL.
47. Robertson, R. E. and Westin, S. S. (2000) *Biotechnology: Information on Prices of Genetically Modified Seeds in the United States and Argentina.* Report GAO/RCED/NSIAD-00-55. US General Accounting Office, Washington, DC. Available on-line at: www.gao.gov. Accessed September 2003.
48. AgWeb from Pro Farmer. (2002) Brazil Beans Losing Their "Clean" Status re: GMOS. Available on-line at: www.afaa.com.au/news/news-757.asp. Accessed September 2003.
49. Dow Jones. (2002) Brazil's Parana May Have More GMO Soybeans Than Thought. Available on-line at: www.afaa.com.au/news/news-757.asp. Accessed September 2003.
50. American Soybean Association. (2003) Brazil's Second Biggest Soy Producing Region May Test Shipments From Other Regions for GMOs. Available on-line at: www.asasoya.org/News/Weekly/Soy031703.htm. Accessed September 2003.
51. Pirovano, F. (2003) Argentina Planting Seeds Annual 2003. Foreign Agricultural Service, GAIN Report AR3016. Available on-line at: www.fas.usda.gov/gainfiles/200304/145885447.pdf. Accessed September 2003.
52. Rafiq, M. (2000) New Frontier in Cotton Production. International Cotton Advisory Committee. Available on-line at: http://www.icac.org/icac/cotton_info/Speeches/Chaudhry/2000/Turkey2000.pdf. Accessed September 2003.
53. Pray, C. E., Huang, J., Hu, R., and Rozelle, S. (2002) Five years of *Bt* cotton in China–the benefits continue. *Plant J.* **31**, 423–430.
54. Rao, C. K. (2001) Illegal *Bt* Cotton in India. Foundation for Biotechnology Awareness and Education, Bangalore, India. Posted to AgBioView. Available on-line at: www.biotech-info.net/illegal_cotton_India.html.
55. Jayaraman, K. S. (2001) Illegal *Bt* cotton in India haunts regulators. *Nat. Biotechnol.* **19**, 12.
56. Iqbal, N. (2002) Pakistan opens doors to GM seed. *Asia Times*, November 15. Available on-line at: www.atimes.com/South_Asia/DK15Df03.html.
57. Genetic Resources Action International. (2001) *Bt* cotton ... through the back door. Available on-line at: www.grain.org/publications/seed-01-12-2-en.cfm.
58. Reuters. (2001) Argentina Finds and Destroys Illegal GM Seeds. Available on-line at: www.gene.ch/genet/2001/May/msg00033.html.
59. Wroclavsky, D. (2002) GMOs Help Argentina Fight Subsidies, Monsanto. Reuters. Available on-line at: http://members.tripod.com/~ngin.121202a.htm.
60. Kryder, R. D., Kowalski, S. P., and Krattiger, A. F. (2000) *The Intellectual and Technical Property Components of pro-Vitamin A Rice (GoldenRice™): A Preliminary Freedom-to-Operate Review.* International Service for the Acquisition of Agri-biotech Applications, Ithaca, NY.
61. Fulton, M. and Giannakas, K. (2000) Agricultural biotechnology and industry structure. *AgBioForum*, **4**, 137–151. Available on-line at: www.agbioforum.org.
62. Graff, G. D., Rausser, G. C., and Small, A. A. (2003) Agricultural biotechnology's complementary intellectual assets. *Rev. Econ. Stat.* **85**, 349–363.
63. May, M. (1998) Agricultural Biotechnology Networks: Transforming Promise Into Practice. Strategic Partnerships to Successfully Commercialize Agricultural Biotech, July 23–24, 1998, Chicago, IL.
64. Thayer, A. M. (2000) Agbiotech Industry is gambling on an information campaign, continued farmer acceptance, and promises for the future. *Chem. Eng. News* **78**, 21–49. Available on-line at: http://pubs.acs.org/cen/coverstory/7840/7840bus1.html.
65. Vidal, J. (2001) Global GM market starts to wilt. *Guardian*, August 28. Available on-line at: www.guardian.co.uk/gmdebate/Story/0,2763,543222,00.html.

66. Hellemans, A. (2003) Consumer fear cancels European GM research. *Scientist* **17**, 52.
67. Pew Initiative on Food and Biotechnology. (2003) US vs EU: An Examination of the Trade Issues Surrounding Genetically Modified Food. Available on-line at: http://pewagbiotech.org/
68. Vasil, I. K. (2003) The science and politics of plant biotechnology–a personal perspective. *Nat. Biotechnol.* **21**, 849–851.
69. Taylor, M. R. (2003) Rethinking US leadership in food biotechnology. *Nat. Biotechnol.* **21**, 852–854.
70. Burton, M., James, S., Lindner, B., and Pluske, J. (2002) A way forward for Frankenstein foods. In *Market Development of Genetically Modified Foods* (Santanielllo, V., Evenson, R. E., and Zilberman, D., Eds.). CAB International, New York, pp. 7–23.
71. Jia, H. (2003) GM labeling in China beset by problems. *Nat. Biotechnol.* **21**, 835–836.
72. Hewitt, H. G. (1998) The fungicides market. In *Fungicides in Crop Protection.* CAB International, New York, p. 13.
73. Lilliston, B. (2000) Decision by Fast Food Giants to Reject Genetically Engineered Potatoes Sends Strong Message to Farmers. Available on-line at: www.biotech-info.net/decisions_fastfood.html.
74. Townsend, M. (2003) Supermarkets tell Blair: we won't stock GM. *Observer*, June 8. Available on-line at: http://observer.guardian.co.uk/print/0,3858,4686465-102279,00.html.
75. Demetrakakes, P. (2000) The GMO balance, processors Are Trying to Gauge the Practical Meaning of the Backlash Against Genetically Modified Crops. Available on-line at: www.foodprocessing.com/Web_First/fp.nsf/ArticleID/MEAT-4L8NVB/.
76. United Nations. (2003) Impact of New Biotechnologies, With Particular Attention to Sustainable Development, Including Food Security, Health and Economic Productivity. Report to the United Nations General Assembly p.4. Available on-line at: http://www.unicttaskforce.org/community/documents/1013213441_sixth_cstd.pdf.

11

Biosafety Issues, Assessment, and Regulation of Genetically Modified Food Plants

Yong Gao

1. Introduction

Human beings have always relied on plants for food, shelter, clothing, and fuel. The demand for plant resources will increase as the world's population grows. In 1900, the global population was approximately 1.6 billion. Now, at the beginning of a new century, this number has surged to 6 billion, and the United Nations estimates that the global population will reach 10 billion by 2030. Modern biotechnology (genetic engineering) can help meet the ever-increasing needs for food and fiber resources by increasing crop yields, decreasing production inputs such as water and fertilizer, and providing pest control methods that are more compatible with the environment.

For centuries, plant breeders have relied on pure breeding, hybridization, and other genetic modification techniques to improve the yield and quality of crops and to provide crops with built-in protection against insects and diseases. The earliest agriculturists performed genetic modification to convert wild plants into domesticated crops long before genetics was understood. As knowledge of plant genetics increased, plant breeders used controlled pure breeding and hybridization of plants with desirable traits to produce offspring with the best traits of the parental plants. These conventional processes are often time consuming, inefficient, and subject to significant practical limitations (e.g., breeding can only be achieved within reproductively compatible species).

The tools of modern biotechnology allow plant breeders to select single genes that produce desired traits and move genes from one plant (or microorganism) to another plant. The process is far more precise and selective than traditional breeding. Many of the traits that scientists want to incorporate with modern biotechnology into crops are similar to those that plant breeders have been trying to incorporate through conventional breeding methods. Examples of these traits include increased yield, enhanced nutrition, improved quality (such as delayed ripening and softening of tomato), the ability to withstand environmental stress (such as drought and salt), resistance to diseases, resistance to pests (such as insects and nematodes), and tolerance to herbicides. The new crop varieties produced by modern biotechnology are commonly called genetically modified (GM) plants, genetically engineered plants, transgenic plants, or

From: *The GMO Handbook: Genetically Modified Animals, Microbes, and Plants in Biotechnology*
Edited by: S. R. Parekh © Humana Press Inc., Totowa, NJ

biotechnology-derived plants. Given the widespread public use, the term GM is used in this chapter although it is somewhat a misnomer because plants generated through conventional breeding methods are also genetically modified.

Since the first commercial introduction of GM plants in 1996, the global planting area of GM crops has increased at an annual growth rate of over 10% *(1)*. Overall, during the 7-year period from 1996 to 2002, the global area of GM crops increased 35-fold, from 1.7 million ha in 1996 to 58.7 million ha in 2002. It is estimated that between 5.5 and 6 million farmers in 16 countries planted GM crops in 2002. Four principal countries accounted for 99% of the global GM crop area. The United States grew 39 million ha (66% of global total GM area), followed by Argentina with 13.5 million ha (23%), Canada with 3.5 million ha (6%), and China with 2.1 million ha (4%).

The principal GM crops grown in 2002 were soybean, corn, cotton, and canola. On a global basis, 51% of the 72 million ha of soybean grown worldwide were GM; 9% of the global 140 million ha of corn were GM; 20% of the global 34 million ha of cotton were GM; and 12% of the global 25 million ha of canola were GM *(1)*. Among the commercialized GM varieties, herbicide tolerance is the most common trait, with insect resistance second. The two most common GM crop–trait combinations are herbicide-tolerant soybean, occupying 36.5 million ha grown in seven countries, and insect-resistant corn, occupying 7.6 million ha and planted in seven countries. More than half of the world's population lives in countries where GM crops are approved and grown *(1)*.

As with all new technologies, there are concerns about unknown effects. Attitude toward the use of modern biotechnology in food and agriculture has become highly polarized, particularly in western Europe. Although GM crops are under rapid development and used over large areas in the Americas, Asia, Australia, and Africa, in western Europe there has been heated debate on whether such technology should be applied to agriculture and food production *(2)*. The critics of GM products question the relative food safety and environmental benefits vs the risks of GM crops compared to crops derived through conventional breeding methods. Critics and some supporters of GM technology also question whether new GM crops receive sufficient governmental regulatory oversight.

In this chapter, the general safety issues associated with GM food plants are described; these include toxicity, allergenicity, nutrition, environmental, and ecological effects. Although this chapter is mainly focused on the potential risks and concerns of plant GM technology, the benefits are also briefly discussed. Then, the scientific principles and approaches used by worldwide regulatory authorities and scientists in the assessment of GM food safety are introduced, highlighting the principle of risk analysis, the concept and application of substantial equivalence, the novel genes, and the toxicity and allergenicity of the novel proteins. Finally, an overview is given on the current international regulatory frameworks on GM plants, in which the efforts of the United Nations in providing a platform for the establishment of a global consensus on the regulation and safe use of agricultural biotechnology is introduced. As well, the regulation and approval processes of GM plants for environmental release and for use as food or animal feed in Argentina, Australia, Canada, China, European Union, Japan, and the United States are reviewed.

2. Toxicity

2.1. Toxicity of Plant Food in General

Plants have been an important source of food for humans since the very beginning of humankind. Humans have learned that some plants can be eaten safely, but others cannot. It was also learned that some plants could only be eaten after cooking. Knowledge gained on the safety of different plants was passed on from generation to generation. Also, through much of human history, humans have been selecting and breeding comestible plants for desirable properties and traits, such as higher yield or better nutrition. As a result, food plants have been changed dramatically in their morphology, growth, performance, and nutritional composition.

With the advances in science and technology, humans have learned more about the chemical composition of plants. It was revealed that food plants are composed of many different compounds and molecules that provide humans with essential macro- and micronutrients and energy. It was also found that, even in some highly domesticated, common food crops, in addition to the various nutrients, there are certain substances that either are toxic or have antinutritional effects to humans and animals; these are called toxicants or antinutrients. Examples of these substances are glycoalkaloids and solanine in potato; erucic acid, phytic acid, and glucosinolates in rapeseed; tomatine in tomato; and trypsin inhibitors in the grains of wheat, rice, and corn. Safety for consumption for some plant foods is relative because safety margins for many natural plant toxicants are quite small *(3)*. For example, a vegetable, *Sauropus androgynus*, which was safely eaten in Borneo and Malaysia, caused severe toxicity and death when consumption increased because of a new perception that it was a "health food" *(4)*.

Besides small compounds, plants may also produce protein toxins. As an example, a family of proteins called γ-thionins is a group of small (5-kDa), highly basic, disulfide-rich proteins found in seeds, stems, roots, and leaves of several common food plants, such as corn and barley. The different members of this family of plant proteins show both sequence and structural homology and are toxic to bacteria, fungi, yeasts, and animal cells *(5)*. In fact, there are structural similarities between γ-thionins of plants and μ-conotoxins, a group of toxins isolated from the venom of the piscivorous sea snail *Conus geographus*. It was reported that plant γ-thionins and sea snail μ-conotoxins share a similar mode of action by blocking sodium channels of the cells *(6)*.

2.2. Potential Alteration in Toxicity of Plants After Genetic Modification

Genetic modification of plants is achieved by the introduction of new genes from one plant, bacterium, or animal species to the desired plant species using recombinant DNA technology. The insertion of the new genes results in either the production of one or more new substances in the plant or a change in the synthesis of existing substances. The effect can be direct and intentional, in which case the introduced genes result directly in the production of proteins or enzymes to obtain a desired phenotype. Alternatively, the introduction of new genes may increase or decrease the expression of the existing proteins or enzymes, which in turn results in the change of other substances in plants. Either of these processes may alter protein profiles within a given crop and therefore require evaluation with respect to their potential toxicity to humans or animals.

The process of genetic modification may also have indirect effects that lead to the production or accumulation in plants of substances that could have potential toxic effects. Insertion of a transgene into a host plant, by the methods of *Agrobacterium*-mediated transformation or biolistic particle bombardment transformation, is a random process with the possibility of changing the regulation of other genes in plants. If the transgene inserts into a gene involved in the regulation of a toxicant, this insertion may lead to either elevated or reduced levels of this toxicant. In another scenario, the product of a transgene can be an enzyme intended to work in a specific metabolic pathway. If this enzyme also shows activity in another metabolic pathway (e.g., a pathway of a toxicant synthesis), it will alter the toxicant level in the plants. This is called a *pleiotropic effect (7)*. However, this effect is not unique to GM plants. It can also be caused by natural or induced mutations (via such methods as irradiation or chemical induction) and by other traditional breeding methods. Trait developers are required by government regulations to monitor toxicants, antinutrients, and other key food components in new GM varieties to ensure that the concentrations of toxicants and antinutrients in new varieties have not increased above the levels of conventional nontransgenic crops.

2.3. Reduction in Levels of Stress-Mediated Toxicants and Fungal Mycotoxins in GM Plants

Plants are susceptible to attack in the field and during storage by various microorganisms and insects. They are also subjected to various physiological stress conditions during cultivation, harvesting, and storage. In response to stress situations, the affected plant tissue produces "unusual" compounds in an attempt to counteract the stress or infection. The unusual compounds produced by plants in response to various exogenous stimuli are generally referred to as *phytoalexins (8)*. Many phytoalexins are toxic to certain microorganisms and animals, and some (toxicants) are hazardous to humans. There is a growing literature body about the abundant levels and varieties of toxic phytoalexins in human foods. These toxicants can cause mutagenicity, carcinogenicity, teratogenicity, neurotoxicity, and visceral organ toxicity in routine laboratory tests *(9)*.

As an example, potato plants respond to infection and stress with increased levels of glycoalkaloids *(10)*. The acute toxic properties of glycoalkaloids include anticholinesterase activity and saponinlike properties that disrupt membrane function and cause nausea, diarrhea, vomiting, and abdominal pain at 1–2 mg/kg body weight, with death occurring in humans at 5–6 mg/kg *(11)*. In addition, high levels of potato glycoalkaloids have been shown to cause birth defects and increased fetal mortality in a number of animal species *(12,13)*. Risk analyses indicated that, although average exposures are low, a small percentage of children may be exposed to much higher levels of glycoalkaloids on an acute basis *(9)*.

Because increases in plant natural toxicants are stress mediated, the prevention of infection, predation, and physiological stress on plants will conceivably reduce the levels of the natural toxicants in the food supply *(9)*. GM crops with insect-resistant or disease-resistant traits are already on the market; GM crops with physiological stress resistance are under development. Compared with their nontransgenic counterparts, these GM crops should have lower levels of stress-induced plant toxicants if microbial infection, insect infestation, or environmental stress conditions occur in the fields.

Besides stress-mediated plant toxicants, mycotoxins in foods produced by fungi that infected the plants have long been recognized as a serious hazard to human and animal health. Fumonisins and aflatoxins are typical examples of noxious mycotoxins. Fumonisins are produced by several *Fusarium* species that cause corn ear rot, which can be found in nearly every cornfield at harvest. Fumonisins can be fatal to horses and pigs and are probably human carcinogens *(14)*. Aflatoxins are another group of notorious corn mycotoxins produced by *Aspergillus flavus* and *Aspergillus parasiticus* that causes corn kernel rot *(15)*. The health and economic impacts of aflatoxins have been greater than that of other mycotoxins in corn because aflatoxins can be passed into milk if dairy cows consume contaminated grain. Aflatoxins have demonstrated potent carcinogenic effect in susceptible laboratory animals and acute toxicological effects in humans. It is known that lepidopteran insects can influence the development of stalk rot and ear rot disease in corn. *Fusarium* ear rot and *Aspergillus* kernel rot are often associated with insect damage to ears or kernels *(16)*.

One of the first successful commercial applications of plant genetic engineering was the development of transgenic *Bacillus thuringiensis* (*Bt*) plants, which incorporate specific genes from the soil bacterium *B. thuringiensis* that express insecticidally active endotoxins (Cry proteins). The Cry proteins are selectively active on certain insect species among the orders of *Lepidoptera*, *Diptera*, and *Coleoptera*. Results of field studies have consistently demonstrated that *Bt* corns had significantly lower incidence and severity of *Fusarium* ear rot and *Aspergillus* kernel rot and produced corn grains with lower levels of fumonisins and aflatoxins *(15,17–19)*. Hammond et al. reported field trials of *Bt* and non-*Bt* corns in five countries (Argentina, France, Italy, Turkey, and the United States) from 1997 to 2001 *(19)*. Their data showed that overall fumonisin levels in *Bt* corn were reduced by 47% to 97% compared with non-*Bt* corn. Windham et al. reported that, when plants were infested with southwestern corn borers, a *Bt* corn hybrid had more than 75% reduction in aflatoxin compared with its non-*Bt* counterpart (5 vs 41 ppb) *(15)*.

3. Allergenicity

3.1. Food Allergy

The term *food allergy* is often overused by the public, as well as by some physicians and scientists, to describe any undesired or bothersome problem related to diet *(20)*. For clarity, human adverse reactions to foods can be classified into two categories: toxic reactions and nontoxic reactions *(20)*. Toxic reactions are caused by toxicants naturally occurring in foods or toxic compounds and contaminants introduced during food production, processing, and handling. Nontoxic reactions are further divided into food intolerance and food allergy. Food intolerance refers to those nontoxic reactions that are nonimmune mediated. One example of food intolerance is lactose intolerance, the inability to digest lactose present in dairy food products.

Food allergy is defined as those nontoxic reactions that are primarily immune mediated. The antigenic molecules giving rise to the immune response are therefore called *food allergens*. The immune-mediated food allergy can be mediated by IgE (immunoglobulin E) or non-IgE mediated. Most known cases of food allergy are IgE mediated. In IgE-mediated food allergy, when first ingested, the allergens may be cleaved to some degree by digestive enzymes, absorbed by the gut mucosa, processed in immunopotent cells, and then presented to the immune system, resulting in the production of allergen-specific IgE. The allergen-specific IgE antibodies circulate in the body and will bind to mast cells throughout the body tissues and basophils in the blood. On renewed contact with the food allergens, the allergens bind to the IgE antibodies, which triggers the degranulation of the mast cells and basophils and production of mediators (such as histamine and leukotrienes), which induce various clinical symptoms *(21)*. Symptoms of IgE-mediated food allergy vary among individuals and are nonspecific. The clinical aspects of IgE-mediated food allergy can be oral (itching and swelling of lips, mouth, or throat); dermal (urticaria, atopic dermatitis, angioederma); gastrointestinal (vomiting, cramps, diarrhea, abdominal pain); respiratory (asthma, rhinitis, bronchospasm, wheezing); and cardiovascular (decrease in blood pressure, anaphylaxis) *(21)*. The symptoms may occur within minutes to days after ingestion of the offending food, and in severe cases, death could occur.

The actual prevalence of adverse food reactions is unknown. A consumer survey indicated that one-third of American households believed that at least one member of their family had food allergies *(22)*. Although the public believes the prevalence of adverse food reactions is quite high, many in the medical community suggest that the true prevalence of adverse food reactions is uncommon. In general, the consensus from the medical literature is that the prevalence of adverse food reactions is approximately 2–8% in infants and children and 1% in adults *(23)*.

3.2. Food Allergens and Their Physicochemical Characteristics

All known food allergens are proteins. Foods contain many proteins, but only a small number of them are known to be allergens. Several hundred food allergens have been identified. Based on patient reactivity, food allergens are classified into major allergens and minor allergens. Allergens to which the majority of patients react (more than 50% of individuals sensitive to the allergen react in IgE-specific immunoassays) are described as *major allergens*; allergens to which a minority of patients react are called *minor allergens (24,25)*. Generally, although not always, major allergens tend to be one of the predominant proteins in the allergenic food source. Among the most documented food allergies, over 90% are caused by eight foods or food groups: peanuts, milk, eggs, soybean, tree nuts, fish, crustacea, and wheat *(26)*.

The surfaces of allergenic proteins that interact with cells of the immune system or specific IgE antibodies are called *epitopes*. Epitopes that react with T cells are called *T-cell epitopes*. Those that react with antibody or antibody-producing B cells are called *B-cell epitopes*. Epitopes are either conformational or linear. Conformational epitopes depend on the tertiary structure of the protein or several amino acid sequences on the protein surface. Linear epitopes depend on the linear sequence of amino acids in a protein. The general belief is that T-cell epitopes are linear, whereas B-cell epitopes are conformational *(27)*. However, there are also exceptions *(28)*. If an epitope is composed of a series of covalently linked amino acids, it is called a *continuous epitope*. An epitope that is composed of two different amino acid sequences, which through their tertiary structure form one epitope, is called a *discontinuous epitope*. The minimum number of amino acid residues is 8 for continuous epitopes *(29)*, whereas discontinuous conformational epitopes have 16 or more *(28)*.

Most known food allergens have molecular weights between 10 and 70 kDa. Although smaller molecules may be immunogenic, the molecular weight (MW) of 10 kDa probably represents the lower limit for the allergenic response *(27)*. The upper limit is probably a result of restricted mucosal absorption of larger molecules. However, some allergens, such as the peanut allergens *Ara h* 1 (MW 63.5 kDa) and *Ara h* 2 (MW 17 kDa), exist in native form as large protein multimers that are 200 to 300 kDa in size *(30,31)*. It is not clear if such large molecules act as allergens directly or are disassociated into smaller allergenic fragments during the digestive process.

Many known food allergens are glycosylated, and the carbohydrate structures of plant food allergens in particular are known to represent important IgE epitopes *(32–34)*. For example, IgE from wheat- and barley-allergic patients recognized a peptide containing an N-linked oligosaccharide. This recognition was lost on deglycosylation from the peptide *(32)*. It should be noted that, even though many food allergens are glycoproteins, glycosylation does not represent a unique property for food allergens because glycosylation is also a feature of many other proteins and enzymes that are not food allergens. In addition, some food allergens are nonglycoproteins.

Many food allergen proteins are resistant to heat. The resistance of food allergens to heat suggests the presence of linear epitopes in the protein molecules. For example, the IgE-binding capabilities of a crude peanut extract and two major peanut allergens, *Ara h* 1 and *Ara h* 2, were unaffected by heating at 100°C for up to 60 minutes *(35)*. The IgE-binding ability of another major peanut allergen, the concanavalin A-reactive glycoprotein, is stable to temperatures up to 100°C over a pH range of 2.8 to 10.0 *(36)*. However, heat-labile food allergens have also been identified; heat treatment induces protein denaturation and the loss of conformational IgE-binding epitopes. For example, the IgE-binding ability of the rice glutelin and globulin fractions as assessed by immunoassay was reduced by 40 to 70% by heating at 60°C for 60 minutes or 100°C for 2 to 10 minutes *(37)*. In fact, some food allergens are quite sensitive to heat denaturation. The allergens in fresh fruits and many vegetables are a good example. Consequently, although many food allergens tend to be resistant to heat treatment, this property is not universal.

Several review articles indicated that many food allergens are resistant to proteolysis and hydrolysis *(27,38,39)*. It is also acknowledged that the amount of information on the stability of food allergens to digestion, proteolysis, and hydrolysis is relatively limited *(27)*. Pepsin and trypsin are the major proteases in human gastric fluid and intestinal fluid, respectively. The pH optima for the activity of the two proteases are drastically different, with pepsin most active at acidic and trypsin at basic conditions. Astwood et al. measured and compared the digestive stability of a group of 16 allergens (most of which are storage proteins) and a group of 9 nonallergenic proteins (all enzymes) in a standard simulated gastric fluid (SGF) *(40)*. Although some food allergens were stable in SGF for the full 60 minutes of reaction, others were rapidly degraded within 30 s, but peptide fragments stable for at least 8 minutes were observed. All the nonallergenic enzymes tested were degraded within 15 seconds without forming any stable peptide fragments. Thus, if allergens are more stable than the nonallergenic proteins, then digestive stability may be a parameter to distinguish allergens from nonallergenic proteins. However, Fu et al. conducted a similar study by including more proteins (23 known allergens and 16 nonallergens) and using both SGF and simulated intestinal fluid test systems *(41)*. The test proteins could be divided into four groups based on functions: storage proteins, plant lectins, contractile proteins, and enzymes. The results of this study did not indicate that food allergens are more stable to digestion in vitro than nonallergenic proteins. A comparison of the digestibility among proteins within the four functional groups showed that food allergens could be more, equally, or less susceptible to SGF and simulated intestinal fluid digestion than nonallergenic proteins with similar cellular functions.

3.3. Potential Alteration in Allergenicity of Plant Food After Genetic Modification

Factors that may alter the allergenicity potential of a plant will depend on the newly introduced or altered genes encoding for the novel or changed proteins. If new genes are introduced, the source of the genes is of paramount importance. If the genes are derived from a known allergenic species, caution should be taken to the likelihood of transfer of allergens from the allergenic species. As an example, soybeans and other legumes are an important source of protein in human and animal diets, but are deficient in methionine, an essential amino acid. Many livestock animals are fed soybean meal diets and require supplements to meet their dietary methionine needs.

To improve the nutritional quality of soybeans, researchers at the US company Pioneer Hi-Bred International developed a line of GM soybean that produces a methionine-rich protein (2S albumin) from Brazil nut. Although it was known at the time that Brazil nuts were allergenic to some consumers, no one had identified which protein in Brazil nuts was the responsible allergen. Pioneer scientists and their collaborators at universities investigated the potential of altered allergenicity in the new soybean. Using blood and skin-prick tests, the researchers determined that at least some persons with a hypersensitivity to Brazil nuts were also allergic to the GM soybean. They concluded that the 2S albumin is probably a major Brazil nut allergen *(42)*. Following the study, Pioneer stopped all field testing and destroyed all plant material and seeds of the GM soybean.

In addition to the direct effect of the novel proteins, the introduction of a new gene may also have an indirect influence on the composition of other endogenous plant proteins or allergens. As a result of transformation, neighboring genes at the site of integration of the insert DNA may be turned off or turned on, resulting in changes in existing proteins. Consequently, the possibility exists that the content of existing allergens of the plant could be elevated or reduced. However, this indirect effect is not something unique to GM plants because similar scenarios apply to plants generated by conventional breeding methods *(21)*. It should be noted that, under government regulation, GM plants have to be analyzed for their content of known allergens, and it must be shown that they do not contain levels any higher than nontransgenic commercial varieties; new varieties created through conventional breeding methods have not been subject to similar regulations.

3.4. Removal or Reduction of Allergens From Plants by Genetic Modification

Many allergens in plants have been identified, and researchers are using modern biotechnology tools (such as antisense technology) to remove allergenic proteins from food plants such as rice and soybean *(43,44)* or other plants *(45)* to reduce human exposure. Continued research and product development in this area will expand the choice of foods available to those who suffer from food-related allergies.

For example, soybean is one of the major allergenic food sources. It was suggested that a protein called *Gly m* Bd 30K/P34 is a dominant allergen in soybeans *(44)*. This protein is a cysteine protease and is present in almost all the domestic soybean varieties and its wild relatives. It causes 65% or more of allergic reactions in soybean-sensitive individuals. A genetically modified hypoallergenic soybean is under development by knocking out the expression of *Gly m* Bd 30K/P34 protein *(44)*. The new variety in development looks promising in that it showed negative reaction with IgE from people who were allergic to soybeans. Eventual commercialization of such a hypoallergenic soybean could have a huge favorable effect on the industry and consumers.

Besides food plants, allergens can also be removed from other plants such as ryegrass, a common grass species for lawn and pasture use. Hay fever and allergic asthma, triggered by grass pollen allergens, affect some 20% of the population in cool, temperate climates. Ryegrass is the dominant source of allergens because of its prodigious production of airborne pollen. *Lol p* 5 is the major allergenic protein of ryegrass pollen, judging from the fact that almost all of the individuals allergic to grass pollen show the presence of serum IgE antibodies reactive to this protein. Moreover, nearly two- thirds of the IgE reactivity of ryegrass pollen has been attributed to this protein. A genetically modified ryegrass was reported that demonstrated downregulation of *Lol p* 5 with an antisense construct *(45)*. The transgenic ryegrass plants showed normal fertile pollen development. Immunoblot analysis of proteins with allergen-specific antibodies did not detect *Lol p* 5 in the transgenic ryegrass pollen. The transgenic pollen showed remarkably reduced allergenicity, as reflected by low IgE-binding capacity of pollen extract as compared with that of the nontransgenic control pollen.

In addition to direct removal of allergens from plants by genetic modification, reduction of stress-mediated allergens can be anticipated in some GM plants, depending on the type of traits modified. A series of plant proteins was actively synthesized when plants were exposed to bacterial or fungal pathogens, insects, mechanical wounding, drought, salt, or low-temperature stress conditions. Plant proteins produced as a plant defense response against these stress conditions are called *pathogenesis-related proteins* or PR proteins *(46)*. Some examples of PR proteins are β-1,3-glucanases in banana; class I chitinases in chestnut; thaumatinlike proteins in cherry, apple, and hazelnut; and lipid transfer proteins in barley *(47)*.

PR proteins are categorized into 14 groups based on similarities in their amino acid sequences, enzymatic activities, or other functional or physiological properties. Many PR proteins are stable at low pH and display considerable resistance to proteases and show sequence homologies to other food allergens. In fact, PR proteins represent an increasingly important category of plant-derived allergens. Of the 14 groups of PR proteins, 7 contain food allergens *(48)*. Because increases in PR proteins are stress mediated, the prevention of infection, predation, and physiological stress will conceivably reduce the levels of the PR proteins in plant foods. Compared with conventional crops, GM crops with insect, disease, or physiological stress resistance traits may produce lower levels of stress-mediated allergenic PR proteins if microbial infection, insect infestation, or environmental stress conditions occur in the fields.

4. Nutrition

4.1. Potential Unintentional Alteration in Nutrient Composition After Genetic Modification

There have been concerns that genetic modification could adversely affect the nutritional quality of foods by altering the levels of existing nutrients. This could be important when a specific GM food is an important source of a certain nutrient. Changes in levels of nutrients could theoretically take place in several ways. First, insertion of genes could conceivably disrupt or alter the expression of normally expressed plant genes. Second, expression of the introduced gene—through protein synthesis—might reduce the availability of amino acids used for synthesis of normal plant compounds. Third, production of normal plant compounds might also be affected if the expressed protein diverted substrates from other important metabolic pathways. Finally, either the expressed protein or altered levels of other proteins might have antinutritional effects *(49)*.

These possible concerns, like those discussed in the sections on toxicity and allergenicity, are related to the randomness of DNA insertion. Again, none of these potential changes is unique to genetic engineering of plants; the same changes can also occur when traditional breeding methods are used. Such changes, in fact, may be less frequent in GM plants because only a limited number of genes are transferred with modern biotechnology methods. The potential alteration in nutrient composition of new GM varieties is addressed through characterization of the inserted gene and compositional analysis of the GM foods (*see* Section 6.2.2). The nutritional characteristic of wholesomeness is also evaluated by animal feeding trials. Numerous studies have been conducted on animal performance and nutrition with approved GM crops, and there have been no differences seen compared with conventional crops *(50–52)*.

4.2. Enhanced Nutritional Value of GM Plant Food

Increasing efforts are under way in the modification of food plants for improved nutrients, such as increasing essential amino acids, altering fatty acid composition, and enhancing mineral and vitamin contents. One prominent example is the development of a "golden rice" to provide vitamin A supplement to humans. Vitamin A deficiency (VAD) is a worldwide health problem. VAD can lead to blindness in children and can weaken children's resistance to diseases such as measles and diarrhea. According to the World Health Organization (WHO), 0.5 million children go blind every year from VAD, 1 million die of VAD-associated diseases, and over 100 million are sickened by VAD.

The greatest incidence of VAD occurs in South and Southeast Asia, where 70% of children younger than 5 years of age are affected. The high occurrence is because of the fact that, in Asia, rice is the staple food, and rice plants do not produce vitamin A or β-carotene (the provitamin A) in the grain. A transgenic rice has been developed by inserting several genes (phytoene synthase, phytoene desaturase, lycopene β-cyclase) from daffodil and a bacterium into the genome of a rice variety. As a result, β-carotene is produced in the rice grain, giving it a golden cast *(53)*. Scientists are cross-breeding golden rice with other elite rice varieties. After regulatory clearance, golden rice seeds will be distributed to poor farmers in developing countries.

5. Environmental and Ecological Effects

5.1. Potential Nontarget Effects of GM Plants

Nontarget effects refer to the unintentional effects of GM plants on organisms living in or around the GM crop field. Some GM plants and their associated agricultural practices are intended to kill specific pests, which are target effects. However, they may have effects on other species that are not the intended targets. The nontarget effects of GM plants could be both direct and indirect. Direct effects are those potential toxicological effects of GM plants on nontarget organisms, such as beneficial insects, birds, aquatic life, worms, and soil microorganisms. Indirect effects are the impacts of GM plants or associated agricultural practices on populations of nontarget species that depend on the target pests for survival or reproduction.

The commercialized insect resistant *Bt* crops have been extensively tested for direct and indirect effects on nontarget organisms in both laboratory and field studies. Their direct toxicological effects on nontarget organisms (such as birds, fish, honeybees, ladybugs, parasitic wasps, lacewings, springtails, aquatic invertebrates, and earthworms) have been minimal *(54)*. It is noted that, in 1999, concerns temporarily rose over nontarget effects of *Bt* crops after Losey et al. *(55)* corresponded to *Nature* on the potential risk of corn pollen expressing *Bt* Cry proteins to the monarch butterfly *Danaus plexippus*. According to this report, when young monarch larvae were given no choice but to feed on milkweed (*Asclepias curassavica*) leaves dusted with pollen from a *Bt* corn hybrid, they ate less, grew more slowly, and had a significantly higher mortality rate than larvae feeding on leaves dusted with nontransgenic pollen.

Ecological risk is a function of exposure (environmental dose) and effect (toxicological response). The US Environmental Protection Agency (EPA) concluded that the potential impact of *Bt* corn pollen on sensitive larvae of *Lepidoptera* was negligible because of factors that limit environmental exposure *(56)*. The amount of pollen dusted onto the milkweed leaves was not quantified in the study of Losey et al. *(55)*. As a result, it is impossible to establish a relationship between pollen exposure and effect from these data. Consequently, the US and Canadian governments, universities, and industries funded a series of studies to investigate further the nontarget effect of *Bt* corn pollen on the monarch butterfly in several US states and Canada *(57–62)*.

Based on the information produced by the collaborative research efforts, Sears et al. conducted a comprehensive risk assessment of the impact of *Bt* corn on monarch butterfly populations *(63)*. They used a weight-of-evidence approach, an approach to risk assessment that has been performed for many nontarget species in relation to pesticides, industrial by-products, and other potential toxicants found in the environment. It requires consideration of both the expression of toxicity and the likelihood of exposure to the toxicant as the basic components for a risk assessment procedure.

In the assessment, information was sought on the acute toxic effects of *Bt* corn pollen and the degree to which monarch larvae would be exposed to toxic amounts of *Bt* pollen on its host plant milkweed found in and around cornfields *(63)*. Expression of Cry proteins, the active toxicant found in *Bt* corn tissues, differed among hybrids, especially in the concentrations found in pollen of different events. In most commercial hybrids, *Bt* expression in pollen is low, and laboratory and field studies show no acute toxic effects at any pollen density that would be encountered in the field. Other factors mitigating exposure of larvae include the variable and limited overlap between pollen shed and larval activity periods, the fact that only a portion of the monarch population utilizes milkweed stands in and near cornfields, and the current adoption rate of *Bt* corn at 19% of North American corn-growing areas. The article concluded that the impact of *Bt* corn pollen from current commercial hybrids on monarch butterfly populations is negligible *(63)*.

It should be pointed out that insect-resistant GM crops that express *Bt* endotoxins have the potential to complement the aims and tools of integrated pest management. Because of the specificity of *Bt* endotoxins, each protein only affects a relatively small set of related insect species, and unrelated nontarget species are unaffected. In contrast, many commonly used conventional insecticides, such as pyrethroids, have been shown to affect a broad range of nontarget species adversely, including natural enemies of insect pests. Replacing these chemicals with plant-incorporated protectants (such as *Bt* crops) should allow natural populations of predators and parasitoids to increase, which would be beneficial for biodiversity and improved control of pest species not directly impacted by the *Bt* crops. Head et al. initiated a set of large-scale, long-term field studies to evaluate the relative impact of a transgenic Cry1Ac cotton (Bollard®) and conventional varieties treated with insecticides on natural enemy abundance *(64)*. It was found that, in fields where specific insecticide was used to control lepidopteran pests, a significant reduction in numbers of various arthropod natural enemies was observed relative to the Cry1Ac cotton fields where no chemical insecticide was needed to control the lepidopteran pests. Populations of predatory bugs, including *Orius* and *Geocoris* species, spiders, and ants, were all significantly decreased by chemical insecticide use relative to the Cry1Ac cotton fields *(64)*.

Besides the insect-resistant *Bt* plants, another major development of commercial GM plants is herbicide tolerance. Plants have been transformed with genes encoding enzymes to allow plants to tolerate the presence of herbicides, predominantly glyphosate (Roundup™) and glufosinate ammonium or phosphinothricin (Liberty™), both of which are considered relatively benign herbicides. Both of these are widely used broad-spectrum herbicides. The planting of herbicide-tolerant GM plants such as soybean, corn, and canola leads to use of more environmentally friendly herbicides, which results in reduced direct toxicological effect to all kinds of organisms. On the other hand, it may potentially result in an indirect effect on some nontarget organisms (such as birds) that rely on the weeds as food and shelter. However, proper farming practices could alleviate this effect.

Dewar et al. demonstrated how the development of appropriate weed management systems, used in conjunction with herbicide-tolerant GM crops (in this case, sugar beet) can result in more weeds and insects (which provide food and shelter to birds and other wildlife) without an adverse effect on the crop yield *(65)*. The results of another study

(66) indicated that conservation tillage is an effective farming system in managing the nontarget effect of herbicide-tolerant crops (*see* Section 5.5).

5.2. Vertical Gene Flow and Weediness Potential of GM Plants

Vertical gene flow is the flow of genetic material from parent plants to their descendants, with the descendants either clones of their parental plants or the result of mating between sexually compatible organisms *(67)*. *Weediness* refers to the phenomenon of plants of any kind growing in the wrong place, causing damage, having no benefit, and suppressing cultivated plant species. It is a relative concept because one plant species can be considered a weed in one place and a useful species in another place.

The potential of vertical flow of transgenes from GM plants to wild relatives receives high attention in the environmental risk assessment of GM plants. Vertical gene flow from GM plants to wild relatives requires successful hybridization to take place between GM plants and their wild relatives. It is known that many important crops, such as wheat, rice, corn, soybean and oilseed rape, hybridize with wild relatives somewhere in the cultivation area *(67)*. The concern is that, if expressed in the genetic background of a weed species, a transgene could potentially change the fitness of the wild relative.

The possibility for increased fitness of transgenic hybrids and backcrosses depends on the nature of the transgenes. For example, weeds containing a transgene that confers resistance to a herbicide would be a nuisance to agriculture, but would have little effect in a nonagricultural environment in which the herbicide is not applied. In contrast, an insecticidal *Bt* transgene in a weed host could alter natural ecology by giving transgenic weeds a selective advantage as the result of natural insect pressure *(68)*.

As an example, oilseed rape (*Brassica napus*) is one of the transgenic crops for which gene flow to wild plants is a concern because it has many wild weedy relatives, such as birdseed rape (*Brassica rapa)* and wild radish *(Raphanus raphanistrum)*, persisting in or near areas of cultivation. Much attention has been paid to the determination of the actual gene flow from transgenic oilseed rape. Transgenic hybrids have been produced between *B. rapa* and transgenic oilseed rape modified with herbicide resistance genes and Bt genes under laboratory conditions *(69–71)*.

Halfhill et al. reported that, in a laboratory experiment, transgenic hybrids were obtained from crosses of six Cry1Ac oilseed rape lines and the weedy *B. rapa*; the hybrids exhibited an intermediate morphology between the parental species *(71)*. The Bt transgene was present in the hybrids, and the Cry1Ac protein was synthesized at similar levels to the corresponding Cry1Ac oilseed rape lines. Insect bioassays confirmed that the hybrid material was insecticidal. The hybrids were backcrossed with the weedy parent *B. rapa*, and only half the oilseed rape lines were able to produce transgenic backcrosses. After two backcrosses, the ploidy level and morphology of the resultant plants were indistinguishable from *B. rapa*. This study indicated that, in an open field of Cry1Ac oilseed rape, insect-resistant weedy *B. rapa* could be produced if proper preventive measures were not taken.

It should be pointed out that transferring a resistance trait for one herbicide (such as glyphosate) to a weedy species does not alter the ability to control the weed with other herbicides or other control measures. Therefore, any potential problem may not be realized if farmers employ other weed control measures at different times of the year. Similarly, transfer of an insect resistance trait from a GM crop to a wild weedy relative will only confer selective advantage to that relative when the target insect is a key factor regulating that particular weed population, which is rare.

5.3. Concerns Regarding Horizontal Gene Transfer From GM Plants to Other Organisms

Horizontal gene transfer (or flow) is the nonsexual or parasexual transfer of genetic material between organisms belonging to the same or different species. Although actual evidence of its occurrence or feasibility (except among bacteria and fungi) is exceedingly rare, the issue is taken seriously in the safety assessment of GM plants. The most relevant food safety issue concerning horizontal gene transfer is the potential consequence of the transfer of an introduced gene from material derived from a GM food to microorganisms in the gastrointestinal tract or mammalian cells in such a way that the gene can be successfully incorporated and expressed and result in an impact on human or animal health.

Marker genes are inserted into GM plants to facilitate identification of genetically modified cells or tissue during development. There are several categories of marker genes, including herbicide resistance genes and antibiotic resistance genes. Examples of the antibiotic resistance marker genes used in plant genetic engineering include those that confer bacterial resistance to ampicillin, streptomycin, kanamycin, neomycin, hygromycin, or the like. Questions are raised on the potential for transfer of antibiotic resistance genes because these genes are the most likely to raise safety problems if they are transferred and expressed in gastrointestinal microflora or mammalian cells.

The WHO held a workshop in 1993 titled "Health Aspects of Marker Genes in Genetically Modified Plants" *(72)*. The Food and Agriculture Organization (FAO) and WHO also organized a joint expert consultation in 2000 on foods derived from biotechnology *(73)*. The workshop and the joint consultation both concluded that there is no authentic recorded evidence for the transfer of genes from plants to microorganisms in the gut or to mammalian cells. The conclusion was based on the judgment that transfer of antibiotic resistance genes would be unlikely to occur given the complexity of steps required for gene transfer, expression, and impact on antibiotic efficacy.

The transfer of plant DNA into microbial or mammalian cells under normal circumstances of dietary exposure would require all of the following events to occur *(73)*: (1) the relevant genes in the plant DNA would have to be released, probably as linear fragments; (2) the genes would have to survive nucleases in the plant and in the gastrointestinal tract; (3) the genes would have to compete for uptake with dietary DNA; (4) the recipient bacteria or mammalian cells would have to be competent for transformation, and the genes would have to survive their restriction enzymes; and (5) the genes would have to be inserted into the host DNA by rare repair or recombination events. There have been numerous experiments aimed at evaluating the possibility of transfer of plant DNA to microbes and mammalian cells. To date, there are no reports that marker genes in plant DNA transfer to these cells *(73)*.

Another concern is regarding horizontal gene transfer from GM plants to soil bacteria. Bacteria exist in soil in large numbers. DNA from crop plants has been shown to remain in soil for several months to several years *(74,75)*, and there is evidence that some of this DNA is protected from soil nucleases by binding to clay or organic components of soil and remains capable of transforming bacteria. The amount of this DNA, however, appears to be extremely small after months to a year *(75)*, with the probability of transformation diminishing considerably with time. It should be noted that the majority of known bacteria are not naturally transformable. Horizontal gene transfer from transgenic plants to bacteria has not been demonstrated under field conditions *(74,76)*. Even with much larger concentrations of DNA, transformation with plant transgenes has only been accomplished at low frequencies and under optimized conditions, that is, when homology to existing DNA in the recipient bacteria occurs, as well as high selection pressure for the horizontal transfer event *(77,78)*. When homology does not occur, horizontal transfer has not been observed, even at extremely low frequencies of less than about 10^9 to 10^{17} *(75,77)*. Therefore, DNA transfer occurs rarely, if at all, from plants to soil bacteria *(54)*.

5.4. Concerns Regarding Insects Developing Resistance to Bt Plants

Insects are a major cause of crop damage. They have proved to be remarkably adaptable and capable of developing resistance to a broad range of pesticides. This is not a new issue to agriculture, and breeders and pesticide developers constantly struggle to stay ahead of evolving plant pests. The insect-resistant *Bt* plants have many environmental advantages over conventional pest control with chemical insecticides. When pressure from pests is high, *Bt* crops produce greater yields. However, there have been concerns that the widespread planting of genetically modified *Bt* plants might hasten the development of insect resistance to *Bt* endotoxins. To prevent or delay the emergence of insect resistance to *Bt* crops, the biotechnology industry, the regulatory authorities, and farmers have worked together to develop and implement insect resistance management (IRM) programs.

An IRM plan specifies the practices aimed at reducing the fitness advantage of alleles that code for resistance, thereby delaying the development of insect resistance to *Bt* crops. It also includes resistance monitoring for the development of resistance (and increased insect tolerance of the protein), grower education, a remedial action plan (if resistance is identified), annual reporting, and communication *(54)*. A scientific basis for an IRM plan requires understanding of pest biology and ecology, population genetics, farmer practices, and the environment. IRM is required by the US EPA and is a component of the seed contracts that biotechnology companies sign with farmers. Since 1995 when the first *Bt* plant was registered for commercialization, there has been no case of *Bt* crops leading to insect resistance.

One strategy of IRM is the use of high dose and planting of a refuge (a portion of the total acreage planted with non-*Bt* seed) to provide sufficient numbers of susceptible adult insects. This strategy is based on the assumption that resistance to *Bt* is recessive and is conferred by a single locus with two alleles resulting in three genotypes: susceptible homozygotes (SS), heterozygotes (RS), and resistant homozygotes (RR). Allele R codes for resistance, and allele S codes for susceptibility. It also assumes that there is a low initial resistance allele frequency, and that there will be extensive random mating between resistant and susceptible adult insects. The high dose provides a high likelihood that resistance will be functionally recessive (i.e., RS individuals will have fitness similar to SS individuals). Under ideal circumstances, only rare RR individuals will survive a high dose produced by the *Bt* crop. Both SS and RS individuals will be susceptible to the *Bt* toxin. A structured refuge of a non-*Bt* portion in a grower's field will provide for the production of susceptible (SS) insects that may randomly mate with rare resistant (RR) insects to produce susceptible RS heterozygotes, which will be killed by the *Bt* crop. This will remove resistant alleles from the insect population and delay the evolution of resistance *(54,79)*.

IRM strategies can be further enhanced by multiple genes. For example, *Bt* crops with two or more *Bt* genes active against the same target pest are under development. For such stacked *Bt* crops, insects would have to develop resistance to two or more toxins simultaneously to survive. Under most circumstances, such resistance is expected to evolve extremely slowly *(79)*.

5.5. Environmental Benefits Caused by Plantation of GM Plants

The United States grows the largest area of GM crops in the world, with 35.7 and 39 million ha of GM crops in 2001 and 2002, respectively *(1)*. The widespread adoption of GM technology in major commodity crops has resulted in significant yield increases, savings for growers, and huge reductions in pesticide use, which brought significant environmental and ecological benefits. Insect resistance and herbicide tolerance traits account for the majority of currently adopted GM crops. A comprehensive study conducted by the National Center for Food and Agricultural Policy in the United States revealed huge economic and environmental benefits because of plantation of GM crops *(80)*.

According to the National Center for Food and Agricultural Policy study, in 2001, 8 GM cultivars adopted by US growers reduced chemical pesticide use by 46 million pounds and increased crop yields by 4 billion pounds, adding an extra $1.5 billion value to the growers. These cultivars include insect-resistant corn and cotton; herbicide-tolerant canola, corn, cotton, and soybean; and virus-resistant papaya and squash. In addition to these 8, many other GM cultivars have been or are under development to control pests. Case study analyses of 32 such cultivars under development showed that they would cut chemical pesticide use by another 117 million pounds per year and increase crop yields by an additional 10 billion pounds, adding another $1 billion net value to the US growers. In addition, the study also pointed out that the next wave of GM cultivars would emphasize prevention of crop diseases and nematodes. Examples of these cultivars are fungus-resistant barley; virus- and fungus-resistant potatoes; and virus- and bacteria-resistant grapes, stone fruit, and citrus. It predicted that the next wave of GM cultivars could result in the reduction of chemical pesticide use in the United States by another 170 million pounds per year.

Besides the large-scale reduction of chemical pesticide use, another example of environmental benefit because of adoption of GM crops is regarding low- or no-till agricultural practices. Most of the land that today makes up America's row-crop farms was vast expanses of grasslands or forests 200 years ago. The cycle of natural plants created a deep layer of litter, which protected the soil from wind and water erosion and temperature extremes. As plants decayed, carbon and other nutrients returned to the soil. Water, instead of running off fields, seeped back into the soil, replenishing groundwater and nearby streams. These areas supported an ecological cycle that was changed radically after settlers first put plows to the soil. Organic matter has been lost, and erosion has taken topsoil.

A comprehensive study conducted by the Conservation Technology Information Center (CTIC) at Purdue University found that within the past decade, however, many farmers in the US have begun to practice conservation tillage and recreate the cycle that once characterized the prairie soils and forests before they were cleared for farming *(66)*. Instead of plowing and disking their fields before planting, many farmers are leaving the residue of the previous crop on the soil surface. This layer of decaying plant material provides protective litter and begins to create conditions that existed before people first began to till the soil.

Conservation tillage, as defined by the CTIC, means any minimal tillage system that leaves the soil surface at least 30% covered by crop residue. Conservation tillage has a number of environmental benefits, including reduced soil erosion; improved moisture content in soil; healthier, more nutrient-enriched soil; more earthworms and beneficial soil microbes; reduced consumption of fuel to operate equipment; the return of beneficial insects, birds, and other wildlife in and around fields; less sediment and chemical runoff entering streams; and reduced potential for flooding *(66)*.

The CTIC study found an association between the introduction of GM crops and a 35% increase in no-till acres in the United States. Because the main reason farmers till their soil is to control weeds, with herbicide-tolerant GM crops farmers can allow weeds to emerge with their crops and then apply herbicide over the top of their crop, removing the weeds without harming the GM crops. This improvement in weed control gives increased confidence that weeds can be controlled economically without relying on tillage. The study found that the no-tillage system has grown more than other reduced tillage systems since 1996 (when GM crops were first introduced commercially), and nearly all the growth has occurred in crops for which the herbicide tolerance trait is available: soybean, cotton, and canola.

According to the CTIC study *(66)*, conservation tillage and the Conservation Reserve Program have reduced soil erosion by 1 billion tons per year, which is a 30% reduction since the early 1980s when traditional plowing methods were more common. Reducing sediment flow into streams and rivers improves water quality and aquatic habitat and eases flooding. Minimizing sediment flow is particularly helpful because soil often carries organic carbon that reacts with chlorine in water treatment systems to create carcinogens that must be filtered and removed. The study credited reduced tillage practices with a savings of $3.5 billion in 2002 in water treatment and storage, waterway maintenance, navigation, fishing, flooding, and lost recreation costs.

Americans now have cleaner and more affordable drinking water because farmers tripled the number of acres they planted with conservation tillage in the past two decades, and the trend will continue as more farmers plant GM crops and convert to no-till farming systems. Because conservation tillage requires fewer trips across the field for weed control, farmers are using 306 million fewer gallons of fuel per year to power their equipment. It also reduces the amount of carbon dioxide released into the air by as much as 1 billion pounds per year over the traditional plowing practices of a generation ago. In addition to building healthier soils, the study concluded that cropland in conservation tillage provides a more benignant environment for wildlife, such as birds and insects. The study noted that wildlife, such as quail, thrives by cutting the time of their daily hunt for food as much as 80% in no-till soybean fields vs traditional plowed soybean fields. Meanwhile, no-till fields have three to six times as many soil-loosening earthworms that help incorporate organic residues, aerate the soil, and improve water filtration.

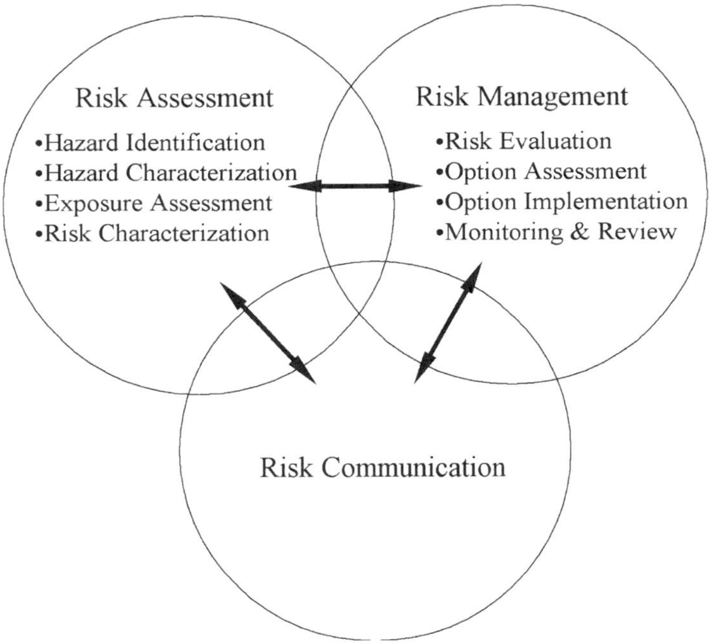

Fig. 1. Structure of risk analysis *(81)*.

6. Scientific Principles and Approaches in the Food Safety Assessment of GM Plants

6.1. The Principle of Risk Analysis

Regulatory authorities base GM food safety and environmental risk analysis on the general principle of risk analysis defined by the Codex Alimentarius Commission (Codex). The Codex was established in 1963 to implement the joint FAO/WHO Foods Standards Program. Codex is an intergovernmental statutory body with a 165-country membership. Its purpose is to protect the health of consumers, to ensure fair practices in food trade, and to promote coordination of all food standards work undertaken by international governmental and nongovernmental organizations. According to Codex's definition, risk analysis has three components: risk assessment, risk management, and communication of risk (Fig. 1).

Risk assessment includes a safety assessment, which is designed to identify whether a hazard (nutritional or other safety concern is present), and if present, to gather information on its nature and severity. The safety assessment of GM foods should include a comparison between the food derived from modern biotechnology and its conventional counterpart, focusing on determination of similarities and differences (*see* Section 6.2). If a new or altered hazard (nutritional or other safety concern) is identified by the safety assessment, the risk associated with it should be characterized to determine its relevance to human health. A safety assessment is characterized by an assessment of a whole food or a component thereof relative to the appropriate conventional counterpart. It should be conducted by taking into account both intended and unintended effects, identifying new or altered hazards, and identifying changes in key nutrients relevant to human health *(82)*. It is noted that diminished risk from natural plant toxicants and fungal mycotoxins and improved nutrition are not considered in contemporary risk assessment of new insect-resistant GM plant varieties. This benefit (not risk) should be taken into consideration if crop protection technologies are to be given a scientifically rational review.

Risk management measures for GM foods should be proportional to the risk, based on the outcome of the risk assessment and, if relevant, taking into account other legitimate factors. It should be recognized that different risk management measures may be capable of achieving the same level of protection regarding the management of risks associated with safety and nutritional impacts on human health and therefore would be equivalent. Risk managers should take into account the uncertainties identified in the risk assessment and implement appropriate measures to manage these uncertainties. Risk management measures may include, as appropriate, food labeling, conditions for marketing approvals and postmarket monitoring. Postmarket monitoring may be an appropriate risk management measure in specific circumstances. Its need and utility should be considered, on a case-by-case basis, during risk assessment, and its practicability should be considered during risk management *(82)*.

Effective risk communication is essential at all phases of risk assessment and risk management. It is an interactive process involving all interested parties, including government, industry, academia, media, and consumers. Risk communication should include transparent safety assessment and risk management decision-making processes. These processes should be fully documented at all stages and open to public scrutiny, yet respect legitimate concerns to safeguard the confidentiality of commercial and industrial information. In particular, reports prepared on the safety assessments and other aspects of the decision-making process should be made available to all interested parties *(82)*. The risk communication process implemented by the Australian government during the genetically modified organism (GMO) approval process is an excellent example (*see* Section 7.2).

The risks associated with foods derived from GM plants are not inherently different from the risks associated with conventional ones. Those most likely to be related to consumption of GM foods concern toxic effects, allergic reactions, and changes in nutrient composition. The process of risk assessment begins with identification and characterization of any potential hazards regarding toxicity, allergenicity, and nutrition of a new GM food. This is followed by an assessment of human exposure to the expressed trait associated with the hazards that have been identified.

6.2. Substantial Equivalence

6.2.1. The Concept of Substantial Equivalence

Because of the difficulties of applying traditional toxicological testing and risk assessment procedures to whole foods, a different approach is required for the safety assessment of foods derived from GM plants. This has been addressed by the development of a multidisciplinary approach for assessing safety of GM foods using the concept of substantial equivalence.

The use of animals for assessing toxicological end points is a major element in the risk assessment of many compounds, such as pesticides. In most cases, the substance to be tested is well characterized, of known purity, and of no particular nutritional value, and human exposure to it is generally low. It is therefore relatively straightforward to feed such compounds to animals at a range of doses several orders of magnitude greater than the expected human exposure levels to identify any potential adverse health effects of importance to humans.

However, animal studies cannot readily be applied to testing the risks associated with whole foods, which are complex mixtures of compounds often characterized by a wide variation in composition and nutritional value. Because of their bulk and effect on satiety, they can usually only be fed to animals at low multiples of the amounts that might be present in the human diet. In addition, a key factor to consider in conducting animal studies on foods is the nutritional value and balance of the diets used to avoid the induction of adverse effects that are not related directly to the material itself. Detecting any potential adverse effects and relating these conclusively to an individual characteristic of the food can therefore be extremely difficult *(82)*.

The concept of substantial equivalence was elaborated within the Organization for Economic Cooperation and Development (OECD) *(83)*. Determining substantial equivalence entails consideration of the trait encoded by the genetic modification; phenotypic characterization of the new food source compared with an appropriate comparator (conventional counterpart) already in the food supply; and compositional analysis of the new food source or the specific food product compared with the conventional counterpart *(84)*. The OECD has agreed that safety assessment based on substantial equivalence is the most practical approach to address the safety of foods and food components derived through modern biotechnology. The concept of substantial equivalence embodies the idea that existing organisms used as food, or as a source of food, can be used as the basis for comparison when assessing the safety of human consumption of a food or food component that has been modified or is new *(83)*.

In 1996, an FAO/WHO consultation endorsed the application of substantial equivalence in the safety assessment of GM foods *(84)*. It recognized that the establishment of substantial equivalence is not a safety assessment *per se*, but that establishing the characteristics and composition of the new GM food as equivalent to those of a familiar, conventional food with a history of safe consumption implies that the new food will be no less safe than the conventional food under conditions of similar exposure, consumption patterns, and processing practices. Substantial equivalence is not intended to be a measure of absolute safety; instead, it recognizes that, although demonstrating absolute safety is an impractical goal, it is possible to show that a GM product is no less safe than a conventional food product.

Three possible scenarios are envisaged as a result of a substantial equivalence evaluation *(84)*:

1. When substantial equivalence has been established for an organism or food product, it is considered to be as safe as its conventional counterpart, and no further safety evaluation is needed.
2. When substantial equivalence has been established apart from certain defined differences, further safety assessment should focus on these differences. A sequential approach should

focus on the new gene products and their structure, function, specificity, and history of use. If a potential safety concern is indicated for the new gene products, further in vitro or in vivo studies may be appropriate.

3. When substantial equivalence cannot be established, this does not necessarily mean that the food product is unsafe. Not all such products will require extensive safety testing. The design of any testing program should be established on a case-by-case basis, taking into account the reference characteristics of the food or food component. Further studies, including animal feeding trials, may be required, especially when the new food is intended to replace a significant part of the diet.

One important benefit of the substantial equivalence concept is that it provides flexibility that can be useful in food safety assessment. It is a tool that helps identify any difference, intended or unintended, that might be the focus of further safety evaluation. Because it is a comparative process for evaluating safety, the determination of substantial equivalence can be performed at several points along the food chain (e.g., at the level of the harvested or unprocessed food product, individual processed fractions, or the final food product or ingredient). Although from a practical point of view substantial equivalence should typically be determined at the level of the unprocessed food product, the flexibility of the concept permits the determination to be targeted at the most appropriate level based on the nature of the product *(49)*.

6.2.2. Application of Substantial Equivalence

Applying the substantial equivalence concept involves a comparison of a GM food with an appropriate conventional counterpart that has an acceptable history of safe food use. The choice of conventional counterpart is crucial to effective application of the concept of substantial equivalence to establish the safety of a GM food. The conventional counterpart used in this assessment should ideally be the near-isogenic parental line. In practice, this may not be feasible at all times, in which case a line as close as possible should be chosen. In addition, an appropriate conventional counterpart must have a well-documented history of use. The amount and type of data required for the assessment of substantial equivalence depends on the plant species, end products, and how the plant is used in food production. The data can be chemical (compositional), morphological, agronomic, or physiological in nature. If the application only covers one specified product, such as oil from canola, the analysis will be narrowed.

When the comparison is made at the plant level, the data required for assessment of substantial equivalence could be divided into three sets *(85)*. The first set should include the kind of data normally used for the characterization of the organism at the appropriate taxonomic classification level (such as subspecies, variety, cultivar) and data on agronomic performance. For plants, this includes the morphological description, flowering period, time to maturity, resistance to disease and stress, reproductive viability, yield, and the like. The second set of data is related to the final product used by the consumers. For example, canola is used to produce vegetable oil for human consumption. In this case, the content of triglycerides and fatty acid composition is the most important data to obtain. The third set of data is on the key nutrients, antinutrients, toxicants, and allergens of a particular plant as a whole. This is the so-called composition analysis. The specific substances to be analyzed vary among crops, but generally they include (1) proximate analysis (the content of moisture, protein, carbohydrates, glycerides, ash, and fiber); (2) protein and amino acids profile; (3) quantitative and qualitative composition of lipids and complete fatty acid profile; (4) composition of carbohydrate fractions; (5) vitamin profile; (6) mineral profile; (7) levels of toxicants and antinutrients; (8) secondary plant metabolites; and (9) major known allergens.

In addition, specific types of modification may introduce or alter specific compounds. These can be assessed on a case-by-case basis. For example, introduction of genes encoding herbicide tolerance may detoxify the herbicide in the plant, thereby generating intermediate metabolites as residues as well as the parent molecule of the herbicide. Safety assessment of such plants requires investigation of herbicide residue and metabolite levels in addition to the general composition analysis. When altered residue or metabolite levels are identified in foods, consideration should be given to the potential impacts on human health using conventional procedures for establishing the safety of such metabolites.

Data collected from a GM plant or food should be compared with that from a conventional counterpart. In many cases, the data on the conventional counterpart is available in databases, handbooks, or other publications. An International Network of Food Data Systems (INFOODS; www.fao.org/infoods) was established in 1984 under the auspices of the UN University to coordinate efforts to improve the quality and availability of food analysis data worldwide. INFOODS has provided leadership and an administrative framework for the development of standards and guidelines for collection, compilation, and reporting of food component data. It is establishing and coordinating a global network of regional data centers directed toward the generation, compilation, and dissemination of accurate and complete data on food composition. It is also the generator and repository of special international databases and serves as a general and specific resource for persons and organizations interested in food composition data on a worldwide basis. The Secretariat of INFOODS has developed the necessary software for the electronic storage of food composition data and its interchange among databases. In addition, the International Life Sciences Institute (ILSI) released an on-line comprehensive crop composition database that is available for public use via the Internet (www.cropcomposition.org).

If analytical data on the composition of a selected conventional counterpart is not already available, the data can be generated along with the data on GM food. The location of trial sites for compositional analysis should be representative of the range of environmental conditions under which the plant varieties would be expected to be grown. The number of trial sites should be sufficient to allow accurate assessment of compositional characteristics over this range. To minimize environmental effects and to reduce any effect from naturally occurring genotypic variation within a crop variety, each trial site should be replicated. An adequate number of plants should be sampled, and the methods of analysis should be sufficiently sensitive and specific to detect variations in key components.

Food composition data are obtained by various analytical methods (such as spectroscopy, chromatography, electrophoresis, mass spectrometry). This analytical approach serves well with the GM products, for which a limited number of foreign genes are transformed, which has been the case for the GM plants introduced on the market so far. Future genetic modification may employ an increasing number of multiple genes, which allows the introduction of new metabolic pathways. This may render the assessment of substantial equivalence more difficult solely by individual component analysis because unforeseen secondary effects may not be identified. New technologies and approaches, such as metabolite profiling *(86)*, nucleic acid or protein microarray *(87)*, and in vitro screening tests of whole food extracts *(88)*, are under development and evaluation and could provide additional future tools for identifying any important differences between GM plants and their nontransgenic counterparts.

It should be noted that, when substantial equivalence has been established, this does not mean that a GM product is identical to the conventional counterpart. Because the comparison does not take all components into account, substantial equivalence only provides assurance that those components most likely to be relevant to the product's safety are present in equivalent amounts. Because the comparative approach links the composition of the new food to existing products with a history of safe use, the new food's impact in the diet can be predicted. Applying the concept allows everything that is the same between the GM food and conventional food to be considered safe. The statistical significance of any observed differences should be assessed in the context of the range of natural variations for that parameter to determine its biological significance *(49)*. Differences identified in the comparison are the focus for further scrutiny involving traditional nutritional, toxicological, or immunological testing or long-term studies, depending on the identified differences. In many cases, the differences are the new genes, the new proteins, and possible other components.

6.3. The New Gene

Risks associated with DNA consumption *per se* are the same for DNA derived from both conventional and GM plants *(89)*. The US Food and Drug Administration (FDA) has concluded that DNA is generally regarded as safe, independent of its source, inasmuch as all DNA is composed of the same four components, which are and always have been constituents in nonprocessed or whole foods *(90)*. Therefore, the safety evaluation of the newly inserted DNA focuses on potential risks associated with acquisition and expression of the specific genetic information inserted into the plant if the DNA were to transfer to human, animal, or bacterial cells.

The origin and nature of all the genetic elements that have been introduced into the GM plants need to be provided by the developer, including structural and regulatory sequences and any remaining parts of vector sequences. The possibility of horizontal gene transfer and its consequences are also considered. If transfer of antibiotic resistance genes were to occur, the potential significance to human health would need to be considered in relation to existing levels of antibiotic-resistant microorganisms in the human gut or other parts of the body. Approval of antibiotic resistance genes has been restricted to those resistance markers for which resistance is already widespread in nature. It is recommended that the gene selected be a resistance marker to an antibiotic that is not used to a significant extent in human or veterinary medicine. It should be noted that some GM crops do not contain antibiotic resistance marker genes, and that new marker genes and selection strategies that reduce the need for antibiotic resistance markers are under development *(25)*.

To provide a clear understanding of the impact on the composition and safety of GM foods, a comprehensive molecular and biochemical characterization of the genetic modification to the plants should be carried out. Information should be obtained on the DNA insertion into the plant genome. The number of insertion sites should be identified. The organization of the inserted genetic material at each insertion site, including copy number and sequence data of the inserted material and of the surrounding region (border sequences), should be characterized clearly to identify any substances expressed as a consequence of the inserted material. Analysis of the expressed products (i.e., proteins) of the new genes should be conducted by standard purification and characterization techniques. Any open reading frames within the inserted DNA or created by the insertions with contiguous plant genomic DNA, including those that could result in fusion proteins, should be identified.

6.4. The New Protein

6.4.1. Toxicity Assessment

Hazard identification requires knowledge of which introduced genes are expressed; the characteristics, concentration, and localization of the expressed proteins; and the consequences of expression. If the GM food differs from its traditional counterpart by the presence of one or a few novel proteins, it is possible to assess the potential toxicity of these proteins in a manner analogous to traditional toxicity testing *(73)*. That is, the assessment is applied to the novel protein itself rather than the whole food. In considering the potential toxicity of a novel protein, it is first important to determine whether it is likely to be present in the food as consumed and thus whether exposure is likely. Once likely human exposure to a novel protein is established, a number of different pieces of information can collectively be used to demonstrate there is a reasonable certainty that no harm will result from that exposure.

An assessment of potential toxicity of a novel protein should consider such factors as the following: (1) whether the novel protein has a prior history of safe human consumption or is sufficiently similar to proteins that have been safely consumed in food; (2) whether there is any amino acid sequence homology between the novel protein and known protein toxins and antinutrients; and (3) whether the novel protein causes any adverse effects in acute oral toxicity testing.

Acute oral toxicity testing is an important component of the safety assessment of novel proteins and is particularly useful when there is no prior history of safe consumption of the protein. Acute tests should be sufficient because—if toxic—proteins are known to act via acute mechanisms, and laboratory animals have been shown to exhibit acute toxic effects from exposure to proteins known to be toxic to humans *(91)*. The acute toxicity tests are performed using purified protein administered at very high dose levels, usually many orders of magnitude above the anticipated level of human exposure. Ideally, the protein to be tested should be that directly purified from the GM plant. However, protein produced from alternative sources (such as recombinant bacteria) is often used because, in certain cases, it is infeasible to obtain sufficient quantities of the purified protein from the GM plants in which the proteins are expressed in very low amount. In such cases, it is essential to ensure that the recombinant protein produced for testing is structurally, functionally, and biochemically equivalent to that present in the GM plants *(82)*.

6.4.2. Allergenicity Assessment

There is no definitive reliable test to predict allergic response in humans to a novel protein; therefore, it is recommended that an integrated, stepwise, case-by-case approach be used in the assessment of possible allergenicity of proteins *(82)*. In 1996, the International Food Biotechnology Council and the Allergy and Immunology Institute of the International Life Sciences Institute (IFBC/ILSI) presented a decision tree approach to the evaluation of the potential allergenicity of introduced novel proteins in GM foods *(26)*. It is a strategy that focuses on the source of the gene, the sequence homology of the newly introduced protein to known allergens, the immunoaffinity of the newly introduced protein with IgE from the blood serum of individuals with known allergies to certain foods and the physicochemical properties of the newly introduced protein *(26)*.

The issue of allergenicity of genetically modified foods was specifically addressed by the 1996 Joint FAO/WHO Consultation on Biotechnology and Food Safety *(84)*. An assessment approach similar to that developed by IFBC/ILSI was advocated. In the 2000 Joint FAO/WHO Consultation on Safety Aspects of Genetically Modified Foods of Plant Origin, the IFBC/ILSI decision tree approach was adapted with modification *(73)*. The decision tree was furthered modified in the 2001 Joint FAO/WHO Expert Consultation on Allergenicity of Foods Derived from Biotechnology (Fig. 2).

The Codex ad hoc Intergovernmental Task Force on Foods Derived From Biotechnology further considered the FAO/WHO 2001 decision tree and came to the conclusion that it was not possible scientifically to arrive at clear yes/no decisions at each and every step in the decision process. The task force therefore recommended a more holistic approach that took into account a broader range of information to be examined in a stepwise and structured manner *(82)*. Based on the Codex draft guidelines, the initial steps in assessing possible allergenicity of any newly expressed proteins are the determination of the source of the introduced protein; any significant similarity between the amino acid sequence of the protein and that of known allergens; and the properties of the protein, such as its susceptibility to enzymatic degradation *(82)*.

6.4.2.1. Source of the Protein

It is important to establish whether the source of the introduced protein is known to cause allergic reactions. Genes derived from known allergenic sources should be assumed to encode an allergen unless scientific evidence demonstrates otherwise *(82)*. Allergenic sources of genes would be defined as those organisms for which reasonable evidence of IgE-mediated oral, respiratory, or contact allergy is available. Knowledge of the source from which the introduced protein is derived allows the identification of tools and relevant data to be considered in the allergenicity assessment. These include the availability of sera for screening purposes; documented type, severity, and frequency of allergic reactions; structural characteristics and amino acid sequence; physicochemical and immunological properties (when available) of known allergenic proteins from that source.

6.4.2.2. Amino Acid Sequence Homology

The purpose of a sequence homology comparison is to assess the extent to which a newly expressed protein is similar in structure to a known allergen. This information may suggest whether that protein has an allergenic potential. Sequence homology searches comparing the primary structure of the newly expressed protein with all known allergens should be done. Searches should be conducted using various algorithms to predict overall similarities. Strategies such as stepwise contiguous identical amino acid

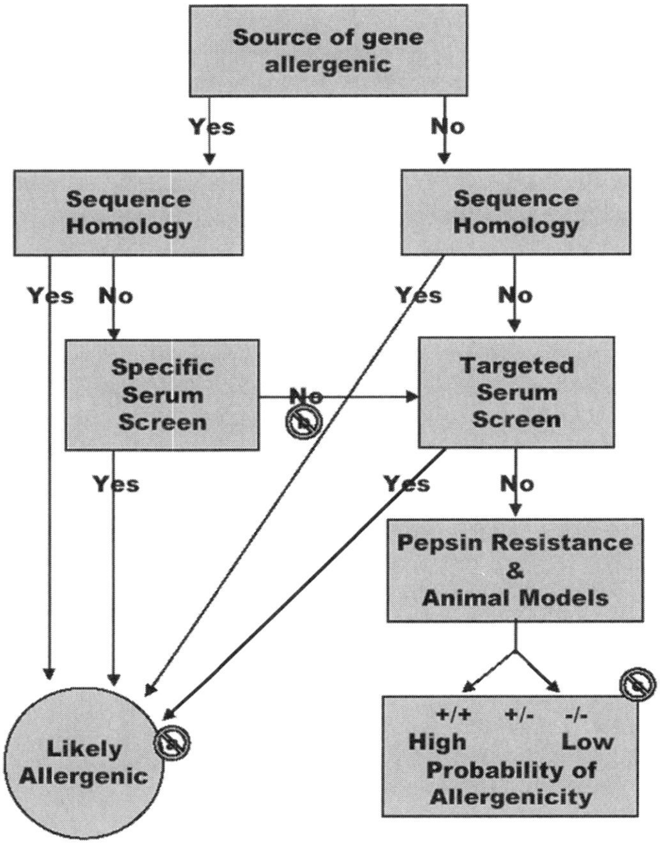

Fig. 2. FAO/WHO 2001 decision tree of assessment of allergenic potential of foods derived from biotechnology *(25)*.

segment searches may also be performed for identifying sequences that may represent linear epitopes. The size of the contiguous amino acids search should be based on a scientifically justified rationale to minimize the potential for false-negative or false-positive results *(82)*. Generally 8 contiguous amino acids are considered the appropriate size for linear epitope identification, although the size of 6 contiguous amino acids is also used *(25)*. It should be noted that the smaller the size of the peptide, the greater the likelihood for a false-positive result. IgE cross-reactivity between the newly expressed protein and a known allergen should be considered a possibility when there is more than 35% identity in a segment of 80 or more amino acids *(25)*.

A negative sequence homology result indicates that a newly expressed protein is not related to a known allergen and is unlikely to be cross-reactive to known allergens. A result indicating absence of significant sequence homology should be considered along with the other test data. A positive sequence homology result indicates that the newly expressed protein must be more closely scrutinized as a potential allergen. If the product is to be considered further, it should be assessed using serum from individuals sensitized to the identified allergenic source *(25,82)*.

6.4.2.3. Pepsin Resistance

It is generally recognized that food allergens tend to resist pepsin digestion (*see* Section 3.2). Pepsin resistance is suggested and commonly conducted as a test in allergenicity assessment of novel proteins *(25,82)*. The resistance of a protein to degradation in the presence of pepsin under appropriate conditions indicates that further analysis should be conducted to determine the likelihood the newly expressed protein is allergenic. However, it should be taken into account that a lack of resistance to pepsin does not exclude the possibility that the newly expressed protein can be a relevant allergen *(82)*. Pepsin resistance test results should be considered along with other test data.

6.4.2.4. Other Considerations

As scientific knowledge and technology evolves, other methods and tools may be considered in assessing the allergenicity potential of newly expressed proteins as part of the assessment strategy. These methods should be scientifically sound and may include specific serum screening; targeted serum screening; and the use of animal models.

For those proteins that originate from a source known to be allergenic (such as soybean) or to have sequence homology with a known allergen, specific serum screening should be performed if sera are available. Sera from individuals with a clinically validated allergy to the source of the protein can be used to test the specific binding to IgE antibodies of the protein in in vitro assays. A critical issue for testing will be the availability of human sera from sufficient numbers of individuals. A minimum of 8 relevant sera is required to achieve a 99% certainty that the new protein is not an allergen in the case of a major allergen. A minimum of 24 relevant sera is required to achieve the same level of certainty in the case of a minor allergen *(25)*. In the case of a newly expressed protein derived from a known allergenic source, a negative result in in vitro immunoassays may not be considered sufficient, but should prompt additional testing, such as the possible use of skin test and ex vivo protocols *(82)*. A positive result in such tests would indicate a potential allergen.

For proteins from sources not known to be allergenic and that do not exhibit sequence homology to a known allergen, targeted serum screening may be considered if such tests are available *(82)*. For targeted serum screening, six groups of source organisms are distinguished: yeasts/molds, monocots, dicots, invertebrates, vertebrates, and "others" *(25)*. A panel of 50 serum samples with high levels of IgE to allergens in the relevant group is used to search for IgE antibodies that are cross-reactive with the expressed protein. If a positive reaction is obtained with one of these sera, the expressed protein is considered an allergenic risk, and further evaluation for allergenicity would typically not be necessary. If a gene was obtained from a bacterial source, no targeted serum screening would be possible because no normal population of individuals are known to be sensitized to bacterial proteins *(25)*.

Despite increasing research efforts to develop suitable animal models that can mimic the induction of an IgE antibody response and the induction of symptoms on a challenge reaction like those observed in human allergic patients, no validated and widely accepted animal model is yet available *(92)*. An ideal animal model should satisfy several important criteria *(27,92)*, such as the following:

1. Sensitization and challenge should occur orally.
2. Adjuvants are not needed.

3. The test animal should produce a significant amount of IgE or other Th2-specific antibody.
4. The test animal should tolerate most food proteins.
5. On a challenge with the allergen, clinical reactions with respect to organ sensitivities should be similar to those seen in humans.
6. Antibody responses should be directed to similar proteins in the allergenic food as found in patient sera.
7. Tests with the model should be relatively easy to conduct and reproducible both in time and in different laboratories.

Several animal models have been reported for assessing on a relative scale the potential allergenicity of proteins; these models include the Brown Norway rat *(93,94)*, mice *(95,96)*, and guinea pigs *(97)*.

7. International Regulatory Frameworks on GM Plants

7.1. The United Nations

Today, many countries have their own regulatory framework for safety assessment, approval, and handling of products derived from GMOs; however, they are not necessarily consistent in the basic approach or methodologies *(98)*. Other countries, especially many developing nations, have not established a competent regulatory system dealing with modern biotechnology. International agencies, particularly FAO, WHO, and OECD, have been very active in providing an international platform for the establishment of a global consensus on the safe use of biotechnology and regulation of foods derived from modern biotechnology. Since 1990, FAO and WHO have convened a series of joint expert consultations to provide advice to member countries and to the Codex Alimentarius Commission on the safety assessment of GM foods. The 1990 joint expert consultation concluded that foods from modern biotechnology were not less safe than those from traditional technology. The 1996 consultation recommended that the principle of substantial equivalence was an important component in the safety assessment of foods derived from GM plants *(84)*.

The 2000 consultation reviewed experience gained in this field since 1996 and discussed overall safety aspects of foods derived from GM plants *(73)*. This consultation acknowledged that applying traditional toxicological studies and risk assessment procedures to whole foods is difficult because foods are complex mixtures of compounds characterized by wide variation in composition and nutritional value. It also recognized the fact that foods have been accepted as safe, although very few have been subjected to toxicological studies. Therefore, for the safety assessment of foods derived from GM plants, the consultation agreed that a different approach is required in which a safety assessment concentrates on the differences from its traditional counterpart after applying the concept of substantial equivalence rather than establishing absolute safety by conducting safety assessment on whole foods.

The 2000 consultation further noted several critical points in conducting safety assessment or in securing safety *(98)*:

1. The safety assessment of GM foods requires methods to detect and evaluate the impact of unintended effects, such as the acquisition of new traits or loss of existing traits.

2. When the change in nutrient levels in a particular plant has an impact on overall dietary intake, it is important to monitor changes in dietary patterns as a result of the introduction of the GM foods and evaluate its potential effects on nutritional and health status of consumers.
3. The transfer of genes from commonly allergenic foods should be discouraged unless it can be documented that the genes transferred do not code for an allergen.
4. The horizontal gene transfer from plants and plant products consumed as food to gut microorganisms or human cells is considered a rare possibility.
5. The genes that confer resistance to drugs important for medical use should be avoided in the genome of widely disseminated GM plants.

The 2000 consultation also identified specific areas required for further consideration, such as allergenicity of genetically modified foods. The January 2001 consultation was subsequently convened to address the allergenicity assessment *(25)*. This consultation elaborated a modified decision tree approach for the assessment for allergenicity of GM foods (*see* Section 6.4.2).

In addition to expert consultations, FAO/WHO also aim to develop harmonized international standards, guidelines, or recommendations for GM foods through the Codex, which in 1999 established the ad hoc Intergovernmental Task Force on Foods Derived From Biotechnology. The conclusions of the joint FAO/WHO expert consultations have been used to advance discussions at the Codex ad hoc task force, and are incorporated into the text of draft guidelines under development by the task force. Four sessions of the task force have been convened in the past four years (2000 to 2003), and a number of guidelines were drafted and refined, such as "Principles for the Risk Analysis of Foods Derived From Modern Biotechnology" and "Guideline for the Conduct of Food Safety Assessment of Foods Derived From Recombinant-DNA Plants" *(82)*.

It should also be noted that, in January 2000, the UN Convention on Biological Diversity, which grew out of the 1992 Earth Summit in Rio de Janiero, met in Montreal, Canada, and announced the Biosafety Protocol (known as the Cartagena Protocol on Biosafety; www.biodiv.org/biosafety). The protocol focuses on transboundary movement of any live GMOs that could harm conservation and sustainable use of biological diversity. It allows a country to require prior notification through an advanced informed agreement from countries exporting GM seeds and living organisms intended for introduction into the environment. Further, it requires that shipments of products that may contain live GMOs, such as bulk commodities for food or feed, be labeled accordingly. The Advanced Informed Agreement provided by the exporter would include written notification of shipment accompanied by an extensive risk assessment. Further, it establishes a biosafety clearinghouse to help countries exchange scientific, technical, environmental, and legal information about live GMOs. The agreement requires governments to provide information on final decisions on the domestic use of a GMO commodity within 15 days of making that decision. The clearinghouse should provide needed transparency on where products have been approved and on regulatory requirements of participating countries. Although primarily an environmental agreement, the Biosafety Protocol will have a significant impact on trade, particularly biotechnology-derived seeds exported for planting purposes. By May 30, 2003, the Biosafety Protocol was ratified by 50 governments, which means the protocol entered into force on September 11, 2003, in accordance with provision of its Article 37.

7.2. Argentina

Argentina regulates GM plants through a combination of GMO-specific legislation and preexisting laws covering seeds. Approving the environmental release of GMOs, and their use in human food or livestock feeds, is the responsibility of the Secretariat of Agriculture, Livestock, Fisheries, and Food (SAGPyA; www.sagpya.mecon.gov.ar), under regulations administered by both SAGPyA and the National Service of Health and Quality Agrifood (SENASA). In 1991, SAGPyA created the National Advisory Committee on Agricultural Biosafety (CONABIA) as a mechanism to provide advice on the technical and biosafety requirements to be met in environmental releases, human food, and livestock feed uses of GMOs. CONABIA's membership is composed of both public and private sector representatives including research institutes, universities, industry, government offices, and professional societies.

Argentine regulation on GMOs is based on the identified characteristics and possible risks of the biotechnological product rather than on the process by which such product is obtained. In other words, it focuses on transgenic crops, in connection to their intended use, and takes into consideration only those aspects in the procedures that might entail hazards to the environment, to agricultural or livestock production, or to public health. The regulatory framework for biotechnology encompasses the contained use, deliberate release (confined field trial), and commercialization of GM crops. Under this framework, specific regulations were developed to establish conditions under which environmental releases of transgenic materials may be conducted and the resulting data reviewed by CONABIA and enforced by SAGPyA (SAGPyA Resolution 656, July 30, 1992; SAGPyA Resolution 837, September 9, 1993; SAGPyA Resolution 289, May 9, 1997). Permits to conduct field trials or precommercial multiplication are granted on condition that a number of biosafety measures should be in compliance. The biosafety measures of a release are determined by (1) the biological characteristics of the GMO subject of the release; (2) the features of the agroecosystem where the proposed trial is to be performed; and (3) compliance with the necessary safety measures, including the technical capacity of the applicant.

The monitoring of field trials is performed through inspections by staff inspectors from the INASE (formerly National Institute of Seeds), SENASA, and SAGPyA. This is to assess *in situ* actual compliance with the biosafety measures under which the authorization was granted, as well as eventually to adopt measures to avoid adverse effects on the adjacent environment if compliance is not secured. Plots are also controlled after harvest with the purpose of reducing the likelihood of gene flow from the GM plants to other organisms.

Before a GM plant and the products (food, feed, or other products) can be granted a commercialization permit, it must fulfill three stages of evaluation:

1. To be granted the flexibilization status. Once a permit for introduction into the environment is granted and biosafety is adequately established, an application for extensive cultivation (called *flexibilization*) can be submitted. Granting this status means that, for future releases, only information on sown area, date of seeding, site of release, and harvest date

is required. Under this status, CONABIA shall mandate monitoring of the harvest and final disposition of the material.

2. To meet the requirements under regulations administered by the SENASA and be approved for food and feed uses. In accordance with SAGPyA Resolution 511/98 ("Safety Assessment of Genetically Engineered Foods and Food Ingredients"), the criteria applied by SENASA to evaluate GMOs for food and feed uses include (1) natural toxicants, (2) toxicants of new expression, (3) homology of the transgene product with known allergens, (4) nutritional modification, (5) nutrient modification and nutritional characterization, (6) modification of macro- and micronutrient bioavailability, and (7) modified food characterization considering its suitability for food and feed uses. Resolution 511/98 has been replaced by Resolution 412/02, which requires information on compositional analysis to prove substantial equivalence of a GM food with its conventional counterpart. It also requires basic data to perform risk assessment concerning its intended use for feed and food. Allergenicity and toxicity of the new protein(s) and the whole product are also to be evaluated.
3. To have a favorable technical report from the National Bureau of Agrifood Markets about the marketing of the GM product to avoid any negative impact on Argentine exports.

In addition, all GM plants must comply with the pre-existing regulations in Argentina pertaining to plant protection (Decree-Law of Agricultural Production Health Defense, no. 6704/66 and its amendments), seeds and phytogenetic creations (Seed and Phitogenetic Creations Law, no. 20.247/73 and its regulatory decree), and animal health (Law of Veterinarian Products, Supervision of the Creation and Commercialization, no. 13.636/49).

7.3. Australia

Food Standards Australia New Zealand (FSANZ, www.foodstandards.gov.au), formerly Australia New Zealand Food Authority, is a binational independent statutory authority that develops food standards for composition, labeling, and contaminants that apply to all foods (including GM foods) produced or imported for sale in Australia and New Zealand. FSANZ works in partnership with Australia's commonwealth, state, and territory governments and the New Zealand government.

The Office of Gene Technology Regulator (OGTR; www.health.gov.au/ogtr/index.htm) in the Health Ministry is responsible for dealings with GMOs other than food issues and only in Australia. The OGTR seeks advice from other agencies, including FSANZ, Environment Australia, Australian Pesticide and Veterinary Medicines Authority, and Australia Quarantine and Inspection Service. Their responsibilities encompass broader issues, such as the environmental release, quarantine, and the registration of insect-resistant or herbicide-tolerant plants. They coordinate with each other on products that require the approval of several agencies.

The OGTR is responsible for evaluation and approval of GM plants before intentional release to the environment. Each license application for a dealing involving intentional release (DIR) of a GMO into the environment is subject to a comprehensive evaluation (risk assessment) process under the Gene Technology Act 2000. The regulator has 170 working days in which to make a decision either to issue or to refuse to issue a license for the dealings proposed in a DIR application. The general evaluation process for DIR license application can be summarized as the following steps:

1. DIR application received and preliminary assessment: OGTR conducts an initial assessment to determine if there is any "significant" risk. If it is determined that the proposed dealings may pose a substantial risk to human health and safety or the environment, public consultation on the application is immediately sought on matters relevant to the preparation of the risk assessment and risk management plan (RARMP). Otherwise, an "early bird" public notification of receipt is made, and a summary on the application is posted on the Web site. Input is sought from prescribed expert groups and key stakeholders.
2. RARMP: A thorough risk assessment is undertaken, and a risk management plan is prepared with proposed license conditions if approval is contemplated. OGTR considers a number of specific issues in preparing the RARMP, including the properties of the parent organism, the effect of the genetic modification, the scale of the proposed release, and the potential for dissemination or persistence of the GMO or its genetic material in the environment and any likely impacts of the proposed dealings on the health and safety of people. Comment is sought from expert groups and key stakeholders. Public consultation is sought on the plan through publications of government gazette, newspapers, and mailouts. RARMP will be finalized taking into account comments and submissions relating to public health and safety and the environment.
3. License decision and conditions: OGTR makes decision on whether to issue a license. The applicant, key stakeholders, and public are notified of the decision, including the license conditions.
4. Monitoring, audit, and reporting compliance by OGTR on the approved DIR.

In Australia, GM foods are regulated by the Food Standard (Standard 1.5.2 in the new code or A18 in the old code, Food Produced Using Gene Technology). Standard 1.5.2 regulates the sale of GM foods in Australia and New Zealand and was incorporated into the Food Standards Code on May 13, 1999, and in an amended form on December 7, 2000. The standard has two provisions: (1) a mandatory premarket safety assessment requirement and (2) a mandatory labeling requirement. Under this standard, the sale of food produced using gene technology is prohibited unless the food is officially included in the standard. However, the standard provides an exemption for those GM foods already on the market and with these three conditions met: (1) the producer of the GM food submitted a comprehensive data package to the Australia New Zealand Food Authority by April 30, 1999, so that a full health and safety assessment could be completed; (2) the GM food was lawfully sold overseas and considered safe by overseas regulatory authorities and; (3) Australia New Zealand Food Standards Council had no evidence indicating that the GM food was unsafe.

In assessing the safety of GM foods, FSANZ requires applicants to submit a complete data package that addresses issues such as the nature and stability of the genetic modification, general safety issues, toxicological issues, and nutritional issues. The legal requirement of the Australian food standard-setting process requires full disclosure and public consultation. All applications for safety assessment of GM foods are subject to public notification and comment. FSANZ seeks public comment on all proposed amendments to Standard 1.5.2 (i.e., GM food approval). In the process of soliciting public comment, FSANZ publishes a draft risk analysis report that provides a background of the application, highlights the issues addressed during the risk assessment, summarizes public comment submitted in response to the notification of application, and deals with legitimate issues raised in public comments.

Standard 1.5.2 has a mandatory labeling provision on GM food and food ingredients when novel DNA or protein is present in the final food and when the food has altered characteristics. However, the following would be exempt from the labeling requirement: (1) highly refined food for which the effect of the refining process is to remove novel DNA or protein; (2) processing aids and food additives except those for which novel DNA or protein is present in the final food; (3) flavors present in a concentration less than or equal to 0.1% in the final food; and (4) food prepared at the point of sale. The labeling exemption also applies to foods that contain up to 1% of GM material, but only if its presence is unintended.

7.4. Canada

Plants in Canada are regulated on the basis of the traits expressed and not on the basis of the method used to introduce the traits. A plant with novel traits (PNT) is defined as a plant variety or genotype that possesses characteristics that demonstrate neither familiarity (defined as the knowledge of the characteristics of a plant species and experience with the use of that plant species in Canada) nor substantial equivalence to those present in a distinct, stable population of a cultivated species of seed in Canada and that have been intentionally selected, created, or introduced into a population of that species through a specific genetic change. PNTs include those derived from both recombinant DNA technology (GM plants) and plants derived through traditional plant breeding and mutagenesis.

The Canadian Food Inspection Agency (CFIA; www.inspection.gc.ca/english/plaveg/pbo/pbobbve.shtml) under the federal department of Agriculture and Agri-Food Canada is the primary agency in assessing, approving, and regulating PNTs for environmental release (both confined field trials and unconfined commercial release), for plant or seed import to Canada, and for livestock feed use. The relevant laws are the Canadian Environmental Protection Act, Seeds Act and Seeds Regulations, Feeds Act and Regulations, and Plant Protection Act. Various specific guidelines have been developed in helping applicants to prepare data packages and to meet the regulatory requirements.

Processing of new applications for PNTs starts with the Plant Biosafety Office (PBO; for environmental release) and Feed Section (for animal feed use) of CFIA and the federal department of Health Canada (for food use petition). The PBO will distribute a copy of the application to the Pest Management Regulatory Agency of Health Canada for evaluation if the material involves altered pesticidal tolerance or altered pesticidal properties (such as *Bt* plants). If the application is for confined field trials, the PBO will send nonconfidential information about each new trial to designated provincial government contacts in those provinces where proposed trials are to be conducted. Secondary agencies and provincial governments have a 30-day turnaround time, and any comments from them are considered by the PBO in the final evaluation of the application.

CFIA's determination of the need for unconfined environmental release authorization and the environmental safety assessment of PNTs will be conducted on a case-by-case basis, founded on familiarity and substantial equivalence. The environmental effects that will be assessed include (1) potential of the PNT to become a weed of agriculture or be invasive of natural habitats; (2) potential for gene flow to wild relatives with hybrid offspring that may become more weedy or more invasive; (3) potential for the PNT to become a plant pest; (4) potential impact of the PNT or its gene products on nontarget species, including humans; and (5) potential impact on biodiversity.

CFIA's guiding principle in livestock feed safety assessment of PNTs is the comparison of molecular, compositional, and nutritional data of products derived from the modified plant to those of traditional counterparts. Once substantial equivalence to an existing feed product can be established, it is expected that no additional safety testing will be required. If similarity or degree of equivalence cannot be established, a more extensive feed safety and efficacy assessment may be necessary.

In Canada, GM foods are regulated under the Novel Food program by Health Canada (www.hc-sc.gc.ca/food-aliment/mh-dm/ofb-bba/nfi-ani/e_division28.html) under the Food and Drugs Act. A premarket notification and safety evaluation is mandatory for "novel food," including GM foods. The manufacturer or importer of GM foods should submit notification to Health Canada. Health Canada shall review the information included in the notification within 45 days after receiving it. If Health Canada determines that additional information of a scientific nature is necessary to assess the safety of the novel food, the director will request in writing that the applicant submit that information. Then, within 90 days after receiving the additional information, Health Canada will assess it and determine whether the novel food is safe for consumption and notify the applicant. Health Canada evaluates the GM food safety based on its "Guidelines for the Safety Assessment of Novel Foods." In line with other international practices, the guideline bases GM plant food safety evaluation on the source and history of safe use of the organisms, the nature of genetic modification, dietary exposure, and data on nutrition, toxicity, and allergenicity.

7.5. China

China has the world's fourth largest area of GM crops after the United States, Argentina, and Canada. China is also a prominent importer and exporter of agricultural products. For example, the total world soybean trade was estimated at 55 million tons in the year 2000 to 2001. China alone imported over 13 million tons, accounting for about 24% of the world soybean trade. China's soybean import mainly comes from the United States, Brazil, and Argentina, where most of the soybean is GM. With China expected to import more agricultural food products to meet the consumption demand of its 1.3 billion population, its regulation of GMOs will have a significant influence on the international trade.

In 1993, China's State Science and Technology Commission issued a "Safety Administration of Genetic Engineering" document that provided general provisions on application, safety evaluation, approval, and management on the laboratory research, field trials, and commercialization of GMOs. In 1996, the Ministry of Agriculture (MOA; www.agri.gov.cn/index.htm) issued "The Implementation of Agricultural Genetic Engineering Safety Measures," which provides a guideline specifically regarding agricultural GMOs.

The MOA has a management office responsible for daily administration and a commission charged with assessing the safety of GMO technology and products. From March 1997 to February 2002, the MOA approved 501 cases of GMOs for field trials or commercialization out of a total of 695 applications. Commercial approval has been granted to 59 applications, which include insect-resistant cotton, shelf-life altered tomato, virus-resistant tomato, virus-resistant sweet pepper, and color-altered petunia *(99)*.

In May 2001, China's State Council (the central government) issued a more comprehensive "Regulation on the Safety Administration of Agricultural Transgenic Organisms," which replaced the 1996 MOA regulation. The new regulation prescribes the official approval of all production and marketing of GMOs in China for environmental release and for import as food or feed use. The regulation also requires mandatory labeling on GMO products. The Biosafety Office of Agricultural Transgenic Organisms (BOATO), under the MOA, is the designated agency for handling GMO applications and approval, although several other ministries, such as Health, Environment, and Trade, coordinate the major issues in GMO regulation with MOA. The evaluation is conducted by a Biosafety Committee of Agricultural Transgenic Organisms with expert members in biological science, food production and processing, health, environmental protection, and quarantine. In January 2002, MOA issued several guidelines on the application and approval processes regarding safety evaluation, import, and labeling of GMOs. They are "Regulatory Measures on Safety Evaluation of Agricultural Transgenic Organisms," "Regulatory Measures on Import of Agricultural Transgenic Organisms," and "Regulatory Measures on Labeling of Agricultural Transgenic Organisms."

Under the new regulation, GMO safety evaluation and regulation is administered in five phases: laboratory research, intermediate experiment, environmental release, production trial, and safety certification. GMO developers need regulatory approval before conducting each phase of development, starting from the phase of environmental release. Requirements for each type of submission are specified in the relevant guidelines.

For the final safety certification on GM plants for commercialization, applicants need to submit a complete data package on the biology of the organisms; genetic modification, characterization, and stability; and food and environmental safety. It also requires a test report from a MOA-designated institution on the GMO and reports of the three trial phases (i.e., intermediate experiment, environmental release, and production trial). Applications should be submitted to the BOATO, which arranges two rounds of evaluations per annum. The submission deadline is March 31 for the first round of evaluation and September 30 for the second round. BOATO will notify the applicant within two months of receipt of the application whether the submission is accepted. BOATO will make its decision 3 months after each of the two submission deadlines.

An import approval is required before a GMO or its product can be moved to China for experimental use, field trial, or use as food or feed. Applications should be submitted to the BOATO, which will notify the applicant within two months on receipt of the application whether it is accepted. BOATO will make its decision within 270 days from receipt date. The required materials include (1) import application forms specified by MOA; (2) a complete data package on the biology of the organisms, genetic modification, genetic characterization and stability, food and environmental safety information; (3) approval document from the exporting country on the intended use of the GMO; (4) experimental data generated in the exporting country on the safety of the GMO to humans and the environment; and (5) safety protection measures that the exporting company intends to take during the process of introduction of the GMO to China.

7.6. European Union

The European Union is built on an institutional system, which is the only one of its kind in the world. Currently, there are 15 countries (member states) in the union: Austria, Belgium, Denmark, Finland, France, Germany, Greece, Ireland, Italy, Luxembourg, The Netherlands, Portugal, Spain, Sweden, and the United Kingdom. There are ten more countries accepted to join the union formally in 2004: Czech Republic, Estonia, Cyprus, Latvia, Lithuania, Hungary, Malta, Poland, Slovenia, and Slovakia. In the European Union, the member states delegate sovereignty for certain matters to the independent institutions that represent the interests of the union as a whole. The three major institutions are the Commission, the Council, and the Parliament; these are flanked by two more institutions, the Court of Justice and the Court of Auditors, and five other European bodies. The Parliament and the Council are the Union's legislative bodies; the Commission is the executive body.

In the European Union, environmental release of GMOs has been regulated since the early 1990s. Until October 2002, Directive 90/220/EEC from the European Parliament and European Council was the main legislation regulating GMOs to be released into the environment (also covering import). This directive was replaced on October 17, 2002, by the new, updated Directive 2001/18/EC (biosafety.ihe.be/TP/TPMenu.html). The revised Directive 2001/18/EC tightens the existing rules on the release of GMOs into the environment. In particular, it introduces (1) the precautionary principle* for the environmental risk assessment of GMOs; (2) the mandatory postmarket monitoring requirement, including long-term effects associated with the interaction with other GMOs and the environment; (3) mandatory information to the public; (4) a requirement for member states to ensure labeling and traceability at all stages on the market; (5) first approvals for the release of GMOs to be limited to a maximum of 10 years; (6) the consultation of the Scientific Committees to be obligatory; (7) an obligation to consult the European Parliament on decisions to authorize the release of GMOs; and (8) the possibility for the Council of Ministers to adopt or reject a Commission proposal for authorization of a GMO by qualified majority.

Part C of Directive 2001/18/EC determines the provisions for the placing on the market of GMOs as or in products. No GMOs intended for deliberate release are to be considered for placement on the market without first having been subjected to satisfac-

*The precautionary principle was introduced in Europe in the early 1970s to provide regulatory authority with a tool for decision making on environmental threats. Based on this principle, regulatory officials can stop specific environmental contamination without waiting for conclusive evidence of harm to the environment. An early environmental application of the precautionary principle involved the prohibition of the purging of ship bilge contents and other wastes into the oceans. This requirement was justified by the paucity of data on the effects of the purged substances. But, the advocation of applying the precautionary principle in GMO safety assessment by the European Union is controversial. Critics argue that the precautionary principle was developed to prevent harm from hazardous environmental practices, not to serve as a food safety standard. Its application in GMO safety assessment could create an impossible burden of proof for food products and ingredients. Overzealous application of precaution, which is urged by the precautionary principle, can do more harm than good through substituting an alternative with a greater risk or through the economic costs associated with rejection of adequately safe foods *(100)*. The United States and many other countries did not adopt the precautionary principle in GMO biosafety assessment.

tory field testing at the research and development stage in ecosystems that could be affected by their use. Market authorizations are only given under a step-by-step approval process and after a case-by-case environmental risk assessment has been carried out. In addition to the environmental risk assessment, this information shall also contain appropriate safety and emergency response and, in the case of products, precise instructions and conditions for use and proposed labeling and packaging. Labeling, at all stages of placement on the market, is mandatory, and it must be stated clearly that "this product contains genetically modified organisms." Directive 2001/18/EC also introduces an obligation to implement a monitoring plan to trace and identify any direct or indirect, immediate, or delayed or unforeseen effects on human health or the environment of GMOs as or in products after they have been placed on the market.

Since the old Directive 90/220/EEC entered into force in October 1991, a number of transgenic crops (such as maize lines *Bt* 176, T25, MON810) have been authorized in the European Union by a Commission decision. Since October 1998, no further authorizations have been granted, and there were 13 applications pending by October 2002. Pending applications will be subject to the provisions of the new Directive 2001/18/EC. Under the new procedure, a company intending to market a GMO must first submit an application to the competent national authority of the member state (rapporteur) where the product is to be placed on the market first. The notification is composed of a technical dossier (including the environmental risk assessment, a plan for monitoring, a proposal for labeling, and a proposal for packaging) and a summary of the dossier (Summary Notification Information Format).

If the national authority gives a favorable opinion on placing the GMO concerned on the market, this member state informs the other member states via the Commission of the European Union. If there are no objections, the competent authority that carried out the original evaluation grants the consent for the placing the product on the market. The product may then be placed on the market throughout the European Union in conformity with any conditions required in that consent. Consent is given for a maximum period of 10 years for the initial consent. After 10 years, a renewal application has to be submitted, which in principle is subject to the same time limit of 10 years.

If any objections are raised after the rapporteur member state informed the rest of the member states, a decision has to be taken at the union level. The Commission first asks for the opinion of its Scientific Committees, composed of independent scientists, highly qualified in the fields associated with medicine, nutrition, toxicology, biology, chemistry, or other disciplines. If the scientific opinion is favorable, the Commission then proposes a draft Decision to the Regulatory Committee composed of representatives of member states for opinion. If the Regulatory Committee gives a favorable opinion, the Commission adopts the decision. If not, the draft decision is submitted to the Council of Ministers for adoption by qualified majority or rejection. If the council does not act within 3 months, the Commission can adopt the decision.

In the European Union, GM foods are regulated under Regulation (EC) 258/97 (biosafety.ihe.be/NF/GMfoods/GMfood_market.html) on Novel Foods and Novel Food Ingredients, which sets out rules for authorization and labeling of GM food products and other categories of novel foods. The authorization procedures for GM foods are slightly different from the procedure used for the environmental release (Directive 2001/18/EC), but the basic rule is similar. In general, the authorization of GM foods is a one-step process if all member states agree with the initial assessment carried out by a rapporteur member state and a two-step process if one or more member states have objections.

The first step is an assessment by the rapporteur member state where the food is to be placed on the market first. The application should consist of a comprehensive summary report and a copy of all study reports necessary to support the data requirements identified in Guidelines 97/618/EC. The application should also include information on the proposed labeling of the GM food. For evaluation at the rapporteur state, the regulation dictates a period of 90 days, which can be extended if further information is required from the applicant. In case of a favorable opinion, the rapporteur member state shall forward the Initial Assessment Report to the European Commission. The Commission forwards it further to the competent authorities of all member states, which have exactly 60 days to evaluate the notification and possibly raise objections. If there are no objections to the application, the first member state can authorize for the whole European Union, and the product can circulate freely in the European Union.

If there are objections by member states, the second step has to be taken. After consulting the Scientific Committee for Foods, the Commission shall draft a Proposal for Commission Decision that will be delivered for opinion to the Standing Committee for Foodstuffs, which is made up of representatives of the Competent Authorities of the different member states. If a qualified majority still cannot be reached, the issue is forwarded to the European Council of Ministers. Once at union level, the time frame for authorization is necessarily extended.

As a deviation from the full authorization procedure, Regulation (EC) 258/97 provides a simplified procedure for foods derived from GMOs, but no longer containing GMOs, that are substantially equivalent to existing foods with respect to composition, nutritional value, metabolism, intended use, and the level of undesirable substances. In such cases, the applicants only have to notify the Commission when placing a product on the market together with either scientific justification that the product is substantially equivalent or an opinion to the same effect, delivered by the competent authorities of a member state. The Commission forwards to the member states a copy of the notification. The product can then be marketed in the entire European Union.

Currently, a proposal of regulation from the European Parliament and Council on GM food and feed is being drafted. The proposed new food and feed regulation covers specifically the products of modern biotechnology and places essentially the same safety evaluation requirements on them wherever they are used in the food chain. The major changes compared with the previous legislation include the following:

1. Feed ingredients are explicitly covered, and the existing Novel Food Regulations will remain in place for non-GM foods.
2. Authorization will not be granted for a single use as either food or feed when such products are likely to be used both as food and feed.
3. Food and feed applications shall be submitted directly to the European Food Authority.

7.7. Japan

Japan's regulation of GM plants before environmental release and use as food or feed is mainly administered by two government departments: the Ministry of Agriculture, Forestry, and Fisheries (MAFF; www.maff.go.jp/eindex.html) and the Ministry of Health, Labor, and Welfare (MHLW; www.mhlw.go.jp/english/topics/food/index.html). MAFF is responsible for safety to the environment and the safety of animal feeds and feed additives of GM plants. Several guidelines are produced by MAFF, such as "Guidelines for Application of rDNA Organisms in Agriculture, Forestry, Fisheries, the Food Industry and other Related Industries," "Guidelines for Safety Assessments of Application of rDNA Organisms in Feed," and the "Guidelines for Safety Assessment of Application of rDNA Organisms in Feed Additives."

Based on the guidelines, GM plants destined for commercial agriculture use must be evaluated in a "simulated model environment" (confined field trial in Japan) prior to submission of the environmental safety assessment for review. At this stage, the GM plant is cultivated under conditions of reproductive isolation to investigate its properties as an agricultural product, including weediness and hybridizing potential with related species. The field trials must be conducted on sites accredited by MAFF, which may be government-owned research farms or sites managed by the developer. The results of field tests are incorporated into the environmental assessment of the potential impact that commercial release could have on the agriculture and ecosystem in Japan. The requirement for prior evaluation in a simulated model environment applies to the domestic cultivation of GM plants as well as to import of such plants that may propagate in the natural environment.

Environmental safety evaluations for commercial release consider the molecular characterization of the GM plant and its reproductive and propagative properties, including potential outcrossing, tendency to weediness, production of toxic substances, and other physiological characteristics. In the case of nonfood organisms, immediate commercialization is possible; otherwise, compliance with MHLW is required.

Safety assessment of GM foods is administered by MHLW. Since 1991, MHLW assessed the safety of foods and food additives produced by recombinant DNA technology on a voluntary basis based on its "Guidelines for Safety Assessment of Foods and Food Additives Produced by Recombinant DNA Techniques." From May 1, 2000, this became a mandatory requirement.

The food safety assessment is built on the principle of substantial equivalence and seeks to compare the GM food product with a traditional counterpart, focusing on the defined differences between the two. The criteria examined are consistent with those applied internationally. When the safety of the GM food cannot be established based on substantial equivalence, additional data may be required. These include studies on acute toxicity, chronic toxicity, mutagenicity, carcinogenicity, and other studies (intestinal tract, immunological or neurological toxicity). GM plant developers send applications to Secretariat of the Food Safety Investigation Council of MHLW, which arranges for the safety assessment review that normally takes about 1 year. This time frame does not include the time that the developer takes to generate additional data requested by the Food Safety Investigation Council. After the evaluation, MHLW will publish a declaration of the results in the official gazettes.

Beginning in April 2001, under a revised Japanese Agricultural Standard law, all GM food products must be labeled. These labeling requirements apply to all products containing detectable recombinant DNA or novel recombinant protein above a threshold of 5%.

7.8. United States

Regulatory oversight of GMOs has been in place in the United States for longer than in most other parts of the world. In 1984, a White House committee was formed under the auspices of the Office of Science and Technology Policy to propose a plan for regulating biotechnology. This plan, published by the Office of Science and Technology Policy on June 26, 1986, as the "Coordinated Framework for the Regulation of Biotechnology," is still in use. It is based on the principles that techniques of biotechnology are not inherently different from traditional plant technologies, and that biotechnology should not be regulated as a process, but rather the products of biotechnology should be regulated in the same way as products of other technologies. Responsibility for implementing the Coordinated Framework falls onto three lead federal agencies: the US Department of Agriculture, the FDA, and the Environmental Protection Agency (EPA).

The US Department of Agriculture, via its Animal and Plant Health Inspection Service (APHIS; www.aphis.usda.gov) is responsible for assessing the effects of GM plants on agricultural systems under the authority of the Plant Protection Act and the National Environmental Policy Act. Given the authority of APHIS to regulate the movement, import, and release of plant pests or potential plant pests under the Plant Protection Act, the primary focus of the APHIS review is to determine whether a plant produced through biotechnology has the potential to become a weed, create plant pests through outcrossing, or otherwise adversely affect natural habitats or agriculture. APHIS has guidelines covering laboratory and greenhouse research and field testing and has developed specific regulations for containment, transport, and field testing of GM plants. The results from initial screenings in laboratory and greenhouse settings typically are included in requests to APHIS for a permit for contained field trials.

Permit requests must be submitted to APHIS at least 120 days before planting and must contain information on the introduced gene, selectable markers, safeguards to prevent dissemination or carryover of plants, and differences between the GM plant and its unmodified parent. In addition, APHIS reviews required information on the agronomic and fitness characteristics of the GM plant. Conditions for the conduct of field trials are tailored to the individual field test requests and are based on the relevant biological and environmental characteristics of individual crops to ensure that the regulated materials do not persist in the environment. It is noted that, for certain crops, notification (shortened version of permit) may be submitted. Approval of notifications requires less time than permit approvals.

APHIS does not impose limitations on the size of field trials. However, for field tests of GM plants expressing plant-incorporated pesticides (such as *Bt* toxins) greater than 10 acres in size, the developer is required to obtain an Experimental Use Permit from the EPA (www.epa.gov/pesticides/biopesticides/pips/index.htm).

Before a GM plant can be produced on a wider scale and sold commercially, the developer must petition to APHIS for a "determination of non-regulated status." APHIS rules contain detailed requirements for the data and information that must be included in a petition for this determination. The type of information considered includes the biology and genetics of the host plant, including methods of reproduction, tendency to weediness, and modes of gene escape; a complete molecular characterization of the transgenic plant; and the environmental consequences of introducing the transgenic plant, including any changes in potential for weediness, impacts on nontarget organisms, or horizontal gene transfer that would be a consequence of the genetic modification. Detailed requirements for the petition can be found in the Code of Federal Regulations 7 CFR Part 340.

On receipt of a petition and following a review for completeness, APHIS publishes a notice in the *Federal Register* and provides a 60-days public comment period for each petition that meets the eligibility criteria. This notification includes a synopsis of the petition and explains the role of other regulatory bodies (EPA and FDA) and the process for submitting comments and obtaining more information. Following its assessment, and if it determines that the plant poses no significant risk to other plants in the environment and is as safe to use as traditional varieties, APHIS publishes a determination of non-regulated status in the *Federal Register*.

For GM plants with pesticidal properties (plant-incorporated protectants), such as *Bt* plants, APHIS coordinates its review with the EPA, which is responsible for the regulation of pesticidal substances under the Federal Insecticide, Fungicide, and Rodenticide Act; Federal Food, Drug, and Cosmetic Act (FFDCA); and Toxic Substances Control Act. In addition to examining data on product characterization (e.g., source of the gene; its expression of the protein; nature of the pesticidal substance produced; modifications to the introduced trait as compared to that trait in nature; biology of the recipient plant; effects on nontarget organisms; exposure; and environmental fate), the EPA also requires data on toxicology, digestive fate, and potential allergenicity of the pesticidal substance.

The EPA reviews the material on a case-by-case basis. A risk–benefit analysis balances potentially adverse environmental effects with proposed environmental and social benefits, such as reducing chemical applications and ensuring an abundant, economical food supply. There are several distinct stages (such as Experimental Use Permit and full registration for commercialization) in the EPA regulatory process, depending on the product involved. At each stage, the agency publishes a notice in the *Federal Register*, allowing interested parties an opportunity to make comments.

In the United States, GM food safety assessment is administered by the FDA (www.cfsan.fda.gov/~lrd/biotechm.html) under the FFDCA. The first biotechnology-derived food to enter the market was the Flavr Savr tomato, cleared by the FDA in 1994 after approval of a food additive petition for the neomycin phosphotransferase protein encoded by a marker gene that had been introduced into the tomato. Since 1994, the FDA has been using a voluntary consultation process to work together with developers, beginning from an early stage in product development, to identify and resolve any issues regarding the food that would necessitate legal action by the agency if the product were introduced into commerce. Examples of such issues include significantly increased levels of plant toxicants or antinutrients, reduction of important nutrients, the presence of new allergens, or the presence in the food of an unapproved food additive. To date, all GM foods and feeds marketed in the United States (about 55 by the end of 2002) have gone through the voluntary consultation program before they have entered the market.

When the developer has accumulated sufficient data that it believes are adequate to ensure that its product is safe and complies with the relevant provision of the FFDCA, the developer submits a safety and nutritional assessment summary to the FDA, which typically includes (1) the purpose of intended technical effect of the modification on the plant, together with a description of the various applications or uses of the bioengineered food, including animal feed use; (2) a molecular characterization of the modification, including the identities, sources, and functions of the introduced genetic material; (3) information on the expressed protein products encoded by introduced genes; (4) information on known or suspected allergenicity and toxicity of expressed products; (5) information on the compositional and nutritional characteristics of the food, including antinutrients; (6) for foods known to cause allergy, information on whether the endogenous allergens have been altered by the genetic modification; and (7) in some cases, the results of comparisons of wholesomeness feeding studies with foods derived from genetically engineered plants and the nonmodified counterpart.

In keeping with the voluntary nature of this process, there are no requirements for public notification in the *Federal Register* or public consultation. In such a case, the FDA does not issue a product approval *per se*, but informs the developer by letter that it has no further questions based on the information presented and reminds the developer of its legal responsibilities. The FDA does publish a list of completed consultations on the Internet (www.cfsan.fda.gov/~lrd/biocon.html).

Although this voluntary consultation process has worked well since its inception in 1994, it has been suggested that the establishment of a mandatory review process would clarify FDA requirements, impose equal rules on all agricultural biotechnology organizations, and provide a framework for a transparent and open regulatory review that would enhance consumer confidence. The FDA has responded to public comments by publishing, on January 18, 2001, a proposed rule to replace the current voluntary consultations with mandatory premarket consultations *(90)*. The proposed new regulations would require GM food and feed developers to provide a "premarket biotechnology notice" to the FDA at least 120 days prior to marketing such products. The information required in the notice generally would be similar to that provided in present consultations. However, the information submitted would be publicly available, although there would be opportunities for submitters to request that certain information be designated as confidential business information. Within 120 days of the filing of the premarket biotechnology notice, the FDA would respond with a letter describing its conclusion about the regulatory status of the food or animal feed.

Consistent with FDA policies for foods developed by other methods, labeling of food derived from biotechnology-derived crops is required only when the biotechnology-derived crops differ significantly in composition, nutritional value, or health effects from their conventional counterparts. These differing products, when offered for sale, could not be represented as equivalent to their conventional counterparts. If the product is approved, labels would be required to declare the compositional differences clearly. All biotechnology-derived products commercialized to date, except for two, have been found substantially equivalent, and because the FDA has concluded that they are as safe as their conventional counterparts, the FDA does not require these products to be labeled differently from their conventional counterparts. The two exceptions are high laurate canola and high oleic acid soybean products, both of which have oil composition that differs from the conventional counterpart *(101)*. This difference requires that the common or usual name be changed to describe the new foods. The FDA has determined that voluntary labeling is permissible and has circulated proposed voluntary labeling guidelines (published on January 17, 2001) designed to ensure that such labeling is truthful and not misleading *(102)*.

8. Concluding Remarks

Feeding an expanding and increasingly urbanized world population is one of the major challenges of the new millennium. Modern agricultural biotechnology, if properly integrated with other technologies for the production of food, can be one of the powerful tools to achieve global food security. Today, over 3 billion people live in countries where GM food and fiber plants are approved, grown, or consumed. The word *safety* appears always to be defined in terms of an absence of harm, danger, or damage; such a definition inevitably limits an absolute assertion of safety for food or any other product or activity because the future cannot be known. In practice, we extrapolate from past and present experience to make what we hope is a reasonable forecast of the future, but it can never be more than that *(2)*.

Agricultural biotechnology can offer human beings tremendous opportunities and benefits. However, like any other technologies, there are certain risks in applying the technology. As human beings, although it is not possible to ascertain absolute safety of any food, including those derived from modern biotechnology, what can be done is to assess and balance the risk and benefit based on human wisdom and knowledge accumulated to date. After all, human beings have always been doing so in every step forward throughout the tapestry of human history.

Acknowledgment

I gratefully acknowledge my colleagues Matt Cahill, Yan-San Chyi, Renjun Gao, Penny Hunst, Gaston Legris, Joel Mattsson, Guillermo Mentruyt, Deb Straw, Nick Storer, Laura Tagliani, and Jeff Wolt for reviewing all or part of the draft of this chapter and for their constructive comments.

References

1. James, C. (2002) Global status of commercialized transgenic crops: 2002. Briefs No. 27. International Service for the Acquisition of Agri-biotech Applications, Ithaca, New York.
2. Moses, V. and Brannan, M. (2001) One hundred percent safe? GM foods in the UK. Collected and collated for Cropgen. Available on-line at: http://www.cropgen.org. Accessed June 2003.
3. Essers, A. J. A., Alink, G. M., Speijers, G. J. A., et al. (1998) Food plant toxicants and safety risk assessment and regulation of inherent toxicants in plant foods. *Environ. Toxicol. Pharmacol.* **5**, 155–172.
4. Spyker, D., Love, L. A., and Brooks, S. M. (1996) An outbreak of pulmonary poisoning. *Clin. Toxicol.* **34**, 15–20.
5. Florack, D. E. and Stiekema, W. J. (1994) Thionins: properties, possible biological roles and mechanisms of action. *Plant Mol. Biol.* **26, 25–37.**
6. Kushmerick, C., Castro, M. S., Cruz, J. S., Block, C., Jr., and Beirao, P. S. L. (1998) Functional and structural features of γ-zeathionins, a new class of sodium channel blockers. *FEBS Lett.* **440**, 302–306.
7. Pedersen, J. (2000) *Application of Substantial Equivalence: Data Collection and Analysis.* Document of Joint FAO/WHO Expert Consultation on Foods Derived from Biotechnology, World Health Organization, Geneva, Switzerland.

8. Wood, G. E. (1979) Stress metabolites of plants—a growing concern. *J. Food Protect.* **42**, 496–501.
9. Mattsson, J. L. (2000) Do pesticides reduce our total exposure to food borne toxicants? *Neurotoxicology* **21**, 195–202.
10. Kuc, J. A. (1973) Metabolites accumulating in potato tubers following infection and stress. *Teratology* **8**, 333–338.
11. Friedman, M. and McDonald, G. M. (1997) Potato glycoalkaloids: chemistry, analysis, safety, and plant physiology. *Crit. Rev. Plant Sci.* **16**, 55–132.
12. Renwick, J. H., Caringbold, W. D. B., Earthy, M. E., Few, J. D., and McLean, A. C. (1984) Neural-tube defects produced in Syrian hamsters by potato glycoalkaloids. *Teratology* **30**, 371–381.
13. Gaffield, W. and Keeler, R. F. (1996) Induction of terata in hamsters by Solanidane alkaloids derived from *Solanum tuberosum*. Chem. Res. Toxicol. 9, 426–433.
14. Munkvold, G. P. and Desjardins, A. E. (1997) Fumonisins in maize: can we reduce their occurrence? *Plant Dis.* **81**, 556–565.
15. Windham, G. L., Williams, W. P., and Davis, F. M. (1999) Effects of the southwestern corn borer on *Aspergillus flavus* kernel infection and aflatoxin accumulation in maize hybrids. *Plant Dis.* **83**, 535–540.
16. Dowd, P. F. (1998) Involvement of arthropods in the establishment of mycotoxigenic fungi under field conditions. In Mycotoxins in Agriculture and Food Safety . (Sinha, K. K. and Bhatagnar, D., eds.), Marcel Dekker, New York, pp. 307–350.
17. Munkvold, G. P., Hellmich, R. L., and Rice, L. G. (1999) Comparison of fumonisin concentrations in kernels of transgenic Bt maize hybrids and non-transgenic hybrids. *Plant Dis.* **83**, 130–138.
18. Bakan, B., Melcion, D., Richard-Molard, D., and Cahagnier, B. (2002) Fungal growth and Fusarium mycotoxin content in isogenic traditional maize and genetically modified maize grown in France and Spain. *J. Agric. Food Chem.* **50**, 728–731.
19. Hammond, B., Campbell, K., Pilcher, C., et al. (2003) Reduction of Fumonisin Mycotoxins in Bt Corn. Poster presented at the 42nd annual meeting of the Society of Toxicology, Salt Lake City, UT.
20. Anderson, J. A. (1996) Allergic reactions to foods. *Crit. Rev. Food Sci. Nutr.* **36(S)**, S19–S38.
21. Penninks, A., Knippels, L., and Houben, G. (2001) Allergenicity of foods derived from genetically modified organisms. In *Safety of Genetically Engineered Crops.* (Custers, R., e.). VIB, Zwijnaarde, Belgium, pp. 108–134.
22. Sloan, A. E. and Powers, M. E. (1986) A perspective on popular perceptions of adverse reactions to foods. *J. Allergy Clin. Immunol.* **78**, 127–133.
23. Sampson, H. A. and Burks, A. W. (1996) Mechanisms of food allergy. *Annu. Rev. Nutr.* **16**, 161–177.
24. Lehrer, S. B. and Salvaggio, J. E. (1990) Allergens: standardization and impact of biotechnology, a review. *Allergy Proc.* **11**, 197–208.
25. Food and Agriculture Organization/World Health Organization. (2001) *Evaluation of Allergenicity of Genetically Modified Foods.* Report of a Joint FAO/WHO Expert Consultation, Food and Agriculture Organization of the United Nations, Rome.
26. Metcalfe, D. D., Astwood, J. D., Townsend, R., Sampson, H. A., Taylor, S. L., and Fuchs, R. L. (1996) Assessment of the allergenic potential of foods derived from genetically engineered crop plants. *Crit. Rev. Food Sci. Nutr.* **36(S)**, S165–S186.
27. Taylor, S. L. and Lehrer, S. B. (1996) Principles and characteristics of food allergens. *Crit. Rev. Food Sci. Nutr.* **36(S)**, S91–S118.

28. King, T. P. (1994) Antigenic determinants: B cells. In *Proceedings of Conference on Scientific Issues Related to Potential Allergenicity in Transgenic Food Crops*. FDA Docket no. 94N-0053. US Food and Drug Administration, Washington, DC.
29. Pernis, B. G. (1992) Antigen processing and antigen presentation. *Prog. Allergy Clin. Immunol.* **2**, 395–399.
30. Burks, A. W., Williams, L. W., Helm, R. M., Connaughton, C., Cockrell, G., and B'Brien, T. (1991) Identification of a major peanut allergen, *Ara h* 1, in patients with atopic dermatitis and positive peanut challenges. *J. Allergy Clin. Immunol.* **88**, 712–719.
31. Burks, A. W., Williams, L. W., Connaughton, C., Cockrell, G., O'Brien, T. J., and Helm, R. M. (1992) Identification and characterization of a second major peanut allergen, *Ara h* 2, with use of the sera of patients with atopic dermatitis and positive peanut challenge. *J. Allergy Clin. Immunol.* **90**, 962–969.
32. Garcia-Casado, G., Sanchez, R., Chrspeels, M. J., Armentia, A., Salcedo, G., and Gomez, L. (1996) Role of complex asparagines-linked glycans in the allergenicity of plant glycoproteins. *Glycobiology* **6**, 471–477.
33. Wilson, I. B. H., Harthill, J. E., Mullin, N. P., Ashford, D. A., and Altmann, F. (1998) Core α-1,3-fucose is a key part of the epitope recognized by antibodies reacting against plant N-linked oligosaccharides and is present in a wide variety of plant extracts. *Glycobiology* **8**, 651–661.
34. Bantanero, E., Crespo, J. F., Monsalve, R. I., Martin-Esteban, M., Villalba, M., and Rodriguez, R. (1999) IgE-binding and histamine-release capabilities of the main carbohydrate component isolated from the major allergen of olive tree pollen, Ole e 1. *J. Allergy Clin. Immunol.* **103**, 147–153.
35. Burks, A. W., Williams, L. W., Thresher, W., Connaughton, C., Cockrell, G., and Helm, R. M. (1992) Allergenicity of peanut and soybean extracts altered by chemical and thermal denaturation in patients with atopic dermatitis and positive food challenges. *J. Allergy Clin. Immunol.* **90**, 889–897.
36. Barnett, D. and Howden, M. E. (1986) Partial characterization of an allergenic glycoprotein from peanut (*Arachis hypogaea* L.). *Biochim. Biophys. Acta* **882**, 97–105.
37. Shibasaki, M., Suzuki, S., Nemoto, H., and Kuroume, T. (1979) Allergenicity and lymphocyte-stimulating property of rice protein. *J. Allergy Clin. Immunol.* **64**, 259–265.
38. Taylor, S. L., Lemanske, R. F., Jr., Bush, R. K., and Busse, W. W. (1987) Food allergens: structure and immunologic properties. *Ann. Allergy* **59**, 93–99.
39. Fuchs, R. L. and Astwood, J. D. (1996) Allergenicity assessment of foods derived from genetically modified plants. *Food Technol.* **50**, 83–88.
40. Astwood, J. D., Leach, J. N., and Fuchs, R. L. (1996) Stability of food allergens to digestion in vitro. *Nat. Biotechnol.* **14**, 1269–1273.
41. Fu, T. J., Abbott, U. R., and Hatzos, C. (2002) Digestibility of food allergens and nonallergenic proteins in simulated gastric fluid and simulated intestinal fluid—a comparative study. *J. Agric. Food Chem.* **50**, 7154–7160.
42. Nordlee, J. A., Taylor, S. L., Townsend, J. A., Thomas, L. A., and Bush, R. K. (1996) Identification of a Brazil-nut allergen in transgenic soybeans. *N. Engl. J. Med.* **334**, 688–692.
43. Matsuda, T. (1998) Application of transgenic techniques for hypo-allergenic rice. In *Proceedings of the International Symposium on Novel Foods Regulation in the European Union—Integrity of the Process of Safety Evaluation*. Berlin, Germany, pp. 311–314.
44. Suszkiw, J. (2002) Researchers develop first hypoallergenic soybean. *Agricultural Research*, US Department of Agriculture. September 2002, pp. 16–17.
45. Bhalla, P. L., Swoboda, I., and Singh, M. B. (1999) Antisense-mediated silencing of a gene encoding a major ryegrass pollen allergen. *Proc. Natl. Acad. Sci. USA* **96**, 11,676–11,680.

46. Van Loon, L. C. and van Strien, E. A. (1999) The families of pathogenesis-related proteins, their activities, and comparative analysis of PR-1 type proteins. *Physiol. Mol. Plant Pathol.* **55**, 85–97.
47. Ebner, C., Hoffmann-Sommergruber, K., and Breiteneder, H. (2001) Plant food allergens homologous to pathogenesis-related proteins. *Allergy* **56(Suppl. 67)**, 43–44.
48. Midoro-Horiuti, T., Brooks, E. G., and Goldblum, R. M. (2001) Pathogenesis-related proteins of plants as allergens. *Ann. Allergy Asthma Immunol.* **87**, 261–271.
49. Organization for Economic Cooperation and Development. (2000) *Report of the Task Force for the Safety of Novel Foods and Feeds.* C(2000)86/ADD1. Organization for Economic Cooperation and Development, Paris.
50. Cromwell, G. L., Lindemann, M. D., Randolph, J. H., et al. (2002) Soybean meal from Roundup Ready or conventional soybeans in diets for growing-finishing swine. *J. Anim. Sci.* **80**, 708–715.
51. Reuter, T., Aulrich, K., Berk, A., and Flachowsky, G. (2002) Investigations on genetically modified maize (*Bt*-maize) in pig nutrition: chemical composition and nutritional evaluation. *Arch. Anim. Nutr.* **56**, 23–31.
52. Taylor, M. L., Hartnell, G. F., Riordan, S. G., et al. (2003) Comparison of broiler performance when fed diets containing grain from Roundup ready NK603), YieldGard × Roundup ready (MON810 × NK603), non-transgenic control, or commercial corn. *Poultry Sci.* **82**, 443–453.
53. Ye, X., Al-Balili, S., Kloti, A., et al. (2000) Engineering the provitamin A (β-carotene) biosynthetic pathway into (carotenoid-free) rice endosperm. *Science* **87**, 303–305.
54. Environmental Protection Agency. (2001) *US Environmental Protection Agency, Biopesticide Registration Action Document:* Bacillus thuringiensis *(*Bt*) Plant-Incorporated Protectants.* October 15, US Environmental Protection Agency, Washington, DC.
55. Losey, J. E., Rayor, L. S., and Carter, M. E. (1999) Transgenic pollen harms monarch larvae. *Nature* **399**, 214.
56. Environmental Protection Agency. (1995) US Environmental Protection Agency Publication EPA731-F-95-004.
57. Sears, M. K., Stanley-Horn, D. E., and Mattila, H. R. (2000) *Preliminary Report on the Ecological Impact of* Bt *Corn Pollen on the Monarch Butterfly in Ontario.* Canadian Food Inspection Agency and Environment Canada, Ottawa, ON, Canada.
58. Hellmich, R. L., Siegfried, B. D., Sears, M. K., et al. (2001) Monarch larvae sensitivity to *Bacillus thuringiensis*-purified proteins and pollen. *Proc. Natl. Acad. Sci. USA* **98**, 11,925–11,930.
59. Oberhauser, K. S., Prysby, M. D., Mattila, H. R., et al. (2001) Temporal and spatial overlap between monarch larvae and corn pollen. *Proc. Natl. Acad. Sci. USA* **98**, 11,913–11,918.
60. Pleasants, J. M., Hellmich, R. L., Dively G. P., et al. (2001) Corn pollen deposition on milkweeds in and near cornfields. *Proc. Natl. Acad. Sci. USA* **98**, 11,919–11,924.
61. Stanley-Horn, D. E., Dively, G. P., Hellmich, R. L., et al. (2001) Assessing the impact of Cry1Ab-expressing corn pollen on monarch butterfly larvae in field studies. *Proc. Natl. Acad. Sci. USA* **98**, 11,931–11,936.
62. Zangerl, A. R., McKenna, D., Wraight, C. L., et al. (2001) Effects of exposure to event 176 *Bacillus thuringiensis* corn pollen on monarch and black swallowtail caterpillars under field conditions. *Proc. Natl. Acad. Sci. USA* **98**, 11,908–11,912.
63. Sears, M. K., Hellmich, R. L., Stanley-Horn, D. E., et al. (2001) Impact of *Bt* corn pollen on monarch butterfly populations: a risk assessment. *Proc. Natl. Acad. Sci. USA* **98**, 11,937–11,942.
64. Head, G., Freeman, B., Moar, W., Ruberso, J., and Turnipseed, S. (2001) Natural enemy abundance in commercial Bollgard® and conventional cotton fields. In *Proceedings of Beltwide Cotton Conferences.* National Cotton Council, Memphis, TN, pp.796–798.
65. Dewar, A. M., May, M. J., Woiwod, I. P., et al. (2003) A novel approach to the use of genetically modified herbicide tolerant crops for environmental benefit. *Proc. Royal Soc. Biol. Sci.* **270**, 335–340.
66. Fawcett, R. and Towery, D. (2002) *Conservation Tillage and Plant Biotechnology: How New Technologies Can Improve the Environment by Reducing the Need to Plow.* Report of Conservation Technology Information Center, Purdue University, West Lafayette, IN.

67. Ammann, K., Jocot, Y., and Mazyad, P. R. A. (2001) Safety of genetically engineered plants: an ecological risk assessment of vertical gene flow. In *Safety of Genetically Engineered Crops.* (Custers, R., ed.. VIB, Zwijnaarde, Belgium, pp. 60–87.
68. Stewart, C. N., Jr., All, J. N., Raymer, P. L., and Ramachandran, S. (1997) Increased fitness of transgenic insecticidal rapeseed under insect selection pressure. *Mol. Ecol.* **6**, 773–779.
69. Metz, P. L. J., Jacobsen, E., Nap, J. P., Pereira, A., and Stiekema, W. J. (1997) The impact of biosafety of the phosphinothricin-tolerance transgene in inter-specific *B. rapa* (*B. napus* hybrids and their successive backcrosses. *Theor. Appl. Genet.* **95**, 442–450.
70. Mikkelsen, T. R., Andersen, B. and Jorgensen, R. B. (1996) The risk of crop transgene spread. *Nature* **380**, 31.
71. Halfhill, M. D., Millwood, R. J., Raymer, P. L., and Stewart, C. N., Jr. (2002) *Bt*-transgenic oilseed rape hybridization with its weedy relative, *Brassica rapa. Environ. Biosafety Res.* **1**, 19–28.
72. World Health Organization. (1993) *Health Aspects of Marker Genes in Genetically Modified Plants.* Report of a WHO Workshop. World Health Organization of the United Nations, Geneva, Switzerland.
73. Food and Agriculture Organization/World Health Organization. (2000) *Safety Aspects of Genetically Modified Foods of Plant Origin.* Report of a Joint FAO/WHO Expert Consultation. World Health Organization of the United Nations, Geneva, Switzerland.
74. Gebhard, F. and Smalla, K. (1999) Monitoring field releases of genetically modified sugar beets for persistence of transgenic plant DNA and horizontal gene transfer. *FEMS Microbiol.* **28**, 261–272.
75. Paget, E., Lebrun, M., Freyssinet, G., and Simonet, P. (1998) The fate of recombinant plant DNA in soil. *Eur. J. Soil Biol.* **34**, 81–88.
76. Schluter, K., Futterer, J., and Potrykus, I. (1995) "Horizontal" gene transfer from a transgenic potato line to a bacterial pathogen (*Erwinia chrysanthemi*) occurs—if at all—at an extremely low frequency. *Biotechnology* **13**, 1094–1098.
77. Nielsen, K. M., Bones, A. M., Smalla, K., and van Elsas, J. D. (1998) Horizontal gene transfer from transgenic plants to terrestrial bacteria—a rare event? *FEMS Microbiol. Rev.* **22**, 79–103.
78. Nielsen, K. M., van Elsas, J. D., and Smalla, K. (2000) Transformation of *Acinetobacter* sp strain BD413(pFG4 nptIII) with transgenic plant DNA in soil microcosms and effects of kanamycin on selection of transformants. *Appl. Environ. Microbiol.* **66**, 1237–1242.
79. Gould, F. (1998) Sustainability of transgenic insecticidal cultivars: integrating pest genetics and ecology. *Annu. Rev. Entomol.* **43**, 701–726.
80. Gianessi, L. P., Silvers, C. S., Sankula, S., and Carpenter, J. E. (2002) *Plant Biotechnology: Current and Potential Impact for Improving Pest Management in US Agriculture, an Analysis of 40 Case Studies.* Report of National Center for Food and Agricultural Policy, National Center for Food and Agricultural Policy, Washington, DC.
81. Food and Agriculture Organization/World Health Organization. (1997) *Risk Management and Food Safety.* FAO Food and Nutrition Paper 65. Report of a Joint FAO/WHO Consultation. Food and Agriculture Organization of the United Nations, Rome.
82. Codex Alimentarius Commission. (2002) *Report of the Third Session of the Codex Ad Hoc Intergovernmental Task Force on Food Derived from Biotechnology.* Yokohama, Japan, Codex Alimentarins Commission, Rome.
83. Organization for Economic Cooperation and Development. (1993) *Safety Evaluation of Foods Derived by Modern Biotechnology: Concepts and Principles.* Organization for Economic Cooperation and Development, Paris.
84. Food and Agriculture Organization/World Health Organization. (1996) *Biotechnology and Food Safety.* Report of a Joint FAO/WHO Expert Consultation. Food and Agriculture Organization of the United Nations, Rome.

85. Pedersen, J., Eriksen, F. D., and Knudsen, I. (2001) Toxicity and food safety of genetically engineered crops. In *Safety of Genetically Engineered Crops.* (Custers, R., ed.). VIB Zwijnaarde, Belgium, pp. 27–59.
86. Draper, J., Beckmann, M., Taylor, J., et al. (2003) Metabolomics as a technology platform to contribute to the assessment of compositional differences between GM and conventionally bred crops. Poster of 2nd International Conference on Plant Metabolomics, Potsdam, Germany.
87. Kuiper, H. (2000) Profiling techniques to identify differences between foods derived from biotechnology and their counterparts. Document of Joint FAO/WHO Expert Consultation on Foods Derived from Biotechnology, World Health Organization, Geneva.
88. Charles, G. D., Linscombe, V. A., Tornesi, B., Mattsson, J. L., and Gollapudi, B. B. (2002) An in vitro screening paradigm for extracts of whole food for detection of potential toxicants. *Food Chem. Toxicol.* **40,** 1391–1402.
89. Beever, D. E. and Kemp, C. F. (2000) Safety issues associated with the DNA in animal feed derived from genetically modified crops. A review of scientific and regulatory procedures. *Nutr. Abstr. Rev. (Ser. B: Livestock Feeds Feeding)* **70**, 175–182.
90. Food and Drug Administration. (2001) Premarket notification concerning bioengineered foods. *Federal Register* 66-12:4706–4738 (January 18, 2001).
91. Sjoblad, R. D., McClintock, J. T., and Engler, R. (1992) Toxicological considerations for protein components of biological pesticide products. *Reg. Toxicol. Pharmacol.* **15**, 3–9.
92. Penninks, A. H. (2001) *Animal Model for Allergenicity Assessment.* Document of the Joint FAO/WHO Expert Consultation on Foods Derived from Biotechnology. Food and Agriculture Organization of the United Nations, Rome.
93. Knippels, L. M. J., Penninks, A. H., Spanhaak, S., and Houben, G. F. (1998) Oral sensitization to food proteins: a Brown Norway rat model. *Clin. Exp. Allergy* **28**, 368–375.
94. Knippels, L. M. J., Penninks, A. H., Spanhaak, S., and Houben, G. F. (1999) Immune-mediated effects upon oral challenge of ovalbumin sensitized Brown Norway rats; further characterization of a rat food allergy model. *Toxicol. Appl. Pharmacol.* **156**, 161–169.
95. Li, X., Serebrisky, D., Lee, S. J., et al. (2000) A murine model of peanut anaphylaxis: T- and B-cell responses to a major peanut allergen mimic human responses. *J. Allergy Clin. Immunol.* **106**, 150–158.
96. Dearman, R. J. and Kimber, I. (2001) Determination of protein allergenicity: studies in mice. *Toxicol. Lett.* **120**, 181–186.
97. Piacentini, G. L., Bertolini, A., Spezia, E., Piscione, T., and Boner A, L. (1994) Ability of a new infant formula prepared from partially hydrolyzed whey to induce anaphylactic sensitization; evaluation in a guinea pig model. *Allergy* **49**, 361–364.
98. Endo, Y. and Boutrif, E. (2002) Plant biotechnology and its international regulation—FAO's initiative. *Livestock Prod. Sci.* **74**, 217–222.
99. Jia, S. and Peng, Y. (2002) GMO biosafety research in China. *Environ. Biosafety Res.* **1**, 5–8.
100. Hathcock, J. N. (2000) The precautionary principle—an impossible burden of proof for new products. *AgBioForum* **3**, 255–258.
101. Chassy, B. M., Abramson, S. H., Bridges, A., et al. (2001) *Evaluation of the US Regulatory Process for Crops Developed Through Biotechnology.* Council for Agricultural Science and Technology (CAST), Issue Paper 19, October, Council for Agricultural Science and Technology. Ames, Iowa.
102. Food and Drug Administration. (2001) Draft Guidance for Industry: Voluntary Labeling Indicating Whether Foods Have or Have Not Been Developed Using Bioengineering. US Food and Drug Administration, January 17, US Food and Drug Administration, Washington, DC.

12

Conclusion and Future Directions

Sarad R. Parekh

1. Introduction

The human population has grown steadily since the advent of agricultural practices several thousand years ago. Through domestication of animals and agricultural practice, it has ensured itself continued sustenance and an ample supply of food. Despite several population control measures, the world population is projected to grow steadily to 9–11 billion by 2050 *(1)*. This has triggered human fears about balancing the growth of population and food supplies to feed the hungry.

During the 20th century, the "green revolution" helped the human population avoid major food shortage problems by keeping food productivity and medical advances ahead of population growth. Realistically, this trend cannot continue to sustain the population forever. Furthermore, conventional agricultural practices have serious limitations because local pricing policies and profitability are no longer effective in the current global marketplace. Added to this are lengthy plant breeding practices, reduced availability of land and water, increased deterioration of soil and water, use of chemicals, and unpredictable weather patterns.

Meeting all aspects of human life for food, disease control, medicine, technology advances, quality of life, and population growth cannot be brought about in a short period of time by conventional plant breeding, especially when some of the important crops are approaching their physiological limits of productivity, and it cannot be accomplished by increasing arable land. In fact, the arable land for food cultivation comprises only 5% of the earth's surface and is decreasing as a result of soil erosion, pathogens, pest epidemics, salinization, overcultivation, and unpredictable weather patterns *(2)*. In addition, increased human and agricultural use, polluted agricultural runoff, and widespread use of agrochemicals are limiting freshwater supplies.

All this has engendered new fears that food shortages may lead to widespread social, economic, and political unrest, making food and medical security the most serious threats to international peace. At the same time, overwhelming evidence shows that those countries with more efficient technologies, higher living standards, stronger economies, and higher food productivity will not be able to provide for the ever-demanding nutrition of the expanding human race, protect the environment, and ensure political stability. The challenge for human survival will depend on developing novel biotechnologies in plant

From: *The GMO Handbook: Genetically Modified Animals, Microbes, and Plants in Biotechnology*
Edited by: S. R. Parekh © Humana Press Inc., Totowa, NJ

and animal science and merging classical breeding practices with the new and more effective breeding practices.

In the last three centuries, the agricultural, food and fermentation, microbiological, and pharmaceutical sectors have became highly industrialized; as a result, public health has improved, incomes have increased, and international trade in food and machine supplies has grown. To meet the current trends in population growth and agricultural production within the finite volume of the earth's capacity, it will be necessary to identify new and better varieties of food *(3)*.

Historically, meeting the demand for better food and medicine was managed through traditional plant and animal breeding programs and medical practices to bring about enormous increases in quality of life and crop productivity. However, the improvement and advances in the past came slowly, in the cases of plant hybridization and animal breeding, because of a very limited gene pool and natural barriers for genetic alteration.

In the 1980s, the advent of recombinant deoxyribonucleic acid (DNA) meant natural barriers could be crossed, and an entire gene pool, whether plant, animal, or microbe, could be transferred. Through various cell-culturing technologies, scientists acquired the ability to engineer the stable introduction of DNA into a number of living systems and produce molecules at commercial levels. Prior to recombinant DNA, biotechnology used mostly microbes, and experiments were carried out in laboratories and industrial locations. Very soon after the advent of recombinant DNA, genetically modified organisms (GMO) products derived from plants and animal cells were generated for commercial use and were released into the environment. Prospects are good that, in the near future, most, if not all, food and pharmaceutical products will be derived from GMOs *(3)*. However, if these products derived from GMOs are ever to become more than mere promises, they must prove their safety, and their value, outside in the real world.

Over 1000 GMOs and their products have been tested in field trials or industrial applications, ultimately finding themselves in natural surroundings. Considering the increasing number of international GMO releases derived from transgenics, it is important not to ignore the issues related to risk *(4)*. So, for this reason, the progress of GMO use and technology has always been, and still is, accompanied by interesting and controversial discussions. The agricultural biotechnology debate has addressed the practice of technology and containment, that is, how to prevent the accidental release of GMOs into the environment.

Despite thousands of laboratory experiments, field trials, and risk assessment conducted inside academic, industrial, and movement laboratories around the world, in which GMOs caused no harm to human health or to the environment, the focus now has turned to the scientific technological, legal, and political impacts of deliberate release of GMOs into the ecosystem *(5)*. The reasons for this shift are straightforward: The technology has advanced to the stage at which what were only promising laboratory developments are now products waiting for clinical trials and commercialization. It will become critical for the industrial practice of GMOs to apply technologies derived from cellular and molecular concepts and to create novel products that will be integrated into all sectors of life.

This book was conceived as a result of the accelerated growth of GMO-derived technologies and the new challenges and opportunities presented by the commercial practice of GMOs. This handbook is written to provide scientific background, advice, and recommendations for successful acceptance of GMO practices for field and commercial use. Included throughout are examples of the following: (1) novel, emerging technological breakthroughs that allow modification and construction of GMOs for use in agricultural or animal models; (2) technological hurdles that must be overcome for successful construction of GMOs; (3) issues to be regulated; (4) historical experiences and information on modified plants and microbes; (5) emerging advances on the testing of GMO-derived products, from small-scale through large commercial launches; and (6) the new concerns related to exotic GMOs and bioterrorism. After careful review of the chapters in this book, I summarize and highlight here a few salient features of GMOs.

2. GMO-Based Bioeconomy

A GMO-based economy agglomerates all industrial organized activities. The two segments of the GMO industry are (1) those that exploit biological resources (crop production, livestock and poultry, aquaculture, forestry, fishing, and horticulture) and (2) suppliers to the bioresources (agriculture chemicals and seeds, life sciences, food and fiber processing industries, pharmaceuticals, and health care) *(6)*.

These sectors have a direct and more vital impact on human welfare and induce change to the existing biological system.

3. GMO Technologies

With the advent of recombinant DNA technology and its fusion with computer science, it is now possible to screen rapidly the entire sequence and genome of living cells. The package of tools, including bioinformatics and metabolic engineering, X-ray crystallography, and imaging, allows scientists to unfold the mysteries of the genetic code and protein functions and offer the potential to "custom design" and clone any living system *(7)*. In the future, this may be the area in which more efforts will be tied to developing unique gene transfer, expression, and production systems, whether microbes, plants, animal, or viruses.

4. GMO Products

4.1. Microbial GMOs

With the evolution in DNA sequencing and bioinformatics, therapeutic proteins and other vaccines have advanced rapidly along the biotechnology research pipeline. Hundreds of new proteins derived from microbes are ready to enter clinical trials and are very likely to reach commercial applications. Regulators already have approved use of antibiotics, GMO-derived human therapeutics such as insulin, interferon, γ-globulin, and interleukins, as well as plant insect control agents. In short, the active engagement and involvement of those who regulate microbial GMO products and activities, combined with the history of the demonstrated proof of concept, will support future evolution of these products *(8–10)*.

4.2. Plant GMOs

Considering the power and the potential of agricultural biotechnology, the "gene revolution" will dominate agriculture into the next millennium. Like the earlier green revolution, plant GMOs are anticipated to accelerate food production, save lives from hunger, save large tracts of land, ensure low food prices, stimulate broad-based economies, and expand world trade for future generations.

4.3. Animal GMOs and Medicine

Animal GMOs have progressed rapidly from the first transgenic mice and bovine embryos produced in vitro. Today, animal breeding companies are using marker-assisted selection to provide early and improved selection of breeding animals. Gene mapping and bioinformatics have revolutionized the process, simplifying the identification of a large number of candidate genes.

The propagation of transgenic animals is accelerated by in vitro production of embryos. The dozen or so companies engaged in the practice of cloning embryos for cattle and sheep have shown that the technology is possible; however, at the moment the cloning is at low frequency and still faces commercial hurdles. Animal production through genetic selection and efficient segregation of the sperm for in vitro fertilization of embryos has been successfully demonstrated. The flow-sorting devices for discriminating male and female have been tested in cattle and swine to produce offspring of the desired sex.

The artificial insemination and embryo transfer industries have proliferated. Cryopreservation of oocytes and the combination of embryo biopsy with sexing are accelerating. Microinjection of DNA into the nuclei of an egg and use of viral vectors have been demonstrated in the creation of transgenic cattle, sheep, and poultry. The intent of such GMO research has been to change animal growth, elicit disease resistance, or derive new therapeutic products *(11)*. The effectiveness of this GMO technology is low in mice models. Therefore, progress has been slow in the commercial use of transgenic technology, except in the case of cows, for example, for which novel biopharmaceutical products are expressed in milk *(11)*.

Although the production of GMO birds and fish is feasible, currently it is not as commercially successful as GMO-derived microbes *(12,13)*. Momentum in this area can be accelerated by more study on the anatomy and physiology of vertebrates, especially avian and marine, for successful GMO practice through viral introduction of DNA into the germline.

Last, in calves and sheep embryo, stem cell culture and nuclear transfer have been successful for breeding new offspring. The cloning of new offspring by this method has been successful; however, the technology is not fully matured for commercial use. Most efforts are under development for growth hormones and growth-promoting proteins for enhancing animal growth *(11)*.

Numerous vaccines are derived by GMOs. Parent duplication and cloning in transgenic animals have enabled application of recombinant vaccines against calf sores and rabies. New GMO-derived biotechnology has also revolutionized diagnostics application and facilitated precision assay procedures. Monoclonal antibodies have been the arsenal in several therapeutic sciences. Based on these GMO technologies, there are opportunities for field testing of animals by clinicians to furnish real-time data that permits rapid analysis and treatment. It is anticipated that use of such GMO-derived technology will play a critical role in the future, especially in the diagnostic laboratory to provide a sound basis for control and treatment of animal and human diseases.

5. Hurdles Related to Application of GMO Technology

The use of GMOs has generated protests, lawsuits, countless conferences, bans, limited trials, environmental impact statements, and unauthorized environmental releases. Buried among all of these issues are the public's fear and its justification for safer introductions of GMOs into the world. However, it should be emphasized that the focus of the public and the government agencies that regulate the sector should be on the products synthesized from DNA segments of GMO and not on the technology. This important distinction, although made in 1987 by the committee on the introduction of GMOs into the environment and conveyed by the National Academy of Sciences, is still not clearly interpreted or assimilated by a GMO-sensitive public *(14)*.

With the shift in the focus from GMOs to the products, another essential step is addressing the public perception that all GMO-derived products are not created equal. For example, vaccines and immunoglobulins for animals are not the same as those for humans, and human gene therapy does not have the same platform as breeding sheep. Microbes are not plants, and microbes that manifest infection on oranges are not the same as microbes that produce toxins. Clearly, some GMO products are diversified and intrinsically safer than others.

The key point here is that GMO products should be assessed on the basis of their individual characteristics and not be categorized and lumped into one generic group. Furthermore, it should also be reemphasized that the harnessing of biotechnology and construction of new strains or traits for plant breeding practice is not a substitute or "miracle cure" toward the limitless problems faced in other areas of biotechnology. Good judgment and safe biotechnology practices are as warranted and important to the discovery and the potential use of GMOs as they have been to the advent of the technology.

6. Current Scenario and Future

Pharmaceutical and biotechnology companies and agronomists have been engaged in upgrading and modifying crops, breeding animals, and mutating organisms to derive an improved gene pool for several decades. Thus, ample regulatory agencies have been established (Food and Drug Administration, Environmental Protection Agency, US Department of Agriculture, National Institutes of Health, Centers for Disease Control and Prevention) all of whom have experience and knowledge from field tests and clinical trials to determine the impact of the commercial use of GMOs.

Throughout this time and with all these past experiments, no significant incidents have occurred or harm has been done to the human population or the ecosystem. This selective breeding has not caused an outbreak of any major crop that became noxious or resulted in monstrous weeds. In addition, there has not been evidence of cross transfer of either microbial or animal genes to humans or agricultural crops *(15)*. Also, transfer of the gene pool for a given trait, such as disease resistance, to a nearby ecosystem has not caused any living system to be more competitive, thus causing ecological damage.

Finally, the challenge for the adaptability of GMOs is overwhelming. Although in a short period of time the harnessing of recombinant technology and the use of GMOs have surpassed expectations, continued success and public acceptance has been slow. Although the use of GMOs can significantly improve the availability of high-quality food and supply breakthroughs in medical science, it still needs to gain economic acceptance. Many of the underdeveloped countries cannot afford expensive biotechnology and GMO-based products.

The solution to this requires engagement of critical agencies. In essence, more balanced emphasis on agricultural, fermentation, and medical biotechnology along with full integration of environmental ecology will be necessary. These issues are well articulated in this handbook regarding what is deemed important in terms of technological advances, delineation of urgent needs for educating the public, how to define the boundaries and the framework of work ethics, and what the opportunities and limitations are. Broad, one-on-one discussion and sweeping changes in this arena will be essential to reach a consensus. Recognizing this, the contributors of this book have made specific recommendations and included topics to simplify the debate over GMOs and to provide clearer scientific representation of facts.

The various chapters in this book document these points concisely. It is hoped that the chapters in this book will provide the right balance between communication and information. It is hoped that the handbook will further strengthen mutual areas of agreement among academic institutions, government agencies, industries, the public sector, and those engaged in commercial biotechnology and GMO-based technologies.

References

1. Budd, R. (1993) 100 years of biotechnology. *Biotechnology (NY)* **11**, S14–S15.
2. Drewes, J. (1993) Into the 21st century. *Biotechnology (NY)* **11**, S16–S20.
3. Budd, R. (1991). Biotechnology in the twentieth century. *Soc. Stud. Sci.* **21**, 415–457.
4. Murray, F. and Gaiannakas K. (2001) Agricultural biotechnology and industry structure. *AgBio Forum* **4**, 137–151.
5. Duvick, N. D.(1999) How much caution in the fields. *Science* **286**, 418.
6. Maloney, R. J. (2000) Opportunity for agricultural biotechnology. *Science* **286**, 615.
7. James, C. (2002) Global Status of Transgenic Crops. ISAAA Brief No. 27. Available online at www.Isaaa.org. Accessed January 20, 2004.
8. Glick, B. and Skof Y. (1986) Environmental implications of recombinant DNA technology. *Biotechnol. Adv.* **4**, 261–277.
9. Klausner, A. (1993) Back to the future: biotech product sales, 1982–1993. *Biotechnology (NY)* **11**, S35–S37.
10. Anonymous. (1995) *UNEP. International Technical Guidance for Safety in Biotechnology.* UN Environmental Program, Nairobi, Kenya.
11. Winter, G. and Milstein, C. (1999) Man-made antibodies. *Nature* **349**, 293–299.
12. Graham, F. L. (1990) Adenoviruses as expression vectors and recombinant vaccines. *Trends Biotechnol.* **8**, 85–87.
13. Ratafia, M. (1987) Worldwide opportunities in genetically engineered vaccines. *Biotechnology (NY)* **5**, 1155–1158.
14. Kessler, D., Taylor, M., Maryanski, J., Flamm, E., and Kahl, L. (1992) The safety of foods developed by biotechnology. *Science* **256**, 1747–1749.
15. Martin, Q. and Zilberman S. (2003) Genetically modified crops in developing countries. *Science* **299**, 900–1001.

Glossary

Activator: A substance or agent that stimulates transcription of a gene or operon.
Aerobe: A living organism that requires oxygen for growth.
Algorithm: A step-by-step process for solving a problem or modeling a process.
Anaerobic: Growth or activity that does not require the presence of oxygen.
Allele: Alternative form of a gene. Alleles of a particular gene occupy the same loci on a pair of homologous chromosomes.
Allosteric regulation: A catalysis-regulating process in which the binding of a small molecule to the site on an enzyme influences the activity of another site.
Annealing of DNA: The process of heating (separating) and slowly cooling (renaturing) of double-stranded DNA to allow the formation of hybrid or DNA-RNA molecules.
Antibiotic: A biological substance that is typically produced by one organism that can restrict the growth of or kill another organism.
Antibody: A protein (specific immunoglobulin) that is synthesized by a B lymphocyte and that recognizes a specific site on an antigen. The basic immunoglobulin molecule consists of two identical heavy chains and two identical light chains.
Antigen: A compound that induces the production of antibodies.
Antiidiotopic antibody: An antibody that has the properties of an antigen.
Antisense DNA-RNA: The sequence of chromosomal DNA that is transcribed or a DNA sequence that is complementary to all or part of functional RNA (mRNA).
Antiserum: Fluid component of the blood that contains the antibodies of an immunized organism.
Attenuated vaccine: A virulent organism that has been altered to make a less virulent form but nevertheless retains the ability to elicit antibody response against the virulent form.
B cells : Lymphocytes derived from bone marrow that produce antibodies.
Baculovirus: Viral agent that infects arthropods, typically insects.
Bacteriophage: A virus that infects bacteria. Also called a phage.
Bioaccumulation: Increasing levels of a chemical agent (e.g., PCP) in high levels in organisms in a food chain.
Biocontrol: Any process using a living organism or system to restrain the growth and development of pathogenic organisms.
Biodegradation: The breakdown of a complex chemical compound to its simple constituents by living organisms.
Bioremediation: A process that uses living organisms to remove contaminants, pollutants or unwanted substances from soil or water.
Bioreactor: A vessel in which cell or cell materials or enzymes are used under controlled conditions to carry out specific biological reactions.
Broad-host-range-plasmid: A plasmid that can replicate in a number of different species.
Cancer: Uncontrolled growth of the cells of tissue or organ in a multi-cellular system.
Capsid: External protein coat or shell of a virus particle that surrounds the core (RNA or DNA) of the virus. The capsid determines the shape of the virus.

Cell line: A cell lineage that can be maintained in culture.

Cell-mediated immunity: The activation of T cells of the immune response to the presence of foreign antigen.

Chimera: Usually a plant or animal that has a population of cells with different genotypes. Sometimes it refers to recombinant DNA molecules that contain sequences from different organisms.

cDNA: A double stranded DNA complement of messenger RNA sequences synthesized in vitro by reverse transcriptase and DNA polymerase.

Chloroplasts: Organelles within plant cells where photosynthesis occurs; they contain small circular DNA molecules that replicate independently.

Chromosome: A physically distinct unit of the genome.

Chromosome walking: A technique that allows the sequential isolation of clones carrying overlapping sequences of DNA, thus allowing large regions of the chromosome to be mapped and sequenced.

Clone: A population of cells or organisms that are genetically identical due to asexual processes, nuclear transplantation or by insertion of DNA molecules that harbor the same sequence as the parent.

Cloning: Incorporating a DNA molecule into a chromosome site or cloning vector.

Cloning vector: Agent that carries the new genes into a host cell where it produces numerous identical copies of itself and the genes it carries. Also called a cloning vehicle, vector or vehicle.

Codon: A set of three nucleotides in sequence in messenger RNA that code for a specific amino acid or a termination signal.

Competence: The ability of an organism to take up DNA molecules, typically plasmids.

Cofactor: An essential component (usually a low molecular weight compound) that is necessary in an enzymatically catalyzed reaction.

Conjugation: Unidirectional transfer of DNA from one cell to another through cell-to-cell contact.

Constitutive synthesis: Continuous synthesis of RNA or protein by a cell.

Copy number Number of molecules (copies) of an individual plasmid or plastid that is usually present in a single cell. Copy number values range from 1 to 50 or more.

Co-segregation: When two genetic conditions appear to be inherited together.

Cosmid: A cloning vector that is a hybrid between a phage DNA molecule and a bacterial plasmid. The cos sites of a λ-DNA are inserted into a plasmid to form a molecule that can be packaged into the λ-phage head through the presence of the cos sites.

Cross: In genetic experiments also called mating of two individuals.

Cryptic plasmid: A plasmid with undetermined functions.

Culture: A population of cell or microbes that are cultivated under controlled conditions.

Cytokinin: A plant hormone that accelerates cell division.

Diploid: A cell that has a set of all pairs of chromosomes.

DNA: The molecule that carries the genetic blue print and information found in most cells, except a few viruses where the hereditary material is RNA (ribonucleic acid).

The information coded on the DNA determines the structure and function of the organism.

DNA probe: A segment of DNA that is labeled (tagged) so that after DNA hybridization, base pairing between the probe and complementary base pair in a DNA sequence can be detected.

DNA Recombination: A laboratory method in which DNA segments from different sources are combined into a single unit and manipulated to create a new sequence of DNA.

Double cross over: Two simultaneous reciprocal breakage and reunion events between two DNA molecules.

Duplex DNA: Double-stranded DNA.

Effector cell: Cells of the immune system that play a part in degrading antigens.

Electroporation: Use of electrical treatment on a cell to induce transient pores through which DNA is taken into the cell.

Encode: To specify, later decoding by transcription and translation, the sequence of amino acids in a protein.

Endotoxin: Cell wall component of Gram-negative bacteria that elicits an immune response and fever in humans.

An Episome: A DNA molecule that may exist either as an integrated part of chromosomal DNA molecule of the host or as an independently replicating DNA molecule (plasmid) free of host chromosome.

Epitope:A specific region on the antigen that is recognized by an antibody.

Exotic species: A species that does not originate from the place where it is found; a nonnative-evolved or introduced species.

Expression vector: A cloning vector that has been constructed in such a fashion that after insertion of the DNA molecule, its coding sequence is properly transcribed, and the RNA is efficiently translated.

Fermentation:A metabolic process where by microbes gain energy from the breakdown and assimilation of organic and inorganic nutrients.

Fed batch process: Growth of cells or microbes in a bioreactor to which nutrients are added periodically to get optimum biological response.

Fusion protein: Two or more coding sequences from different genes that have been cloned together and which after translation act as a single polypeptide sequence. Also known as a hybrid protein.

Gamete: A reproductive cell having a haploid set of chromosomes that is capable of fusing with a similar cell of the opposite sex to yield a zygote—also called a sex cell.

Genes: Physical units of heredity. Structural genes, which make up the majority, consist of DNA segments that determine the sequence of amino acids in specific polypeptides. Other kinds of genes exist. Regulatory genes code for synthesis of proteins that control expression of the structural genes, turning them off and on according to circumstances within the microbe.

Gene cloning: Procedure employed whereby specific segments of DNA (genes) are isolated and replicated in another organism.

Gene pool: The total sum of genes in a breeding population.

Genetic code: The linear sequence of the DNA bases (adenine, thymine, guanine and cytosine) that ultimately determine the sequence of amino acids in proteins. The genetic code is first "transcribed" into complementary base sequences in the messenger RNA molecule which in turn is "translated" by the ribosomes during protein biosynthesis.

Genetic map: The Lenora array of genes on a chromosome based on recombination frequencies. Also called a linkage map.

Genetic recombination: When two different DNA molecules are paired, those regions having homologous nucleotide sequences can exchange genetic information by a process of natural cross-over to generate a new DNA molecule with a new nucleotide sequence.

Germ line cells : The most primitive or earliest stage of development; referring to tissues or cell lineage producing gametes.

GRAS: An acronym for "generally regarded as safe." In the United States, this designation is given to modified foods or drugs that have been used for a considerable period of time and have a history of not causing illness to humans through extensive toxicity testing. More recently certain host organisms for recombinant DNA experimentation have been given this status.

Glycosylation: Covalent attachment of sugar or sugar-related molecule to proteins or polynucleotides.

Host: A cell that maintains a cloning vector.

Hybridoma: The product of fusion of myeloma cells with an antibody producing lymphocyte. This cell combination (hybridoma) can continue to divide in cell culture and secrete a single type of antibody.

Insecticide: Living organism or chemical that kills insects.

Infectious agent: A living cell, virus, organism or parasite that proliferates in plants or animals and causes diseases.

Immune response: A process that includes the synthesis of an antibody in response to the presence of a foreign antigen.

Integration: Insertion of DNA molecule into a chromosomal site.

Intron: A DNA sequence, found within the coding region of most eukaryotic genes, that interrupts the code for the gene product. The full gene sequence is initially transcribed into heterogeneous nuclear RNA and then the intron sequences are removed by cutting and splicing to give the final mRNA molecule which is then translated at the ribosome to give the protein product.

Interspecific protoplast fusion: Method for recombining genetic information from closely related but non-mating cultures by removing the walls from the cells.

In vivo gene therapy: The delivery of gene(s) to a tissue or organ of an individual to alleviate a genetic disorder.

Locus: A site on the chromosome where a specific gene is located.

Mendelian: Refers to a trait that shows a simple dominant/recessive pattern of inheritance.

Metabolic engineering: A scientific discipline that integrates the principles of biochemistry, chemical engineering and physiology to enhance the activity of a particular metabolic pathway.

Microinjection: A process of introducing DNA or other compounds into a single cell with a fine microscopic needle.

Monoclonal antibody:A single type of antibody that is directed against a specific epitope (antigenic determinant) and is produced by a hybridoma cell.

Mutation: Genetic lesion or aberration in a DNA sequence that results in permanent inheritable changes in the organism. The strains that acquire these alterations are called mutant strains.

Mutant: A living cell that differs from the parent because it carries one or more genetic changes in its DNA (e.g., variant).

Native protein: The naturally occurring form of a protein.

Operon: A cluster of genes that are coordinately regulated.

Phenotype: The observable characteristic of an organism produced by the interaction of the genes and the environment.

Plasmid: An autonomous DNA molecule capable of replicating itself independently from the rest of the genetic information.

Polymerase chain reaction: A technique for amplifying a specific segment of DNA using a mixture of thermostable DNA polymerase enzyme, deoxyribonucleotides, and a primer sequence. The reaction is carried out in multiple cycles that involve assembly and synthesis of specific sequences of DNA to produce millions of copies of the desired DNA strands. Also called PCR.

Primer: A short oligonucleotide that hybridizes with a template strand. A primer provides a 3'-hydoxyl end for the initiation of nucleic acid synthesis.

Primary metabolites: Simple molecules and precursor compounds such as amino acids and organic acids that are involved in pathways that are essential for life processes and the reproduction of cells.

Promoter: A segment of DNA to which RNA polymerase attaches. It usually lies upstream of (5') a gene. The promoter sequence aligns the RNA polymerase so that transcription will initiate at specific site.

Protoplast: A bacterium, yeast or plant cell that has had its cell wall stripped either enzymatically or chemically.

Recombinant: An individual who has genes on a chromosome that results from one of more cross over events.

Recombinant DNA: Assembly of hybrid DNA molecules in vitro which are derived from different sources.

Replicon: A genetic material having the sequences that specify the initiation and control of the process by which DNA is precisely duplicated.

Ribosome: The subcellular structure that contains both RNA and protein molecules. It mediates the translation of mRNA into protein. Ribosomes contain both large and small subunits.

Scale up: The transfer of a process such as fermentation from a small scale (flask) to a large scale (stirred tank).

Secondary metabolites : Complex molecules derived from primary metabolites and assembled in a coordinated fashion. Secondary metabolites are usually not essential for an organism's growth.

Selection: A system for either isolating or identifying a specific organism in a mixed population.

Selective pressures: The influence of environmental factors on the ability of the organism to compete with surrounding organisms for reproductive success.

Somatic cells: Any cell of a multi-cellular organism, except the gametes and the cells from which they develop.

Stem cells: A precursor cell that undergoes division and gives rise to a different lineage of differentiated cells.

Subcloning: Splicing part of cloned DNA molecule into different cloning vectors

Strain: A microorganism or multicellular organism that is a genetic variant of a standard parent stock.

Subspecies: Population of cells sharing certain characteristics that are present in other populations of the same species.

Syndrome: Occurrence of some features that collectively makes up the symptoms of a specific disorder or disease.

Transcript: An RNA molecule that has been synthesized from a specific DNA template.

Transcription: The process whereby a molecule of RNA is synthesized by RNA polymerase using a DNA strand as a template.

Transduction: Transfer of DNA from one cell to another by means of virus or bacteriophage.

Transfection: The transfer of DNA to a eukaryotic cell.

Transformation: Introduction of DNA and its integration form one organism into another via uptake of naked DNA.

Transgene: A gene from one source that is incorporated into the genome of another organism.

Transgenesis: The introduction of gene(s) into animal or plant cells that leads to the transmission of the input gene (transgene) to successive generations.

Transgenetic animal: A fertile animal that carries introduced genes in its germ line.

Transgenic plants: A fertile plant that carries introduced genes in its germ line.

Transposon: A genetic segment that can be randomly inserted into the chromosome, exit the site and relocate at another chromosomal site.

Variant: An organism that is genetically different from the original wild type organism. Also called a mutant.

Vector: A DNA molecule employed to introduce foreign DNA into the host.

Vertical transfers: The passage of DNA material from one organism to another through the germ line (e.g., sexual mechanisms or in bacteria by replication and cell division).

Virus: Typically a smaller particle than bacteria. It contains DNA but is unable to reproduce outside the host cell.

Index

U

V